Doing Research in Sound Design

Doing Research in Sound Design gathers chapters on the wide range of research methodologies used in sound design. Editor Michael Filimowicz and a diverse group of contributors provide an overview of cross-disciplinary inquiry into sound design that transcends discursive and practical divides.

The book covers Qualitative, Quantitative and Mixed Methods inquiry. For those new to sound design research, each chapter covers specific research methods that can be utilized directly in order to begin to integrate the methodology into their practice. More experienced researchers will find the scope of topics comprehensive and rich in ideas for new lines of inquiry.

Students and teachers in sound design graduate programs, industry-based R&D experts and audio professionals will find the volume to be a useful guide in developing their skills of inquiry into sound design for any particular application area.

Michael Filimowicz is Senior Lecturer in the School of Interactive Arts and Technology (SIAT) at Simon Fraser University. He has a background in computer mediated communications, audiovisual production, new media art and creative writing. His research develops new multimodal display technologies, exploring novel form factors across different application contexts including gaming, immersive exhibitions, telepresence and simulations.

Sound Design

The Sound Design series takes a comprehensive and multidisciplinary view of the field of sound design across linear, interactive and embedded media and design contexts. Today's sound designers might work in film and video, installation and performance, auditory displays and interface design, electroacoustic composition and software applications and beyond. These forms and practices continuously cross-pollinate and produce an ever-changing array of technologies and techniques for audiences and users, which the series aims to represent and foster.

Series Editor
Michael Filimowicz

Foundations in Sound Design for Linear Media
A Multidisciplinary Approach
Edited by Michael Filimowicz

Foundations in Sound Design for Interactive Media
A Multidisciplinary Approach
Edited by Michael Filimowicz

Foundations in Sound Design for Embedded Media
A Multidisciplinary Approach
Edited by Michael Filimowicz

Sound and Image
Aesthetics and Practices
Edited by Andrew Knight-Hill

Sound Inventions
Selected Articles from *Experimental Musical Instruments*
Edited by Bart Hopkin and Sudhu Tewari

Doing Research in Sound Design
Edited by Michael Filimowicz

Sound for Moving Pictures
The Four Sound Areas
Neil Hillman

For more information about this series, please visit: www.routledge.com/Sound-Design/book-series/SDS

Doing Research in Sound Design

Edited by
Michael Filimowicz

Routledge
Taylor & Francis Group

LONDON AND NEW YORK
A FOCAL PRESS BOOK

First published 2022
by Routledge
2 Park Square, Milton Park, Abingdon, Oxon OX14 4RN

and by Routledge
52 Vanderbilt Avenue, New York, NY 10017

Routledge is an imprint of the Taylor & Francis Group, an informa business

British Library Cataloguing-in-Publication Data
A catalogue record for this book is available from the British Library

Library of Congress Cataloging-in-Publication Data
Names: Filimowicz, Michael, editor.
Title: Doing research in sound design / edited by Michael Filimowicz.
Description: New York : Routledge, 2021. | Series: Sound design |
Includes bibliographical references and index.
Identifiers: LCCN 2020053028 (print) | LCCN 2020053029 (ebook)
Subjects: LCSH: Sound design—Research.
Classification: LCC MT64.S68 D65 2021 (print) |
LCC MT64.S68 (ebook) | DDC 781.3—dc23
LC record available at https://lccn.loc.gov/2020053028
LC ebook record available at https://lccn.loc.gov/2020053029

ISBN: 978-0-367-40490-1 (hbk)
ISBN: 978-0-367-40489-5 (pbk)
ISBN: 978-0-429-35636-0 (ebk)

DOI: 10.4324/9780429356360

Typeset in Times New Roman
by codeMantra

Contents

Contributors

Currently the Senior Consultant for Undergraduate Education at UC Riverside, **Vanessa Theme Ament** has been a sound professional in many films including *Die Hard*, *Platoon*, *Predator*, *Malice*, *A Goofy Movie*, *Edward Scissorhands* and *Sex, Lies, and Videotape*. Her film scholarship includes Foley and sound effects, film music and musical films. As a recipient of the Virginia Ball Fellowship for Creative Inquiry at Ball State University, Vanessa directed the student film documentary *Amplified: A Conversation with Women in Film Sound*. The author of *The Foley Grail*, and co-author (along with Academy Award Winner David E. Stone) of *Hollywood Sound Design and Moviesound Newsletter: A Case Study of the End of the Analog Age*, she is presently completing *Divergent Tracks: How Hollywood, New York, and the San Francisco Bay Area Revolutionized Film Sound*. In summer of 2018, She designed and taught the Massive Open Online Course Mad About Musicals for Turner Classic Movies and hosted two evenings of musicals with Ben Mankiewicz. Vanessa received her PhD in Moving Image Studies from Georgia State University. Vanessa was previously the Endowed Chair in Telecommunications at Ball State University.

Núria Bonet is a Lecturer in Music at the University of Plymouth. She completed a PhD on sonification methods in creative practice under the supervision of Prof. Eduardo Miranda. She received Bachelor's and Master's degrees in Music and Acoustics at the Universities of Manchester and Edinburgh. She has published on the topic of sonification and the history of Catalan instruments; she is also a published composer and performer.

Roberto Bresin is Professor in Media Technology at KTH Royal Institute of Technology, Stockholm, Sweden. Bresin is the head of the Sound and Music Computing group. His expertise is in design and analysis of expressive control of sound models used in music performance, sound in interaction, interactive sonification in performing arts, sports and robotics. He is also a member of the steering committee of the Sound and Music Computing Network.

Brian Bridges is a composer and lecturer based in Derry/Londonderry, Northern Ireland, where he is Research Director for music, drama, film and heritage at Ulster University. Much of his work is inspired by connections between embodied and ecological concepts in theories of music, sound and broader cognition, and creative practices and technologies. His creative output includes installations, electroacoustic and acoustic music, with particular interests in spatial and microtonal music. He is a member of the Dublin-based Spatial Music Collective and is represented by the Irish Contemporary Music Centre. He also serves on the editorial board of *Interference: a Journal of Audio Culture* and the *Journal of the Society for Musicology in Ireland*, and is a past president of ISSTA: the Irish Sound, Science and Technology Association.

Kristin Carlson is an Assistant Professor in the Creative Technologies Program at Illinois State University, exploring the role that computation plays in embodied creative processes. She has a history of working in choreography, computational creativity, media performance, interactive art and design tools due to her background in movement, technical theater, interaction design and programming. Kristin is a researcher with the movingstories: Tools for Digital Movement, Meaning and Interaction research partnership, exploring the cognition of movement experience and designing movement applications for creativity support tools. She publishes in the fields of cognitive science, computational creativity, movement and computing and electronic art.

Stuart Cunningham is a Senior Lecturer in Computer Science and Information Systems at Manchester Metropolitan University (UK) and has been working in the UK University sector for over 17 years in a variety of academic, research and managerial roles. His research interests include: audio compression; emotional interaction with technologies and computers, especially 'affective audio'; and sonic interaction. Dr. Cunningham is a Fellow of the British Computer Society, Chartered IT Professional, Member of the Institution of Engineering and Technology, Fellow of the Higher Education Academy and a Visiting Professor at Wrexham Glyndŵr University.

Nicolas Donin is Senior Researcher at the Institut de Recherche et de Coordination Acoustique/Musique, Paris, where he leads the research group in interdisciplinary musicology (Analysis of Musical Practices) as part of STMS Labs. His work addresses twentieth and twenty-first century musics at the crossroads of music history, analysis and social sciences. He has authored and co-authored over 100 papers across several disciplines and languages. He has edited and co-edited reference collections such as *L'Analyse musicale, une pratique et son histoire* (Geneva, 2009) and *Théories de la composition musicale au XXe siècle* (2 vols., Lyons, 2013).

Kjetil Falkenberg is an Associate Professor in Sound and Music Computing at KTH Royal Institute of Technology in Stockholm, Sweden, and Docent in Media Technology. He received his PhD in Speech and Music Communication at KTH in 2010. After a post-doc period in Japan and an associate professor position at Södertörn University he returned to KTH in 2016. Falkenberg leads research in sonification, sonic interaction design, and sound and music communication, with particular focus on health, special needs, new musical interfaces, and active listening.

David Fierro is an electronic engineer, holder of a master's degree in Musical Computing, and a PhD student at the Paris 8 University. His research interests focus on the neuronal response to acoustic stimulus, brain-computer interfaces, sound design and sound spatialization. He also works as a digital artist, sound designer and a multiplatform developer. His published work focuses on sound design, virtual reality, neuro acoustics and sound spatialization. Currently working in the fields of neuro acoustics, artificial intelligence and virtual reality.

Michael Filimowicz is Senior Lecturer in the School of Interactive Arts and Technology (SIAT) at Simon Fraser University. He has a background in computer mediated communications, audiovisual production, new media art and creative writing. His research develops new multimodal display technologies, exploring novel form factors across different application contexts including gaming, immersive exhibitions, telepresence and simulations.

Emma Frid obtained her PhD in Sound and Music Computing from KTH Royal Institute of Technology in January 2020. Her thesis was entitled "Diverse Sounds – Enabling Inclusive Sonic Interaction," and focused on how Sonic Interaction Design can be used to promote inclusion and diversity in music-making. She is currently conducting postdoctoral research at the STMS Lab, IRCAM/Sound and Music Computing Group, KTH Royal Institute of Technology, focusing on inclusive Sonic Interaction Design and Accessible Digital Musical Instruments (ADMIs).

Hans-Peter Gasselseder has studied and taught Psychology, Communication Science and Musicology/Dance Science at the University of Salzburg (Austria) and at Aalborg University (Denmark). Currently, Gasselseder is an Associate Lecturer at St. Pölten University of the Applied Sciences and gives regular guest lectures at Hong Kong Baptist University. His published work focuses on the cognitive psychology of immersive experience, effects of sound and music in film and video games, as well as technical implementations ranging from virtual orchestration to state-of-the-art ambisonics and six-degrees-of-freedom recording methods for virtual reality. Currently he is preparing his PhD thesis "Re-Dramatizing the Game Orchestra" at Aalborg University.

Nicolai Jørgensgaard Graakjær is Professor of Music and Sound in Market Communication in the Department of Communication and Psychology at the University of Aalborg, Denmark. His research interests span from musicology, sound studies, market communication and media studies to social psychology. His publications include *Analyzing Music in Advertising* (Routledge, 2015), *Sound and Genre in Film and Television* (MedieKultur, 2010, co-editor), *Music in Advertising: Commercial Sounds in Media Communication and other Settings* (Aalborg University Press, 2009, co-editor) as well as papers in, among others, *Critical Discourse Studies*, *Visual Communication*, *European Journal of Marketing* and *Popular Music and Society*.

Visda Goudarzi is a music technologist working at the intersection of audio and human-computer interaction. She is an Assistant Professor of Audio Arts and Acoustics at Columbia College Chicago and works as the principal investigator (PI) for the Austrian funded project COLLAB (Collaborative Creativity as a Participatory Tool for Interactive Sound Creation) at Institute of Electronic Music and Acoustics (IEM) in Graz, Austria. Prior to Columbia College she worked as an artistic and scientific researcher at IEM Graz, Center for Computer Research in Music and Acoustics (CCRMA) at Stanford University, and Vienna University of Technology. Her research interests include auditory interfaces, interactive and participatory design, sound and music computing, live coding and data sonification. Visda holds a PhD in Sound and Music Computing, a Master's degree in Music, Science, and Technology, and a Master's degree in Computer Science. Visda composes and performs live electronics as a member of intra-sonic electronic duo.

Kate Hennessy an Associate Professor specializing in Media at Simon Fraser University's School of Interactive Arts and Technology. She is an anthropologist with a PhD in Anthropology from the University of British Columbia and an MA in the Anthropology of Media from the University of London, School of Oriental and African Studies. As an anthropologist of media and the director of the Making Culture Lab, her research explores the impacts of new memory infrastructures and cultural practices of media, museums and archives in the context of technoscience. Her highly collaborative multimedia and artworks investigate documentary methodologies to address Indigenous and settler histories of place and space. Hennessy is a founding member of the Ethnographic Terminalia Collective, which has curated exhibitions and projects at the intersection of anthropology and contemporary art since 2009.

Maria Kallionpää is an internationally active composer and pianist, currently working as an assistant professor at the Hong Kong Baptist University, and as a composer in residence of the Mixed Reality Laboratory of the University of Nottingham. During 2016–2018 she worked as a postdoctoral fellow at the University of Aalborg, her artistic research focusing on gamification as a composition technique. Furthermore, as a winner of the Fabbrica Young Artist Development Program of Opera di Roma, Kallionpää was commissioned for an opera that was premiered at Teatro Nazionale Rome in 2017. In collaboration with her colleague Markku Klami, Kallionpää composed the first full-length puppet opera produced in the Nordic Countries (premiered in 2018). She was a laureate of Académie de France à Rome in 2016. Kallionpää got her PhD in composition at the university of Oxford in 2015.

Since 2008, **Maximilian Kock** has held a professorship in audio production (Bachelor's and Master's degree program: media production and media technology) at Ostbayerische Technische Hochschule (OTH) Amberg-Weiden in Bavaria. His area of expertise is sound design, audio production and psychoacoustics. He lives in Munich.

Vincent Meelberg is Senior Lecturer and researcher at Radboud University Nijmegen, the Netherlands, Department of Cultural Studies, and at the Academy for Creative and Performing Arts in Leiden and The Hague. He studied double bass at the Conservatoire of Rotterdam and received his MA both in musicology and in philosophy at Utrecht University, and an MSc in sound design at Napier University Edinburgh. He wrote his dissertation on the relation between narrativity and contemporary music at Leiden University, Department of Literary Studies. Vincent Meelberg has published books and articles about musical narrativity, musical affect, improvisation and auditory culture, and is founding editor of the online *Journal of Sonic Studies*. His current research focuses on the relation between sound, interaction and storytelling. Beside his academic activities he is active as a double bassist in several jazz and improvisation ensembles, as well as a sound designer.

Adam Melvin is a composer and lecturer in Popular and Contemporary Music at Ulster University, Derry/Londonderry, Northern Ireland. A great deal of both his compositional and research practice is concerned with interrogating the relationship between music, sound, site and the visual arts, particularly the moving image. He has received numerous international performances and broadcasts of his music; his research has been published in *The Soundtrack*, *Short Film Studies* (Intellect), *The Palgrave Handbook of Sound Design and Music in Screen Media* and in Routledge's *Foundation in Sound Design* series. He is a member of the Dublin-based Spatial Music Collective and is represented by the Irish Contemporary Music Centre.

Nicolas Misdariis is a research director and head of IRCAM STMS Lab / Sound Perception & Design group. He graduated from an engineering school specializing in mechanics (1993), he got his Master's thesis in acoustics and his PhD in synthesis/reproduction/perception of musical and environmental sounds. He recently defended his HDR (Habilitation à Diriger des Recherches) on the topic of Sciences of Sound Design. He has been working at IRCAM as a research fellow since 1995 and contributed, in 1999, to the introduction of sound design. During that time, he developed research works and industrial applications related to sound synthesis and reproduction, environmental sound and soundscape perception, auditory display, human-machine interfaces (HMI), interactive sonification and sound design.

Leo Murray is a Senior Lecturer in sound at Murdoch University, Australia. He spent ten years as a broadcast engineer in the UK before moving into teaching and researching in sound. The focus of his research is around the practice and theory of sound, particularly for film, television and interactive media. He is interested in how and why soundtracks are created in the ways that they are, and how they function and is the author of *Sound Design Theory and Practice: Working with Sound*. He also continues to work as sound recordist and sound editor.

Sandra Pauletto is an Associate Professor at KTH Royal Institute of Technology, Stockholm, Sweden, where she leads research in sound computing and media production. Currently she is the principal investigator of the project *Sonic Interaction Design to support Energy Efficiency Behaviour in the Household* funded by the Swedish Energy Agency.

Frank Pecquet was born and resides in Paris, France. Composer (PhD from University of California, San Diego), musicologist (PhD from the University of Paris), author of works in acoustic and electronic music and articles on contemporary music and sound anthropology. Teacher in computer arts at the University of Paris I Pantheon-Sorbonne (Ecole des Arts de la Sorbonne) and researcher in the Art Creation Theory and Aesthetic (ACTE) laboratory. Currently working in the field of sound design, acoustical ecology and interactive music composition.

Stephen Roddy is a Research Fellow based in the Electronics and Electrical Engineering Department at Trinity College Dublin. He holds a PhD in Sonification, the systematic generation of sound for the communication of information about a data source to a listener. His research explores multimodal strategies for representing and interacting with complex datasets, in human-computer interaction and creative multimedia contexts. Stephen employs both empirical research methods and creative arts–based methods. His empirical work has been widely disseminated through peer-reviewed journal publications, book chapters and conferences. His artistic outputs, including fixed-media electronic music, data-driven music and generative music systems, are performed frequently at international events.

Emma Rodero is the Director of the Media Psychology Lab in the Department of Communication at Pompeu Fabra University, PhD in Communication, PhD in

Psychology, Master in Pathology of Voice, and Master in Psychology of Cognition. She obtained a Marie Curie fellowship (European Union) to conduct research in the US about cognitive processing of sound messages using psychophysiological techniques. She is author of 12 books and 70 scientific papers about voice, radio and sound. Rodero usually teaches Public Speaking and Advertising on Radio at UPF. She has over a decade of experience in the radio industry. She is currently a voice-over artist and has received awards for radio documentaries and dramas.

Prophecy Sun is an interdisciplinary performance artist, movement, video and sound maker, mother and current Jack and Doris Shadbolt Fellow in the Humanities at Simon Fraser University. Originally from Vancouver, she is currently living in Nelson on the unceded territories of the Ktunaxa, Syilx and Sinixt peoples. Informed by her creative studies in vocal and movement improvisation, her practice celebrates both conscious and unconscious moments and the vulnerable spaces of the in-between in which art, performance and life overlap. Her recent research has focused on ecofeminist perspectives, co-composing with sound, objects, matter, extraction and surveillance technologies, and site-specific engagements along the Columbia Basin region and beyond. She performs and exhibits regularly in local, national, and international settings, festivals, conferences and galleries and has authored several peer-reviewed articles, book chapters and journal publications.

Ian Thompson is a Senior Lecturer in media, specializing in sound design, at University of Greenwich, London, having previously worked as a digital media producer on projects for BBC radio, and as a sound engineer. He currently leads courses in sound design and creative film-making technologies across undergraduate and post-graduate film production programs. His research interests cover acoustic ecology, spatial audio and sound design for virtual production. He is also an artist, specializing in sound (but not exclusively), and a musician. He is currently writing a book, *Audio for Artists*, for publication by Routledge in 2023.

Miles Thorogood is an artist/engineer at the University of British Columbia with research in the practice and theory of sound design and interactive digital art. His research seeks to identify formal models of creativity by investigating aspects of human perception and design process in order to encode creative structures for computer-assisted technologies in art-making environments. Miles specializes in quantitative and qualitative methods from Music Information Retrieval, Human Computer Interaction and Artificial Intelligence toward cutting edge research in the development of novel computational models and systems for artistic creation. Research contributions include new knowledge in the fields of soundscape studies, affective computing, music information retrieval and media arts. The research has been featured as interactive museum exhibits, installations and performances.

The interactive installation and performance works frame the research in creative practice that brings meaningful contexts of experience and environment to the foreground, using algorithmic processes combining art-making, audio and visual media, databases, artificial intelligence and physical and network computing.

Jonathan Weinel is a London-based artist, writer and researcher whose main expertise is in electronic music and computer art. He is the author of *Inner Sound: Altered States of Consciousness in Electronic Music and Audio-Visual Media* (Oxford University Press, 2018). In 2012 Jon completed his AHRC-funded PhD in Music at Keele University regarding the use of altered states of consciousness as a basis for composing electroacoustic music. His work operates within the nexus of sound, psychedelic culture and immersive computer technologies. His electronic music and audio-visual compositions have been presented at a variety of international festivals. Jon is a Lecturer in Games Technology (VR) at the University of Greenwich. He is a Full Professional Member of the British Computer Society (MBCS), a member of the Computer Arts Society (CAS) specialist interest group and a co-chair of the EVA London (Electronic Visualisation and the Arts) conference.

Laura Zattra is a scholar of twentieth and twenty-first century music with special emphasis on science and technology studies, electroacoustic music, musical collaboration, philology and analysis. She is Adjunct Professor of the History of Electroacoustic Music and the History of Sound Design in Cinema in Bachelor and Master Programs in Italy and abroad. Research Fellow at IRCAM in Paris and IreMus (Paris-Sorbonne), co-editor-in-chief of the journal *Musica/Tecnologia* (Firenze University Press) and founder of teresarampazzi.it website. Author of, among others, *Live-Electronic Music. Composition, Performance and Study* (with F. Sallis, V. Bertolani and I. Burle, 2018), and *Studiare la Computer Music. Definizioni, analisi, fonti* (2011).

Volume Introduction

Michael Filimowicz

This volume gathers chapters on the wide range of research methodologies used in sound design today. Sound design is usually defined as distinct sets of related fields of practice, e.g., sound design for film and video, or for installation and performance, or for auditory display and sonification, games and so on. Research in sound design often occurs in somewhat siloed scholar communities which draw on the research traditions of various sciences and academic disciplines, though not always so since some sound design conferences and journals are often marked by a strong inter- and multi-disciplinary inclusivity of scholarship.

For example, humanities might be drawn on in sound studies, perceptual psychology is utilized in audio quality evaluation, human-computer interaction provides a frame for auditory display design, philosophy informs sound art aesthetics, computer science and mathematics are used in music software design and music theory is a background for generative compositional systems. This volume provides an overview of cross-disciplinary inquiry into sound design that transcends discursive and practical divides. In doing so, sound design comes more clearly into view as a well-defined area of research that can tap into a diversity of methods and academic as well as industry literature for realizing its goals and artefacts.

The three *Foundations in Sound Design* volumes which started the Routledge Sound Design series presented a comprehensive breadth-first approach toward defining a new conception of sound design that brings together many areas of practice and research that are usually separated in institutional contexts. Sound design is conceived and taught differently and somewhat more narrowly in established academic units such as film and video, new media, audio production and music technology, games, HCI and performance-oriented departments.

Advanced and in-depth research methodology topics in Linear, Interactive and Embedded media systems are presented here to allow for more rigorous cross-pollination of research methods across the practice areas articulated in the three

DOI: 10.4324/9780429356360-1

1

predecessor volumes. One pressing need this volume immediately addresses is that there are no comprehensive and organized textbooks for all the graduate programs in sound design that have proliferated in recent decades. There are of course research methods books, such as the volumes by Creswell and his co-authors. However, books such as these do not speak very directly to the media, design and technology areas associated with sound design.

Qualitative, quantitative and mixed-methods inquiry are covered in the following chapters. For those new to sound design research, each chapter covers specific research methods that can be utilized directly in order to begin to integrate the methodology into their practice. More experienced researchers will find the scope of topics comprehensive and rich in ideas for new lines of inquiry. Each chapter introduces high level methodological concepts as well as practical outlines for applying methods in concrete examples, so that readers can immediately dive into the research practices. Students and teachers in sound design graduate programs, industry-based R&D experts and audio professionals interested in exploring how they can make explicit in discourse the tacit insights from their practice will find the volume to be a useful guide in developing their skills of inquiry into sound design for any particular application area.

Chapter 1 – "Sound Design Thinking" by Vincent Meelberg – begins the volume by drawing connections between sound design practices and the concepts associated with design thinking. As a practice that combines scientific and creative concepts and tools, sound design invites, through the nature of its materials, a high degree of reflectivity by thinking *through* sound, rather than simply about sound.

Chapter 2 – "Meaning-making and Embodied Cognition in Sound Design Research" by Stephen Roddy and Brian Bridges – examines sound design through the lens of more recent developments in embodied cognition which seek non-reductive explanations of our experiences with sound. The chapter explores how embodied cognition approaches have come in wider use within disciplines that sound design often draws upon, such as psychoacoustics, computing, cognitive science and music.

Chapter 3 – "Mapping Space and Time in the Soundtrack: Embodied Cognition and the Soundtrack's Spatiotemporal Contract" by Adam Melvin and Brian Bridges – discusses the multisensory crosstalk between sonic, visual and even haptic dimensions of cinematic experiences. Drawing upon a multidisciplinary set of theorists, they present a new framework for understanding embodied cognition in sound design.

Chapter 4 – "Measuring People's Cognitive and Emotional Response: Psychophysiological Methods to Study Sound Design" by Emma Rodero – presents psychophysiological methods that measure listeners' responses to sound in real time. The specific advantages of these methods for analysis are examined in connection with the communication potentials of audio messages.

Chapter 5 – "Watching and Evaluating Simultaneously: An Approach to Measure the Expressive Power of Sound Design in a Moving Picture" by Maximilian Kock – covers techniques aimed at empirically measuring audiences' emotional responses while undergoing audiovisual experiences. This study offers a robustness of insights, given the large number of participants involved, using the *emoTouch* application designed for capturing emotional data.

Chapter 6 – "Using a Semiotic Approach to the Practice of Sound Design" by Leo Murray – continues the investigation of qualitative methods by showing how semiotic analysis can be used to hone in on the granular details of listeners' interpretation of sound stimuli. Semiotics offers a flexible approach to the wide variety of audiovisual contexts as it explicates the fundamental 'building blocks' in the production of meaning.

Chapter 7 – "Audi(o) Branding and Object Sounds in an Audio-visual Setting: The Case of the Car" by Nicolai Jørgensgaard Graakjær – connects general semiotic method to that of close reading in particular. Such an approach is applied to a highly rigorous account of sound's role in shaping branded experiences.

Chapter 8 – "Designing and Reporting Research on Sound Design and Music for Health: Methods and Frameworks for Impact" by Kjetil Falkenberg and Emma Frid – details the gamut of methods for conducting sound design research in the context of health, accessibility and disability. The chapter highlights strategies for increasing the impact of such research which to date have mainly been applied in the areas of music therapy and music medical interventions.

Chapter 9 – "Practice-led and Interdisciplinary Research: Investigating Affective Sound Design" by Jonathan Weinel and Stuart Cunningham – considers practice-led and interdisciplinary methods in sound design. The reciprocality between theory generation and artefact production produces new opportunities for the production and dissemination of new knowledge.

Chapter 10 – "Research-creation as a Generative Approach to Sound Design" by Prophecy Sun, Kristin Carlson and Kate Hennessy – furthers the discussion of theory's relation to practice by grounding research creation strategies in a presentation of specific art installation works. Attention is given to the generative use of everyday technologies to foreground personal immediateness, chance and improvisation while also finding value in the inevitable technical failures.

Chapter 11 – "The Soundscape Approach: New Opportunities in Sound Design Practice" by Ian Thompson – relates soundscape practices going back to the World Soundscape Project to more recent standardizations of soundscape discourse by the International Organization for Standardization. ISO 12913 codifies an approach for investigating people's perceptions of soundscape beyond traditional concerns around noise mitigation, and offers new research opportunities for sound designers.

Chapter 12 – "Developing a Sound Design Creative A.I. Methodology" by Miles Thorogood – investigates the wicked problems associated with integrating creative A.I. technologies into creative tasks. Methods related to music information retrieval systems in computationally assistive contexts are outlined to show the complexity of the computing requirements for entertainment media.

Chapter 13 – "Sonification Research: Overview and Emerging Topics" by Sandra Pauletto and Roberto Bresin – starts a sequence of several chapters that treat one of the most active research areas in sound design today, that of auditory display and sonification. They consider sonification research in its many facets as they may apply to the differing interests of scientists, teachers, athletes, artists and designers.

Chapter 14 – "Sound Design in the Context of Sonification" by Visda Goudarzi – focuses on parameter mapping themes in sonification research. Methods from user-centered design are used in the creation of new sonification tools as applied to data. The chapter is grounded in case studies which take into consideration the full context of auditory display.

Chapter 15 – "Creating and Evaluating Aesthetics in Sonification" by Núria Bonet – articulates the aesthetic dimensions of sonification, where design success has to integrate perspectives from auditory perception, data and musical structures. The *Data-Mappings-Language-Meaning* framework is presented as a way of ensuring that aesthetic considerations are foregrounded at every step of the sonification process.

Chapter 16 – "Revealing Industry Culture: A Cultural Ethnographic Approach to Postproduction Sound" by Vanessa Theme Ament – is the first of two chapters looking at sound designers themselves. Ament interrogates industry culture through the lens of reflective practice. The working cultures of sound designers are not often analyzed rigorously through suitable observational methods, which can marginalize some professional voices.

Chapter 17 – "Sound Design as Viewed by Sound Designers: A Questionnaire About People, Practice and Definitions" by Laura Zattra, Nicolas Donin, Nicolas Misdariis, Frank Pecquet and David Fierro – extends the discussion of sound design's professional sphere in a probing study of knowledge acquisition of sound designers across Europe. Database cartography and online questionnaires are used to answer questions such as: Who is a sound designer? How do they learn sound design? How do they design sound?

Chapter 18 – "Qualitative, Pragmatic and Hermeneutic Inquiries in Electro-acoustic Compositions: Selected Case Studies" by Maria Kallionpää – explores the situation of the contemporary composer who has to invent the form anew for each composition. Case study methodology is used to show how composition and music analysis work in tandem during the production of new works.

Chapter 19 – "The Bold and the Beautiful Methods of Sound Experience Research: An Introduction to Mixing the Qualitative and Quantitative Study of the Subjective Experience of Sound and Music in Video Games" by Hans-Peter Gasselseder – presents an in-depth exploratory sequential mixed methods account of the experience of immersion and game play in interactive sound design. Qualitative, quantitative and experimental approaches are integrated into a research design to triangulate insights across data types.

Acknowledgment

The chapter summaries here have in places drawn from the authors' chapter abstracts, the full versions of which can be found in Routledge's online reference for the volume.

1

Sound design thinking

Vincent Meelberg

1 Introduction

Sound design is a practice that is as much art as it is science. Sound design involves academic study as well as creativity. Other forms of design also have this mix of art and science, but the obvious difference is that sound design works with sound rather than with other materials. And because of the specific characteristics of sound, such as being both temporal, spatial and vibrational, aspects such as communication, relationality and interaction become perhaps even more important in sound design that in other design disciplines that work with materials other than sound.

It could even be said that sound design is a practice in which such aspects are constantly reflected upon. The practice of sound design involves the thinking and rethinking of the manners in which relations are formed, interactions are stimulated and communication is established. And this thinking is done through working with sound. In this sense, sound design is not primarily a practice that thinks about sound. It does not focus specifically on what sound is, or wat it can be or on how sound should be used in particular situations. Nor is it a practice that only tries to come up with solutions to problems using sound. Instead, sound design can be considered a practice that explores phenomena through sound. Sound is used as a means to reflect on issues related to interaction, communication and relationality. Put differently: sound design is a form of sonic thinking.According to Bernd Herzogenrath (2017), sonic thinking is a thinking with and by means of sound, not a thinking about sound. Herzogenrath asserts that sound is not a knowledge about the world, coming to you only in retrospective reflection, but a thinking of and in the world, a part of the world we live in, intervening in the world directly. And this is exactly sound design's potential: to examine, reflect on, and intervene in the world by means of sound.

In this chapter I will explore the possibilities of approaching sound design as a form of sonic thinking. First, I will discuss the role design thinking plays in sound design.

DOI: 10.4324/9780429356360-2

Design thinking denotes "[...] the ways of framing, approaching and addressing challenges that characterize design" (Dalsgaard 2014, p. 144). According to Nigel Cross (2011), design thinking constitutes a third paradigm of inquiry besides science and the arts. As opposed to other academic disciplines, a problem is not first formulated and then answered. Rather, Peter Dalsgaard (2014) explains, "[t]he design problem is not given, it is developed through the first stages of designerly inquiry" (2014, p. 150). He furthermore adds that, "[a]t its core, design is an interventionist discipline that seeks to bring about change by developing and staging artifacts and environments that alter how we perceive and act in these volatile conditions" (2014, p. 148). This conception of design thinking resembles sonic thinking, in that both are interventionist in nature.

Next, I will focus on sonic thinking and sketch its possibilities. I will address questions such as: How can one think in a sonic manner? What does it mean to think sonically? What kinds of knowledge can sonic thinking generate? I will then relate the outcomes of this inquiry to the practice of sound design that I developed earlier in this chapter in order to conceptualize sound design as a practice of sonic thinking. Finally, I will outline the potentiality of approaching sound design as sonic thinking. I will assert that this approach may provide new possibilities to study those phenomena that sound design generally tends to deal with: interaction, relationality and communication.

2 Design Thinking

Design thinking is a term that refers to the activities specific to the act of designing. As I explained in the introduction, Peter Dalsgaard (2014) uses the term design thinking to denote the ways of framing, approaching and addressing challenges that characterize design. Design thinking thus is not a purely mental activity, but an act in which the entire body is involved. Design thinking is not limited to the planning stages of design, but happens throughout the entire design process, from inception to the completion of a project.

One of the reasons why thinking is incorporated into the entire design process is because, in design, theory-practice and reflection-action are intertwined, Dalsgaard explains:

> Designers' actions yield input for ongoing reflection and their reflections in turn shape ongoing actions to resolve design problems or open up new design opportunities. Designers draw on theories and preconceptions to scaffold their inquiries into design problems and these theories and preconceptions can be transformed, enriched or discarded over the course of time on the basis of how well they scaffold design practice. Theory and practice are thus closely interrelated in design.
> (2014, p. 145)

Through practice ideas and design strategies are tested, and during this practice new ideas may emerge. This is why design is also characterized by emergence and interaction: "Throughout the process, the design space—i.e., the arena in which the designer acts—undergoes changes. This ongoing development is influenced by reciprocal interaction between designers, stakeholders and the various components of the design space" (ibid.).

Designers interact with the people who have commissioned the design, with those who will use the design, with the materials the design will consist of and with environment the design is supposed to function in.

It is because of its interaction with the environment that design is situated and systemic, Dalsgaard explains: "The fundamental concern in design is overcoming a 'hermeneutical gap' between the existing situation and the product of the design process and between designers' current understandings and the crystallization of ideas and concepts embodied by the product itself" (ibid.). Each situation calls for its own solution, and it is impossible to apply fixed procedures to a design problem. As a result, design is experimental in that solutions to design problems can only be found during the design process itself, trying out whether certain option will or will not work.

Design thinking is a search for design solutions, which at the same time is "[…] a way of revealing the design situation" (Löwgren and Stolterman 2004, p. 22). By engaging with an environment through design thinking the characteristics of this specific environment will be uncovered, which is necessary in order to find appropriate design solutions. Referring to Donald Schön, Jonas Löwgren and Erik Stolterman (2004) assert that designers ask questions of the situation through actions or "design moves" rather than words. As a consequence, design is simultaneously reflection-in-action and reflection-on-action; the problem and the solution have to evolve in parallel. Design is a conversation with the situation and at the same time a practice of experimentation where designers "[…] have to be good 'listeners' and 'readers' of the situation" (Löwgren and Stolterman 2004, p. 21). Löwgren and Stolterman therefore conclude that a designer is a researcher "[…] exploring the reality that constitutes the design situation" (2004, p. 30).

Cross (2011) adds that design thinking is abductive, i.e., design thinking suggests that something may be. It is an act of producing proposals or conjectures. Designers make proposals for solutions that cannot be derived directly from the problems, and the particular design that is the result of the process of design thinking is just one of many possible ways deal with the design problem. Referring to Schön as well, Cross maintains that "[…] design is an interactive process based on posing a problem frame and exploring its implications in 'moves' that investigate the arising solution possibilities" (2011, p. 23). There is hardly ever just one "correct" answer to a design problem.

In order to cope with a complex design process, Löwgren and Stolterman (2004) explain, a designer needs to externalize the actual design thinking through representations such as sketches, drafts and models. Cross adds that "[s]ketching provides a temporary, external store for tentative ideas; supports the 'dialogue' that the designer has between problem and solution" (2011, p. 12). As design thinking entails an ongoing conversation between designers and the situation, representations function as tools for thinking. Via external representations designers enter into a dialectical process between the design situation and possible solutions.

Designers thus think via the creation of external objects. As a consequence, design thinking is a form of thinking though the materials used to create these objects. In this sense design thinking resembles artistic research, which also is a thinking through the material, as Krien Clevis (2016) asserts. In artistic research, too, situations and states of affairs are questioned and investigated through the creation of artefacts. In doing so, questions concerning what the material does and what it brings about are addressed. Clevis (2016) calls this practice "the making of material arguments." She points out that art can create worlds by using specific materials, and that a place can be transformed into a new place by using the power of the imagination as well as the performativity of the materials used in the artwork. Again, this is similar to what is done in design thinking, which, as I explained in the Introduction, is an interventionist discipline that creates artefacts in order to incite change and transformation as well.

The parallels between artistic research and design thinking do not stop here. Henk Borgdorff (2012) asserts that knowledge in art is pre-reflective, non-conceptual and non-articulated, other than via the production of art. Schön (1983), for his part, points out that much of our knowing, including design knowledge, is ordinarily tacit. It is implicit in the way we do things. Our knowing is in our action, and often it is very difficult, if not impossible, to put this knowledge into words

Maurice Merleau-Ponty arrives at a similar conclusion in his description of the manner in which an organist plays his instrument:

> There is no space for any "memory" of the position of the stops, and it is not in objective space that the organist in fact is playing. In reality his movements during rehearsal are consecratory gestures: they draw affective vectors, discover emotional sources, and create a space of expressiveness as the movements of the augur delimit the templum.
>
> (1962, p. 145–146)

The knowledge the organist has regarding his musical performance is conveyed via this very performance, and this knowledge does not come from memory or the brain, at least not primarily. Rather, the knowledge is in the very gestures that the organist makes. And one of the main aims of artistic research is to make this knowledge explicit, to understand and to communicate the knowledge enclosed in artistic experience and practice.

Again, the similarities between artistic research and design thinking are obvious. Artistic research helps implicit, tacit artistic knowledge to become shared and discussed by others, to make the implicit explicit in a convincing and clear manner. Design thinking, in turn, is reflection-in-action. Designers are "[...] reflective practitioner[s], with the ability to act and the ability to reflect in and on [their] actions" (Löwgren and Stolterman 2004, p. 64). And when someone reflects-in-action, Schön (1983) suggests, they become researchers in a practice context. During the process of design they construct new theories that pertain to the specific case they are working on. In conventional design situations these theories often remain implicit. Design thinking may become actual research when these theories are articulated and communicated in other ways than the resulting design itself.

Sound design practices generally also involve design thinking. Sound design can be regarded as a form of reflection-in-action as well. Sound design includes a mode of thinking that happens though the materials used to create sound designs, and the material sound designers work with is sound. As a result, design thinking in sound design is a thinking that happens through sound.

3 Sonic Thinking

What does it mean to think through sound? It is not merely thinking about sound, what sound is, or can be and what one can do with sound. Yet, this is also what sound designers do. Thinking through sound, however, encompasses more than that. Thinking through sound, or sonic thinking, is a thinking with and by means of sound, not a thinking about sound. Sonic thinking, Bernd Herzogenrath (2017) asserts, is a thinking of and in the world, as sound is a part of the world we live in, intervening in the world directly. Sounds, as events, can cause other events. For instance, sounds can change the character of an environment, the manner in which an environment is experienced by its inhabitants. Also, sounds can create vibrations in other materials, including the bodies of humans.

Sounds are events that have the potentiality to intervene, but they are also themselves effects, Christoph Cox (2017) explains. Sounds are the results of physical causes. Sounds thus have a causal relation with the event that produced them, but are also distinct from these causes. Consequently, sound is at the same a memory, a result of an event that has already happened. But sound is also a memory because it has the capacity to recall the characteristics of the world it has traveled through (Carlyle 2017). Sound changes because of the environment in which the sound is produced, and thus carries traces of that environment.

Furthermore, sound implies a future. To hear a sound, Aden Evens (2017) suggests, is to hear a motion. This motion is full of stories and promises, as it began at a certain point in time and continues to move for an indeterminate period of time.

This motion cannot but be interpreted by those who are able to perceive it. The future of sound, Evens (2017) asserts, is heard in its tensions, tensions that never fully resolve but only create more and less sympathetic vibrations and resonances, as well as expectations that will be either met or not. And, as I explained elsewhere (Meelberg 2006; 2019), it is the interplay of tension and resolution that is responsible for the narrative potentiality of sound.

Sonic thinking, Cox (2017) points out, follows this flow of motions, of vibrations, of tensions and resolutions created by what we call sound. This implies that sonic thinking is temporal, as sound is a temporal phenomenon itself, being a flow of motions, a sequence of vibrations. Furthermore, sonic thinking is spatial, because sounds need space in order to exist. Moreover, sound changes the space it sounds in, just as sound carries characteristics of that space and is thus changed by the space in which it sounds.

This suggests that sonic thinking is situated. Just as design thinking, sonic thinking is specific to the particular environment and time in which this thinking takes place. The materiality of the situation in which a specific noise propagates is crucial, Holger Schulze (2017) maintains. Therefore, sonic thinking involves a description and analysis of the specific material and physical situation, which, Schulze (2017) explains, includes the architecture, technologies and designs in which the act of sonic thinking takes place.

Next, Schulze (2017) asks how to articulate this situation is such a way that it does justice to its aural dimensions instead of resorting to either visual metaphors or purely descriptive terminology as used in acoustics. He finds a solution in Barry Blesser and Linda-Ruth Salter's notion of aural architecture. Aural architecture, Blesser and Salter (2007) explain, refers to the properties of a space that can be experienced by listening. It is thus experience that is central in aural architecture. Blesser and Salter (2007) furthermore point out that all sensory experience is influenced by culture. It is therefore not sufficient to quantify the elements involved in sensory experience in order to properly understand them. Similarly, cultural factors need to be taken into consideration when trying to understand aural architecture.

One of the main concepts that Blesser and Salter identify in aural architecture is acoustic arena. Acoustic arena is "[…] the area where listeners can hear a sonic event (target sound) because it has sufficient loudness to overcome the background noise (unwanted sounds)" (2007, p. 22). Blesser and Salter point out that the acoustic arena is the experience of what they call a social spatiality, "[…] where a listener is connected to the sound-producing activities of other individuals" (2007, p. 25). Besides being social, an acoustic arena is also determined by cultural factors such as conventions concerning the way we are supposed to behave acoustically:

> In each situation, both collectively and individually, those who occupy or live within a space have the prerogative to manipulate the size and shape of their

acoustic arenas. Open the door, and you are now inside the acoustic arena for the activities taking place in the other room; close the windows, and you are no longer in the arena of children playing on the street. Turn up the volume of your entertainment system, and you are now beyond the acoustic arena of your telephone. Shout, and your arena overpowers the arenas of others nearby.
(2007, p. 25)

An acoustic arena in fact is a cultural phenomenon, as it is "[...] a region where listeners are part of a community that shares an ability to hear a sonic event" (Blesser and Salter 2007, p. 22). Its existence depends on cultural conventions, while an acoustic arena also codetermines the manner in which its inhabitants act culturally.

Related to acoustic arena is the concept of acoustic horizon, which Blesser and Salter (2007) define as the maximum distance between a listener and source of sound where the sonic event can still be heard. The acoustic horizon delineates an acoustic arena. An acoustic arena, Blesser and Salter (2007) explain, is centered at the sound source. An acoustic horizon, on the other hand, is centered at the listener. Thus, every sonic event has an acoustic arena, while every listener has an acoustic horizon.

The main acoustic property of a space is reverberation. Reverberation influences the size of an acoustic arena, because it may increase the maximum distance between listener and sound source where the sonic event can still be heard. According to Blesser and Salter listeners are affected by reverberation, but it is impossible to specify what the listeners' affective response to reverberation will be "[...] without examining the social context; this aspect of reverberation in aural architecture is culturally relative" (2007, p. 62). The experience of reverberation, too, is culturally determined.

From a cultural and social perspective, reverberation is important for other reasons as well. It is through reverberation that a space effectuates an interactive experience, or as Blesser and Salter (2007) call it, an acoustic dialogue, with its inhabitants. And entering into an acoustic dialogue, reflecting on this dialogue, and making this dialogue explicit are acts of sonic thinking.

Finally, sonic thinking is corporeal. Sonic thinking is feeling as much as it is a cognitive activity. Sonic thinking follows the flow of vibrations that sound consists of, sensing the reverberations of a space. Consequently, sonic thinking includes the feeling of these vibrations and reverberations, a feeling that is done via the body.

Sonic thinking thus should not ignore the fundamentally corporeal character of auditory experience, Schulze (2017) insists. The bodies of listeners function as material receivers, amplifiers and interpreters of sound, and the challenge is to articulate these experiences of sound-as-resonance that are situated and corporeal. Because of its corporeal aspect, sonic thinking cannot simply be reduced to language, a challenge that sonic thinking shares with artistic research. One of the solutions that artistic research developed is to combine language with nonverbal expressions such as actual artworks. Sonic thinking has to do something similar to convincingly express itself: combining language with actual sounds. Or rather, following Mary Caton Lingold, Darron Mueller and Whitney Trettien (2018): merging the practice of creating sounds with an informed critical inquiry of this practice.

4 Conceptualising Sound Design as a Practice of Sonic Thinking

Sound design incorporates design thinking, and thus is a form of reflection-in-action as well as reflection-on-action. Moreover, it is a reflection that involves sonic thinking, a thinking through sound that is situated and corporeal. This implies that the actual aural experiences of sound designers themselves play a crucial role in the design process.

Even though experience is inherently subjective, this does not mean that they thus do not carry any scholarly or heuristic value. Radical empiricism, for instance, which was first developed by William James (1912), holds that "[…] everything that is experienced is equally real. Among the things we experience are relations between things; so relations are real, with the same status as the things that stand in relations" (Chemero 2009, p. 141). Everything we experience is real, and this holds for relations we experience in particular. As sound is inherently a relational phenomenon, the relations we experience as being established by sounds while listening are also real.

Each experience starts with perception. Perception is feeling and a bodily change, Steven Shaviro explains:

> The way that I receive a perception, or apprehend its "sensa," is the way that my body changes, or has changed. Perception or excitation, action or bodily changes, and emotion or response, are all one and the same event. It is only in subsequent reflection that we can separate them from one another.
>
> (2009, p. 58)

Initially, perception is nothing but a bodily change, or a set of bodily changes. Perception induces bodily reactions in an observer, which Gilles Deleuze (2003) calls 'affect.' At first this affect has no meaning or signification, because it is entirely physical. While affect is nothing but an energetic movement within the body, the intensity of this movement provokes reflection on this sensation, an interpretation of this affective reaction. As a result, each perception is simultaneously a reflection-in-action, as perception without reflection is just a feeling of bodily changes.

Phenomena thus are first felt and grasped as modes of feeling, before they can be cognized and categorized. As a result, Shaviro (2009) points out, feeling is a basic condition of experience. So it is not cognition but feeling that is the basis of all experience.

Penny McCall Howard (2013) shares this view on perception. She asserts that listening, too, is a form of embodied perception and an extension of feeling. Iain Findlay-Walsh (2017), for his part, considers listening as a kind of embodied thinking-feeling as well. Following Mark Grimshaw and Tom Garner (2015), Findlay-Walsh suggests that the meaningfulness of sound emerges from "[...] a cloud of virtual elements, ideas and possibilities, which include memory, emotion, expectation and the physical particularities of the individual listener's ears and body" (2019, p. 32). Interpreting sounds as meaningful is a process in which both cognitive and physical elements play a crucial role.

Yet the basis of experience, auditory or otherwise, Shaviro (2009) stresses, and following Alfred North Whitehead, is feeling:

> For Whitehead, the questions of how we feel, and what we feel, are more fundamental than the epistemological and hermeneutical questions that are the focus of most philosophy and criticism (including Kant's *Critiques*). This emphasis on feeling leads, in turn, to a new account of affect-laden subjectivity. Most broadly, Whitehead's affect theory places aesthetics — rather than ontology (Heidegger) or ethics (Levinas) — at the center of philosophical inquiry. Aesthetics is the mark of what Whitehead calls our concern for the world, and for entities in the world.
>
> (2009, p. 47)

Feeling is the manner in which we, as human subjects, perceive and interact with the world. Feeling and affect constitute the basis of our ability to function in the world, and any inquiry into perception, interaction or the establishing of relations should therefore be centered around feeling. As a discipline that is concerned with perception, interaction and relationality as well, sound design should first and foremost focus on feeling, too.

Sonic thinking already acknowledges the importance of feeling. Sonic thinking stresses that auditory experience is corporeal. Sounds are felt before they are interpreted and qualified by listeners. Consequently, in sound design, considered as a form of sonic thinking, it is more important to focus on what sound does, or can do, than on what sound is.

As sound design also includes design thinking, which is a thinking through the materials used in design, it makes sense to define the materiality of sound in terms of what it does. According to Cox, the materiality of sound is "[...] its texture and temporal flow, its palpable effect on, and affection by the materials through and against which it is transmitted" (2011, p. 149). Paul Simpson adds that "[...] the sound itself is precisely sound's materiality, its body, its timbre, and about the resonance these produce" (2009, p. 2559). Sound, as material, is both vibration and the effect these vibrations have on other materials, including human bodies. As a consequence, perceiving this material means feeling their vibrations.

Thus, sound design, conceptualized as a practice of sonic thinking, involves a thinking through vibrations, vibrations that establish relations between entities, between entities and spaces, and even entities and feelings. It is a practice that is situated and focuses on relations. Sound design uses sound as a means to think through design problems that will only fully reveal themselves during the design process itself. Sound design thinking implies an ongoing conversation between sound designers and the situation in which sound functions as a tool for thinking.

Above I explained how in design thinking representations such as sketches, drafts and models are often used to externalize the thinking process. In sound design that involves sonic thinking this externalization is done through the creation of sounds, and this is how sound can function as a tool for thinking. Here, sounds function as what Itiel Dror and Stevan Harnad (2008) call cognitive technology that that allows sound designers to distribute some of their cognitive functions onto material supports, in this case sounds. Dror and Harnard assert that cognitive technology can have profound effects on "[...] how we think and encode information, on how we communicate with one another, on our mental states, and on our very nature" (2008, p. 1). The use of sound in sound design influences the manner in which we think about issues such as communication, thinking, interaction, and relationality.

Distributed cognition is the scientific discipline in which the use of cognitive technology is theorized. In this discipline "cognition" is not limited to a single brain. Instead, cognitive processes are delimited by "[...] the functional relationships among the elements that participate in it" (Hollan, Hutchins and Kirsh 2001, p. 175). This means that cognitive processes may be distributed among several people involved in a cognitive task. Furthermore, cognitive processes may involve the coordination between internal — mental — and external — as cognitive technology — structures. Distributed cognition thus is concerned with the manners in which cognitive activity is distributed across internal human minds, external cognitive artefacts, and groups of people, and how it is distributed across space and time (Zhang and Patel 2006).

Jiajie Zhang and Vimla Patel (2006) point out that external representations, which are forms of cognitive technology, provide knowledge and skills that are unavailable from internal representations. And this is very similar to what sonic thinking claims: to provide knowledge and insights through sounds that cannot be attained otherwise. This is how sonic thinking works in sound design: through external sonic representations sound designers are able to enter into a dialectical process between the design situation and possible solutions, a process that may generate new knowledge.

5 The Epistemic Possibilities of Approaching Sound Design
as Sonic Thinking

As I mentioned in the Introduction to this chapter, the practice of sound design involves the thinking and rethinking of the manners in which relations are formed, interactions are stimulated and communication is established. In this chapter I argued that it is through sonic thinking, in particular, that these reflections take place. Through the use of external sonic representations new insights may be gained that cannot be attained otherwise.

Sonic thinking enables the reflection on issues such as relationality, communication and interaction, because sound is vibration. Vibration, Walter Gershon (2013) explains, is patterned oscillation. And it is this patterned movement that causes resonance. Resonance, in turn,

> [...] is theoretically and materially consequential. Theoretically, if everything vibrates, then everything — literally every object (animate and inanimate), ecology ("natural" or "constructed"), feeling, idea, ideal, process, experience, event — has the potential to affect and be affected by another aspect of everything. It is the ability of one's self and/or not-self's affect (object/not-object, ecology/not-ecology, etc.) to effect in a multidirectional fashion.
>
> (Gershon 2013, p. 258)

Sound, as vibration, has the potentiality to let other things, including people and their thoughts, resonate. Thus, sound is relational in that it makes other things resonate with it. And since everything has the potentiality to resonate, sound is able to relate to, interact with and affect anything. As a result, Gershon proposes, "[...] if everything sings and resonates, then sound serves as both a strong theoretical site for conceptualizing what might 'count' as 'data' in qualitative research and how such methodologies might function in practice" (2013, p. 257). Because of its resonating qualities, sound has epistemological value, and it is this value that is exploited in sonic thinking.

One area that may benefit from this kind of sonic thinking in sound design, or sound design thinking, is ethnography. Above I explained that sound can be considered a memory, a trace, a resonance, of an event that has passed. John Drever (2002) argues that soundscape composition can be used as a form of ethnography, as a way to both uncover and articulate knowledge regarding the way of life of communities or social groups. Tullis Rennie points out that both soundscape composition and ethnography involve "[...] a highly sensual and subjective form of data-gathering based around observation, listening and engaging" (2014, p. 119). Rennie furthermore asserts that by employing an ethnographic approach to field recordings, the sounds themselves may act as a guide for the reflexive generation of abstract soundscapes. Listening to field recordings and reassembling and editing these recordings into soundscapes enable sound designers/ethnographers, or "ethnodigital sonologists" (Rath 2018), to reflect on the practices that are recorded, by addressing questions such as: What are the sounds doing to the people involved in these practices? How do these sounds affect me, as an outside listener? In which ways do the sounds establish relations between the participants and the practices? Which aspects of these practices can be highlighted by editing the recordings in particular ways? How is the acoustic arena established during these practices? Which acoustic horizons can be identified? How can the acoustic horizon be altered through editing the recordings, and what does that teach us about this horizon? In doing so, both the actual practice of creating the soundscape and listening to the end result, i.e., the soundscape itself, has epistemological potential and may create new knowledge and insights.

Sound is not only related to the past, but also implies a future. As I explained above, sound is full of stories and promises, as it began at a certain point in time and continues to move for an indeterminate period of time, creating tension and expectations that are either met or not. Sound may therefore also be used to explore possible futures. This is similar to what Iain Findlay-Walsh asserts regarding listening to sound artworks:

> A listener's engagement with the work lets them participate in a possible somewhere else — somewhere real yet fictional. This suggests a potential usefulness for field recording and sound art as an access point for embodying and inhabiting alternative spaces, experiences and perspectives.
>
> (2019, p. 33)

Creating sound artworks implies the construction of alterative worlds and stories, even when field recordings are used. The particular ways in which these recordings are edited and reassembled may result in the creation of aural experiences that no longer concern the present, but alternative realities or possible futures instead. And again, it is both during the creation of these works and while listening to the end result that sonic thinking, in this case about alternative realities and possible futures, may take place.

As sound is not only temporal, but spatial as well, sound design thinking may also be employed in exploring issues related to space, place and environment. Michael Gallagher, Anja Kanngieser and Jonathan Prior (2017) propose that through what they call expanded listening, attention can be directed to "[...] any and every kind of kinetic oscillation, generating insights into the interrelations and flows between humans, animals, objects, technologies, materials, infrastructures, and environments" (2017, p. 3). Expanded listening, Gallagher, Kanngieser and Prior (2017) explain, concerns the responsiveness of bodies encountering sound, while these bodies can be anything of any and every kind. This kind of listening "[…] provides an additional channel of knowledge, producing insights into scale, materiality and landscape morphology that are not available through other ways of knowing" (2017, p. 7). This is very similar to sonic thinking, in that it is through listening to sounds that these insights and knowledge are gained. More specifically, the resonances produced by sounds can "[…] promulgate the spatial dynamics of landscape, revealing spatial contours as well as various material qualities of landscape surfaces — particularly how surfaces may influence the reception of sounds through reflection and absorption" (2017, p. 7). And through the manipulation of the recordings of these sounds and resonances specific features can be highlighted and reflected upon.

The above examples suggest that knowledge is not only gained by listening to field recordings and other sounds. The epistemological potential of sonic thinking also, and perhaps even predominantly, lies in the explicit manipulation of sounds and recordings, as Felix Gerloff and Sebastian Swesinger are also keen to point out:

> Sonic structures or models of course carry information about their cultural contexts themselves, but can also function as epistemic tools, figures of thought or attention guides for an investigation of culture in general. In a kind of reverse writing or oscillation between the empirical research and our thought process we then again try to convey these models sonically. This means not just to present specific sounds but to stylise these audible forms in a somewhat exaggerated way to achieve the kind of conciseness necessary for them to be understood within this novel format of an audio paper. So while recorded soundscapes provide material for the audio paper it is not so much about representing their original cultural context through them, but about expressing our insights, considerations, hypotheses etc. sonically.
>
> (2016, p. 91)

Sounds are themselves informative, but the real epistemological potential lies in the manipulations that can be performed on them. Through the creation of sound designs features of the sounds and their resulting resonances may be brought to light that would have remained obscured otherwise. Moreover, just as language uses rhetorical devices in order to convincingly convey an argument, sound design can be used as a rhetorical as well as epistemological tool to make a point sonically: it is a form of sonic thinking.

Sonification, the technique of rendering sound in response to data and interactions (Hermann, Hunt and Neuhoff 2011), is an example of a practice that uses sound design in order to convey a point in a sonic manner. David Worrall points out that "[d]ata sonification is the acoustic representation of informational data for relational non-linguistic interpretation by listeners, in order that they might increase their knowledge of the source from which the data was acquired" (2019, p. 25).

This is indeed compatible with the practice of sonic thinking as elaborated above: new knowledge and insights, in this case of data that itself is not necessarily sonic, are gained through the listening to sounds that act as representations of the data. And it is not only through listening to the sonification itself that knowledge is gained. The creation of the representations, the sonifications, themselves is a process of sonic thinking during which new insights are found as well.

Finally, sonic thinking can act as an epistemological lens to research the practice of sound design itself. Such research would typically include a focus on how sound designers use sound as cognitive technology in order to solve design problems, and investigate the role intuition, feeling, perception and affect play in their decisions. Also, questions such as the manner in which acoustic arenas and horizons influence sound design practices could be addressed. In this way new insights could be gained about the ways sound designers think through the vibrating and resonating material they work with.

References

Blesser, B., and Salter, L-R. 2007. *Spaces Speak, Are You Listening? Experiencing Aural Architecture.* Cambridge, MA: MIT Press.

Borgdorff, H. 2012. *The Conflict of the Faculties: Perspectives on Artistic Research and Academia.* Leiden: Leiden University Press.

Carlyle, A. 2017. Memories of Memories: Remembering and Recording on the Silent Mountain. In: Herzogenrath, B. (ed.) *Sonic Thinking: A Media Philosophical Approach.* New York: Bloomsbury, pp. 65–82.

Chemero, A. 2009. *Radical Embodied Cognitive Science.* Cambridge, MA: MIT Press.

Clevis, K. 2016. Time|Place|Memory: Artistic Research as a Form of Thinking-Through-Media. In: Herzogenrath, B. (ed.) *Sonic Thinking: A Media Philosophical Approach.* New York: Bloomsbury, pp. 23–40.

Cox, C. 2017. Sonic Thought. In: Herzogenrath, B. (ed.) *Sonic Thinking: A Media Philosophical Approach.* New York: Bloomsbury, pp. 99–110.

Cross, N. 2011. *Design Thinking: Understanding How Designers Think and Work.* London: Bloomsbury.

Dalsgaard, P. 2014. Pragmatism and Design Thinking. *International Journal of Design* 8, pp. 143–155.

Deleuze, G., 2003. *Francis Bacon: The Logic of Sensation* (trans. D.W. Smith). London: Continuum.

Drever, J. 2002. Soundscape Composition: The Convergence of Ethnography and Acousmatic Music. *Organised Sound* 7, pp. 21–27.

Dror, I., and Harnad, S. 2008. Offloading Cognition onto Cognitive Technology. In: Dror, I, and Harnad, S. (Eds.), *Cognition Distributed: How Cognitive Technology Extends Our Minds.* Amsterdam: John Benjamins Publishing, pp. 1–23.

Evens, A. 2017. Digital Sound, Thought. In: Herzogenrath, B. (ed.) *Sonic Thinking: A Media Philosophical Approach.* New York: Bloomsbury, pp. 281–308.

Findlay-Walsh, I. 2017, Sonic Autoethnographies: Personal Listening as Compositional Context. *Organised Sound* 23, pp. 121–130.

Findlay-Walsh, I. 2019. Hearing How It Feels to Listen: Perception, Embodiment and First-Person Field Recording. *Organised Sound* 24, pp. 30–40.

Gallagher, M., Kanngieser, A. and Prior, J. 2017. Listening Geographies: Landscape, Affect and Geotechnologies. *Progress in Human Geography* 41, pp. 618–637.

Gerloff, F. and Swesinger, S. 2016. Sonic Thinking: Epistemological Modellings of the Sonic in Audio Papers and Beyond. *Interference Journal* 5, pp. 89–102.

Gershon, W.S., 2013. Vibrational affect: Sound theory and practice in qualitative research. *Cultural Studies <—>- Critical Methodologies* 13, no. 4, pp. 257–262.

Grimshaw, M., and Garner, T. 2015. *Sonic Virtuality: Sound as Emergent Perception.* New York: Oxford University Press.

Hermann, T., Hunt, A. and Neuhoff, J. 2011. Introduction. In: Hermann, T., Hunt, A. and Neuhoff, J. (Eds.) *The Sonification Handbook.* Berlin: Logos Verlag, pp. 1–6.

Herzogenrath, B. 2017. Sonic Thinking—An Introduction. In: Herzogenrath, B. (ed.) *Sonic Thinking: A Media Philosophical Approach.* New York: Bloomsbury, pp. 1–22.

Hollan, J., Hutchins, E. and Kirsh, D. 2001. Distributed Cognition: Toward a New Foundation for Human-Computer Interaction Research. In: Carroll, J.M. (ed.) *Human-Computer Interaction in the New Millennium.* New York: ACM Press, pp. 75–94.

James, W. 1912. *Essays in Radical Empiricism.* New York: Longman Green and Co.

Lingold, M.C., Mueller, D. and Trettien, W. 2018. Introduction. In: Lingold, M.C., Mueller, D. and Trettien, W. (Eds.) *Digital Sound Studies.* Durham: Duke University Press, pp. 1–25.

Löwgren, J., and Stolterman, E. 2004. *Thoughtful Interaction Design: A Design Perspective on Information Technology.* Cambridge, MA: MIT Press.

McCall Howard, P. 2013. Feeling the Ground: Vibration, Listening, Sounding at Sea. In: Carlyle, A., and Lane, C. (Eds.) *On Listening.* Axminster: Uniformbooks, pp. 61–66.

Meelberg, V. 2006. *New Sounds, New Stories: Narrativity in Contemporary Music.* Leiden: Leiden University Press.

Meelberg, V. 2019. Imagining Sonic Stories. In: Grimshaw-Aagaard, M., Walther-Hansen, M.and Knakkergaard, M. (Eds.) *The Oxford Handbook of Sound and Imagination*, Volume 1. New York: Oxford University Press, pp. 443–457.

Merleau-Ponty, M. 1962. *Phenomenology of Perception* (trans. C. Smith). London: Routledge & Kegan Paul.

Rath, R.C. 2018. Ethnodigital Sonics and Historical Imagination. In: Lingold, M.C., Mueller, D. and Trettien, W. (Eds.) *Digital Sound Studies.* Durham: Duke University Press, pp. 29–46.

Rennie, T. 2014. Sonic Autoethnographies: Personal Listening as Compositional Context. *Organised Sound* 23, pp. 121–130.

Schön, D. 1983. *The Reflective Practitioner: How Professionals Think in Action.* London: Temple Smith.

Schulze, H. 2017. How to Think Sonically? On the Generativity of the Flesh. In: Herzogenrath, B. (ed.) *Sonic Thinking: A Media Philosophical Approach.* New York: Bloomsbury, pp. 217–242.

Shaviro, S. 2009. *Without Criteria: Kant, Whitehead, Deleuze, and Aesthetics.* Cambridge, MA: MIT Press.

Simpson, P., 2009. 'Falling on deaf ears': A postphenomenology of sonorous presence. *Environment and Planning A* 41, pp. 2556–2575.

Worrall, D. 2019. *Sonification Design: From Data to Intelligible Soundfields.* Cham: Springer.

Zhang, J., and Patel, V.L. (2006). Distributed cognition, representation, and affordance. *Pragmatics & Cognition* 14, pp. 333–341.

2

Meaning-making and Embodied Cognition in Sound Design Research

Stephen Roddy and Brian Bridges

1 Introduction

Sound design research is a young and growing field that combines methods and approaches from a broad range of areas. Although this multidisciplinarity is a strength, it also brings some challenges. Psychoacoustics, computing, cognitive science and music are four areas core to sound design research. Throughout their development, each of these fields has had to reckon with the problem of meaning-making, the cognitive process by which people interpret and assign meaning to perceived and mental phenomena to various degrees. We distinguish, here, between two broad approaches to meaning-making: *embodied*, those which account for how physical, perceptual and sensorimotor dimensions shape and constrain cognition; and *disembodied*, those which offer a purely mental description of cognition and do not involve the body.

In general terms, disembodied models of cognition suggest that meaning-making is a purely mental phenomenon that involves the rule-based computation of abstract symbols on a mental layer that is isolated from the physical/biological world (Putnam 1967). This approach has been criticized by thinkers like Searle (1980), Harnad (1990) and Dreyfus (1965) for its insufficient account of how abstract mental symbols are assigned their meaning. The key argument made by these thinkers is that a disembodied mind would not be capable of meaning-making: what Harnad terms the *symbol grounding problem*. Embodied models of cognition acknowledge how the physical, neural and perceptual apparatus of the human body shape and constrain cognition, and hold that mental symbols are grounded in bodily experiences (Lakoff and Johnson 1980). Meaning-making is critical to sound design, and the creation of meaningful sonic symbols is a key focus for the discipline. On each project, sound designers work to ensure that the sounds they create convey the appropriate meaning and can be properly interpreted by a listener. It is in this context that we

DOI: 10.4324/9780429356360-3

believe care must be taken to adopt models of cognition which can account for meaning-making and allow us to better understand how meaning-making functions in sound design research contexts. However, disembodied models of cognition have historically been prevalent in many of the fields related to sound design, and a trend towards embodied models of cognition has only arisen in more recent times. The following sections of this chapter will examine how disembodied models came to the fore within fields which have influenced sound design, and how they were eventually challenged by embodied models of cognition. Although disembodied approaches have often been superseded in these fields, we argue that sound design research may need to address assumptions inherited from these fields whilst they were at an earlier stage of development.

1.2 Disembodied Roots

Dualism is the metaphysical belief that reality is composed of two parts, the mental and the physical. *Mind-body dualism*, chiefly associated with Descartes' "Cogito ergo Sum" (Descartes 1641), is the further and related belief that each human being is composed of two unique and independent parts: a material, publicly observable human body of physical dimensions, and a privately experienced mind of thoughts, emotions and perceptions. This seemingly irreconcilable distinction is often termed the *mind-body problem*. It has persisted in one form or another in Western cultures, shaping historical and contemporary thought (Uttal 2004). Responses to Cartesian dualism have often involved the adoption of *essentialist* positions on the mind-body problem, reducing reality to either purely *materialist* or purely mental, or *idealist,* descriptions and therefore eliminating the role of the human body in constituting and mediating cognition. According to Samuel Todes (2001): "[t]he human body is the material subject of this world" (88). Todes' statement describes an integrated triadic relationship where body, world and subjectivity are mutually constitutive of one another. Removing one element renders the other two elements meaningless, as there can be no body without a world, and no subjectivity without a body, etc. This integrated relationship between body, subjectivity and world is a key factor in embodied models of cognition which differentiates them from disembodied models which tend to draw a hard distinction between mind and world (Varela et al. 1991; Lakoff and Johnson 1999).

Although associated with Descartes, some form of mind-body dualism has permeated Western culture since at least the work of Aristotle (Heinaman 1990), and some have presented evidence for roots that go back to the Paleolithic era (Uttal 2004). Damasio (1994) has been particularly critical of Cartesian dualism, and Todes (2001, 23) argues in a similar vein that mind/body dualism precludes a cohesive understanding of how mind, body and world interrelate. Similar sentiments are echoed in Varela et al.'s (1991), foundational work on *embodied cognition,* where mind-body dualism is implicated in the distinctions between materialism and ideal-ism, as well as rationalism and empiricism, and further linked to the computational theory of mind discussed here below.

1.3 Psychophysics and Psychoacoustics

We can begin to understand how dualism has influenced research methodologies in auditory perception by considering developments in the field of *psychophysics*. Psychophysics and *psychoacoustics* (the study of hearing via psychophysical meth-ods) do not directly deal with meaning-making or describe how a listener assigns a meaning to some sonic symbol. Instead, they quantify relationships between physical stimuli and their perceptual correlates. This field of study has enriched our understanding of human perception and led to innovation in the fields of auditory display and sound localization (Walker and Nees 2011; Middlebrooks et al. 2000).

Psychophysics was pioneered by Gustav Fechner (1860), who built upon Ernst Weber's (1846) earlier observation that the just noticeable difference in a stimulus is proportional to the initial stimulus value to propose a logarithmic relationship between the intensity of a stimulus and its perceptual result. Stevens (1957) revived and updated the Fechner-Weber laws. Through a process of *magnitude estimation*, he had experimental participants estimate the intensity of a series of stimuli assign-ing value measurements they felt to be appropriate in the absence of any set scale. *Psychophysical scales* that govern the relationships between stimulus and response could be inferentially drawn by comparing results across large sets of repeated experiments. Stevens' *power law* was developed in this way, and proposed that sub-jective sensation is a function of the magnitude of the physical stimulus raised to a constant power, with the exponent specific to a given magnitude (e.g., the brightness of a light, loudness of a sound). *Cross-modal matching* (Stevens 1975) was the for-malization of the relationships between these psychophysical scales as experimental participants would match the perceived intensities of stimuli in one modality (e.g., loudness) with a second modality (e.g., brightness).

Morabito and Della Rocha (2010) point out that Fechner had originally gone to great lengths to address and overcome the problem of Cartesian dualism in his meth-odology. He attempted to build *a common system of quantification* that addressed the interconnection between mind and body, rather than focusing on the relation of mental phenomena as distinct from physical phenomena. Stevens departed from Fechner's approach, opting instead to quantify mental and physical phenomena as distinct. From a sound designer's point of view, Stevens' approach is concerned with making sounds that are perceptually relevant and intelligible for a listener, and this is a critical factor for sound design. At the same time, we also need to keep in mind that we cannot assign a meaning to a sound using a framework like Stevens': we only make the sound's meaning clearer or less clear to the degree that we account for relevant psychoacoustic constraints. The *meaning* of the sound has to come from somewhere else.

Key criticisms of early psychoacoustics research focus on shortcomings in both the philosophy and research methods of the field. Presenting experimental subjects with unrealistically simplistic and isolated stimuli in laboratory contexts led to research results that did not stand up in real-world contexts (Neuhoff 2004). This, in turn, led researchers to *ecological* approaches to psychoacoustics, which aimed to study sounds in their natural listening environments, where they are dynamic, complex and heard concurrently. In the ecological approach to psychoacoustics, questions of meaning-making began to move back to the fore (Walker and Kramer 2004).

2 Computer Science and Cognitive Science

As discussed previously, the concept of meaning-making is central to sound design as a crucial concern of the field is effectively conveying correct or intended information to a listener. *Cognitive science*, the interdisciplinary study of thought and mental phenomena, has contributed much to our current models of meaning-making. In the late 1940s, growing dissatisfaction with *behaviorist* approaches to psychology led scientists who were interested in the human mind to seek answers from elsewhere. They turned to new and exciting developments in the field of computer science for answers.

Cognitive science grew out of developments at the first Hixon Symposium on Cerebral Mechanisms and Behaviour in 1948, where John Von Neumann, and also Warren McCulloch and Walter Pitts, drew striking comparisons between the computer and brain and central nervous system, while psychologist Karl Lashley challenged the prevailing behaviorist attitudes and called for a new approach to studying the mind (Gardner 1985). In fact, there were a number of key developments in the early days of computer science that suggested computing processes were similar to human cognitive processes. One early example came in 1936 with the Universal Turing Machine (Turing 1936), a hypothetical device that could simulate the logic of any algorithm using four simple rules. Also that year, the Church-Turing thesis lent formal definition to the concept of the algorithm (Church, 1936; Turing 1936). Shannon (1938) used Boolean logic to represent states in, and solve problems with, electromechanical relay switches, and suggested this approach could model cognitive processes. Similarly, while demonstrating how networks of neurons can be modelled with logic, McCulloch and Pitts (1943) concluded that mental activity could be thus modelled and that these logical transformations could be run on a Universal Turing Machine. In 1945 Von Neumann introduced the concept of a 'stored program' that could be recalled from memory as required (Von Neumann 1945). This made it theoretically possible for a machine to program and reprogram itself, a concept that bore a striking analogy to human-like thought (Aspray 1990). Wiener (1948) introduced the field of cybernetics, arguing that machines which exhibited feedback could be described as striving towards goals because they modified their own behavior to achieve an objective. Finally, Turing (1950) proposed the Turing test, which stated that if a user could not distinguish the responses of a machine from those of a human, then that machine could be said to be capable of thought. Breakthroughs in computer science during this period therefore seemed to be making the subjective and private cognitive processes ignored by behaviorists (psychologists concerned with behavior rather than thinking) publicly observable and objectively verifiable.

The first generation of cognitive science was heavily influenced by these developments and adopted a computational model of the mind. In 1961, philosopher and mathematician Hillary Putnam (1967) proposed the *Classical Computationalist Theory of Mind* (CCTM). He argued that the human mind was an information processor and that thought was a form of computation. Mental content, thoughts and perceptions were rendered as symbols, and thinking was conceived of as the rule-based processing of those symbols. We see this position developed in cognitive science by Jerry Fodor (1975) and Newell and Simon (1976), and by the 1980s, the CCTM was the prevailing model of cognition (Lakoff and Johnson 1999, Chapter 6).

Alongside these developments, the role of the computer was evolving. In 1945 Vannevar Bush, who was instrumental in the Manhattan Project, wrote *As We May Think* (Bush 1945). In it, he presented a radical reimagining of the role of the computer in society, laying the groundwork for a range of innovations including personal computing, hypertext and the Internet (Zachary 1999). This would, in time, usher in a new era of computing, driven by Englbart's introduction of hypertext and the mouse (Baecker 2008), Licklider's envisioned forerunner of the Internet, the ARPANET (Lukasik 2010) and Sutherland's (1964) *Sketchpad*, a prototype of the graphical user interface with 3D modelling and a 'light pen' controller. The mainstreaming of these developments began in the 1980s, and with the advent of personal computing and the wealth of new computing technologies being developed and adopted, attention was increasingly paid to the development of new modes of interaction.

The *home computing boom* saw the widespread consumer adoption of computing hardware and this led to the birth of *human-computer interaction* (HCI) a field of research focused on understanding how users interact with computers in order to develop better systems for interaction. In 1983, the publication of *The Psychology of Human-computer Interaction* (Card, Newell and Moran 1983) helped to define the research agenda for the field. However, as Hürtienne (2009) notes, at the forefront of this movement was Allen Newell, who had played a pivotal role along with Herbert Simon in developing the CCTM, and now looked to HCI to further support and find an application for this theory. Card, Newell and Moran (1983) introduced the *Model Human Processor*, representing the user as a biological machine of integrated perceptual, motor and cognitive systems. This model was used to estimate how long a user might take to perform specific cognitive tasks by accounting for the perceptual and cognitive constraints, including measures of visual and auditory capacity along-side perceptual and conceptual 'processor cycle' times. As such, these early HCI researchers paid little attention to aspects of cognition that were not encompassed in the CCTM, effectively designing the user out of the system by developing tech-nologies that were designed solely for the users' "rational" information processing faculties, and so limited in their usability (Bannon and Bødker 1989).

The failure of these approaches to result in more usable and useful systems even-tually led to a broader *embodied turn* in HCI as researchers adopted a user-centric approach (Bannon 1995). These approaches acknowledged the role of the human body in meaning-making and placed empirical testing of real users above the theo-retical projections from generalized models (Dourish 2004). By the mid-late 2000s, HCI researchers were beginning to consider how embodied cognition principles could be used to guide the design of more meaningful user interactions (Imaz and Benyon 2007; Hurtienne and Blessing 2007).

2.1 Computer Music

The early development of *computer music* was driven by the advances in computer science noted above. As a result, early experiments in computer music inherited disembodied modes of interaction. In 1957, Max Mathews, working at Bell Labs, developed the first significant music software platform, the aptly named MUSIC (Doornbusch 2017); though only capable of generating monophonic melodies (its successor MUSIC-II, could manage four-part polyphony (ibid.), and programming with this software was laborious and computationally expensive work. Still, by the release of MUSIC-V, a system of dedicated subroutines called *unit-generators* made the compositional process manageable and more user-friendly (ibid.). Computer music was becoming a viable creative project and a core group of early enthusiasts of Mathew's work carried the MUSIC-N family of languages to more than a dozen academic research centers, eventually contributing significantly to the development of computer music as a formal research field. In spite of their influence and portabil-ity, MUSIC-N-type languages had their drawbacks too. With the use of orchestra (.orc) files to define synthesis routines and score files (.sco), to define musical pat-terns, they continued the tradition in Western art music of overlooking the contri-bution made to a musical work by the embodied human performer. Wishart (1996) reasons that as Western art music evolved, the focus of composers shifted from creating and organizing musical performances to creating and organizing written scores. This reduced the rich multidimensional musical spectra to just three primary dimensions: pitch, duration and timbre, a small subset of the many possible dimen-sions of sonic experience. Worrall (2010, 2013) argues that this reductive approach to music is informed by the CCTM and that modern music technologies are built around this same disembodied framework. This is compounded by the fact that, until recent releases of Csound, MUSIC-N-type languages did not run in real-time and so could not support interactivity in live performance.

Another early development at Bell Labs, which ran counter to the initial disembodied turn of computer music, was the GROOVE – Generated Real Time Operations On Voltage-controlled Equipment – system (Mathews and Moore 1970). GROOVE was the first interactive computer-controlled analog synthesizer to run in real-time. While relatively simplistic by modern standards, it was foundational in the context of musical interaction, where it opened the question of how a designer can best map physical gestures to sound in a computer music context. It represented a first step towards integrating physical gestures into computer music production, contrasting with the initially disembodied mode of MUSIC-N.

Nonetheless, the original MUSIC-N model remained influential, proliferating on the commercially accessible personal computing platforms of the 1980s via Barry Vercoe's introduction of Csound, an extended version of Mathew's MUSIC-11 (Vercoe 1986) that would work on any PC capable of running C. Other computer music languages and platforms were developed, including the visual, patching-based Max/MSP and Pure Data (Puckette 1996), and other, more purely synthesis-focused platforms, such as Native Instruments' Reaktor. The text-based but object-oriented SuperCollider and ChucK would follow in 1996 and 2002 respectively. Many of these technologies have been adopted for sound synthesis and music generation tasks in a number of fields beyond their original domain, which may draw attention to their implicit conceptual underpinnings. For example, in the field of *auditory display*, tools originating from MUSIC-N have been adapted for the task of sonification (Hermann et al. 2011) although, as discussed above, they have been criticized, along with popular DAW-based approaches, for failing to account for embodied aspects of sound during sonification design (Worrall 2010; 2013). Nonetheless, much recent research and commentary (Paine 2009; Fyans and Gurevich 2011) within the field of computer music has been devoted to questions of embodied cognition, gesture and interactivity in live performance.

2.2 Western Music Theory

Sound design has historically been mediated by technology, with the capture, reproduction and editing of sounds encompassing a variety of technologies whose working methods have found their way into later technologies; for example, the use of tape recording as a metaphor in DAWs. Similarly, our conceptualization of music frequently carries the conceptual baggage of previous generations of practices, from the Pythagorean monochord to the pipe organ, and from both of these to the MIDI standard's use of *twelve–tone equal temperament*: the equal division of the octave to obtain consistent intonation when transposing, obtained at the expense of adherence to integer harmonic ratios (Keislar 1987).

The formalization of musical notes is said to have originated in the West with Pythagorean thought, birthing the integer ratio-based Pythagorean tuning system (Sethares 2005; Barbour 1953, 1). This represents an early attempt to relate the physical and the perceptual attributes of music, though it also introduced the distinction that a particular mathematically-based formalization was the most important. The Pythagorean scale, built upon transpositions of a perfect fifth in 3/2 ratio, produced major thirds based on powers of three or less (*3-limit*) with the form 81/64; a construct which occurs in a very distant position within the harmonic series, and does not usually occur within perceptual experience of harmonic timbres: Parncutt and Hair (2018) have described Pythagorean tuning as being based on "ratios of psychologically implausible large numbers". A simpler form of the major third (which occurs within the harmonic series), the 5/4 ratio, was ruled out by Pythagoreanism because it introduced a new prime (five) as a component multiple within the scale definition. In contrast, later theorists, Didymus in the first century B.C.E. and Claudius Ptolemy in the second century C.E., discussed the use of the 5/4 major third, based on a *5-limit* (ratios expressed as powers of 5 or less) approach, aligning with those found in the earlier parts of the harmonic series (Benson 2006, 160). Thus, one of the foundations of Western music theory prioritizes one of many alternative formalisms at the expense of perceptual experience: the Pythagorean preference for stopping at the early prime three was an apparently arbitrary preference for this particular formalism at the expense of simpler ratios and tuning, i.e., those closer to that of the early harmonic series (Bridges 2012, 41–43). In his 1618 *Compendium on Music*, Descartes reaffirmed the distinction between perceptual experience and formalism, even as he attempted to make sense of musical experience using mathematical quantification.

> [I]n the *Compendium* he attempted to demystify the power of music and substitute for it a geometry of sensation [...] By considering the object as an entity separate from its perception on the basis of its measurable mathematical and physical properties, Descartes automatically excluded the qualitative element from perception.
>
> (Augst 1965, 122)

This account further suggested that *mental* and *emotional* elements of music were basically unworthy of analysis, with a conception of musical responses as based on simple reflexes (ibid. 130–32). Though not as influential as his other work, Descartes' musical ideas have nonetheless been influential, finding parallels in the work of theorists including Mersenne and Rameau (Jorgensen 2012). More broadly influential has been Kant's *transcendental idealism*, which holds that we experience the appearances of things rather than the things in themselves, and so concepts like space, time and causality are mental constructs we use to organize our experience of these representations. This has been criticized for disregarding the relationship between the physical body and world by reducing bodily experience to a purely imaginary process (Todes 2001, 95). It retains, maybe even intensifies, the Cartesian mind/body distinction with a model of the human isolated from the real, objective, world.

Schopenhauer (1883) built upon Kant's ideas by pointing out that the human body is the only object a subject can be aware of both externally, through subjective perceptions (*representation*), and internally, through the experience of embodiment (*will*). Because purely instrumental music was apparently not mediated by imitations of the phenomenal world, in the way that painting and sculpture are, he thought of it as a direct expression of *will*, one that could allow the listener to overcome the everyday suffering that typified existence (Schopenhauer 1883). In a way, Schopenhauer reaffirmed the importance of bodily experiences in his philosophy of music, but the old Cartesian model still prevailed with private bodily experience (will) locked away *inside* a world of mental phenomena (representation), and music offered as a temporary reprieve from this unbearable situation.

Schopenhauer's thought had a profound effect on Wagner (Darcy 1994), and, later, Schoenberg (1941). Schopenhauer's view of music as an expression of objective reality was shared by Wagner (Wagner 1879), who counted his first reading of *The World as Will and Representation* as the most important event of his life and the inspiration for *Tristan and Isolde* (Wagner 1875). In fact Wagner's 'Tristan Chord', a chord based on the augmented fourth, sixth and ninth, in an influential view, is seen as challenging the established modes of functional (i.e., syntactical) tonal harmony (Kurth 1920), although there is now debate on the degree of its novelty (Taruskin 2008). Wagner's readiness to employ chromatic constructs may disrupt the idealism of functional harmony, 'grounding' it within the confines of motifs (*leitmotiv*) which serve thematic concerns. However, in his 1849 essay, *The Artwork of The Future*, he described his compositional philosophy in terms that closely foreshadowed Schoenberg's future work (Stein 1960); in fact, Schoenberg asserts (1941) that there was a formal logic shared between both Wagner's approach and his own, drawing parallels between the *leitmotiv* and the basic 12-tone set (series) as reifying an integrating compositional structure (even if later commentators, notably Taruskin (2008), have expressed some skepticism around this particular parallel).

Even as both composers attempted to forge their own novel musical languages, Wagner and Schoenberg can be seen (at least rhetorically) as striving for an assumed universal and logically-structured aesthetic. Wagner's tendency towards *universalism* eschews Western music's previous prioritization of abstract or 'absolute' music's focus on functional harmony, replacing it with a new, totalizing form of the *Gesamtkunstwerk* (complete artwork) incorporating music, drama and associated forms, whilst maintaining some claims to Western music's formalistic ambitions through permutational and narrative-driven modifications of various *leitmotif*. Perhaps more significantly, Schoenberg's universalism saw him famously claim that his new *serialist* approach would replace the common practice of Western tonality to the extent that, in this teleological progression of musical technique and aesthetics, he had made a "discovery that [...would] ensure the supremacy of German music for the next hundred years" (Schoenberg, quoted in Taruskin 2008, 316). We argue that this musical universalism, later criticized by Meyer (1956), was a symptom of Cartesian dualism inherited from Schopenhauer (and, indeed, via the broader intellectual heritage of Western music), whereby there is an assumed fundamental, objective structure inherent within a musical language, which is then asserted through particular works and musical experiences. Furthermore, these universal structures, such as Schoenberg's series, can be seen as acting as the sole relevant source of structure, to the exclusion of personal embodied experience.

Webern and Boulez, of the Darmstadt School, inherited and extended Schoenberg's approach in pursuit of a 'democratic' serial music (Grant 2005) that while providing a counterbalance to traditional hierarchical structures in Western music, still poses problems for the perception and cognition of musical structures (McAdams 1989; Lerdahl 1992). Lerdahl initially focuses on the more complex case of *integral serialism* (the application of serial approaches to structures other than pitch) of Boulez's (1954) postwar *Le Marteau sans Maître*, whilst McAdams addresses serialism in its classical, pitch-based mode, but both conclude that serialism provides additional challenges to a listener when parsing music. In their assessment, it is almost as if what Lerdahl (1992, 115) terms the 'cognitive opacity' of serialism renders significant aspects of the musical structure into a less accessible formalist domain (even if he concedes this is not necessarily a criticism of it as a compositional tool). Serialism is therefore, in the final analysis, considered by Lerdahl to be a useful tool for structuring creative work, but is not considered a satisfactory explanation for the perception and cognition of this structure; the meaning, again, must be derived from elsewhere. Although these musical developments were not uncontested (and the twentieth century provides many counter-examples, including experimentalism, stochastic approaches to composition, open form and minimalism), serialism was particularly influential in many educational institutions for music composition (Gann 1998), setting a dualistic tone and framing for many of the musical developments preceding the latter part of the twentieth century.

2.3 Elektronische Musik and musique concrète

By the late 1950s two very distinct schools of electronic music had established themselves: the French *musique concrète* school and the German *Elektronische Musik*. Where *musique concrète* adopted an almost idealistic viewpoint inherited from Husserl (Kane 2007), *Elektronische Music* was motivated by a materialist understanding of the world borrowed from the field of acoustics (Eimert 1957; Dunn 1992). The early work of *musique concrète* pioneer Pierre Schaeffer set a positivist tone that would define the work of both schools and exerts a powerful influence on research in electronic music and sound production to this day (Dunn 1992). It is worth noting that Godøy (2006) has more recently points out that Schaeffer's original framework of typological and morphological categories for sounds are embodied to the degree that they are all built around sound-producing physical gestures. However, in Schaeffer's own lifetime, at least some of his methods and approaches were considered to be positivistic, and even reductive of the role of the body in music.

Inspired by the positivistic methods pioneered in Webern's approach to serialism (Gloag 2012, 40), the *Elektronische Musik* movement grew out of the electronic music studio of Cologne's Nordwestdeutscher Rundfunk (NWDR), which opened its doors in 1953. Originally helmed by Herbert Eimert, it distinguished itself from Schaeffer's *musique concrète* early on with its focus on "synthesis from first principles", with Eimert declaring that electronic music should be synthesized from the bottom up using electronically generated signals only (Eimert 1957). The culture that grew up around *Elektronische Musik* was defined by positivist approaches that concealed the role of the human body in determining aspects of the mind. Eimert and Meyer-Eppler elevated the rational in their approach to music, idolizing the mathematical, statistical and formal to the point of eliminating the human subject from music altogether (Pace 2009). Eimert (1957) advocated for the parameters of pitch, amplitude and duration as objectively inherent to the note itself and dismissed the "living flesh and blood" of the musical note, which unfettered would eventually lead to "the abrupt images and naked sensations of Expressionism". He compounded this positivistic attitude by reducing the role of the human body in both performance and composition to that of a machine which was better replaced by tape technology. Eimert believed this new music was purely objective and maintained that the sine tone was the fundamental unit of music perception, seeing it as the logical continuation of Webern's tripartite serialism through electronic means, and Meyer-Eppler, a physicist who advocated stochastic approaches to composition rooted in computer science and information theory, had similar views (Meyer-Eppler 1949). It would be some time before we see this position softening somewhat in Karlheinz Stockhausen's (1956) *Gesange der Jungliche,* which combined the methods of *Elektronische Musik* with the *musique concrète* techniques Stockhausen picked up during his time with Pierre Schaeffer.

By the early 1940's Pierre Schaefer's *musique concrète* introduced new compositional approaches specific to the medium of recorded sound. They were heavily influenced by Husserl's *phenomenology*, which broke with the Cartesian tradition to cast the body as the lived center of an experience defined by our capacities for physical movement and sensory perception. His *epoché*, or phenomenological bracketing, recommended suspending judgement about objects of perception to better understand the objects themselves. Schaeffer (1966), in his *Traité des objets musicaux*, adapted this idea in his 'reduced listening' which encouraged "listening to the sound for its own sake as a sound object by removing its real or supposed source and the meaning it may convey" (Landy 2007). Such sounds, theoretically decoupled from their sources, Schaefer termed *object sonore* (sound objects). Schaefer intended *objet sonore* to be pure sonic substances, free of reference to any meaning beyond their own perceptual appearance, that could be easily manipulated and assembled in musical structures should they prove to be suitable *objet musicaux* (Teruggi 2007; Worrall 2013). It was with these *objet musicaux* that one could compose a new kind of music fit for the newly emerging medium of recorded sound.

For Husserl, the body is an intermediary phenomenon that bridges the Cartesian gap, providing the locus of reality for the disembodied transcendental ego that owns it (Carman 1999); however, although he gives the body a greater role in his philosophy than his predecessors, he is still bound by the old Cartesian framework. This shortcoming would later be criticized by Heidegger (1927) and Merleau-Ponty (Carman 1999). Through the *objet sonore* and his reduced listening, Schaeffer transferred Husserl's Cartesian dualism into the philosophy of electronic music, in many senses reducing the role of the body in composition, listening and performance (Kane 2007). He does this by focusing on the *objet sonore,* an isolated phenomenal object, stripped of its bodily mediated meaningful context and conceptualized instead as a substance of pure sonic potential. Schaeffer would go on to expand his *musique concréte* program with Pierre Henry through the *Groupe de Recherche de Musique Concrète* (GRMC) at the French Radio Institution. According to researcher Hugues Vinet and composer François Bayle, the foundational techniques that Schaeffer developed for working with sound have become standards that define audio recording/manipulation technologies and techniques employed in studios around the world to this day (Teruggi 2007). The sound editing models associated with these technologies are here seen as essentially atomistic, decontextualized and non-ecological. R. Murray Schafer (1993) would later cast these developments in a negative light, describing recorded sound decoupled from its original source as 'schizophonic', favoring a phenomenologically-driven and ecologically grounded approach to, and understanding of, sound that was at odds with the initial conception of *musique concrète*. Later commentators would point out that positivistic attitudes adopted during this period laid the groundwork for the removal of the human body from live performance of electronic and computer music (Fyans and Gurevich 2011; Paine 2009; Worrall 2013).

3 Conclusion

As established early on, disembodied frameworks of cognition do not explain how listeners interpret or assign meaning to a given sound. Without an understanding of this meaning-making process, the efficacy of sound design research may be limited.

Each of the fields explored here experienced a transformation as their methods and theories matured and shifted to account for broader embodied and ecological contexts. In computer science, the second wave of HCI shifted its focus "from human factors to human actors" (Bannon 1995), while roboticists and AI researchers were similarly embracing embodied cognition principles in their work (Steels and Brooks 1995). The second generation of cognitive science focused on embodied cognition and how embodiment shapes cognitive functions like meaning-making (Lakoff and Johnson 1999). Ecological psychoacoustics (Neuhoff 2004) left the lab to examine hearing in real-world environments. Musicologists extended their frameworks to account for embodied models of meaning-making in *musique concrète* (Godøy 2006), *Elektronische Musik* (Chagas 2008) and computer music (Fyans & Gurevich 2011; Paine 2009; Worrall 2013). It is in the interest of sound design research to learn from the advances made in these fields by applying the theories and techniques that have been developed to address meaning-making through embodied cognition.

The meanings which we are concerned with in sound design are created at the intersection between multiple modes of perception and interaction. Embodied frameworks can explain meaning-making in multiple domains of vision and interaction, as well as sound, making them uniquely useful to sound design as the practice is often paired with some visual (e.g., film) or interactive (e.g., video games or AR/VR) element. Understanding how these different modes can be related is another, more pragmatic, imperative behind the adoption of frameworks from embodied cognition. At a more fundamental level, this placing of embodied 'presence', part of Todes' triad, at the center of sound design relationships also calls on us to consider again where potentially embodied correlates may lie within structures which we may previously have regarded as formal. Placing the body, whether as a conceptual framework, or as a physical entity with agency, at the center of the act of designing sound, should help us to map the terrain of meaningful sound design much more quickly and efficiently.

References

Aspray, William. 1990. *John von Neumann and the origins of modern computing*. Vol. 191. Cambridge, MA: MIT Press.

Augst, Bertrand. 1965. "Descartes's Compendium on Music." *Journal of the History of Ideas* 26, no. 1: 119–132.

Baecker, Ronald M. 2008. "TIMELINES Themes in the early history of HCI – some unanswered questions." *Interactions* 15, no. 2: 22–27.

Bannon, Liam J. 1995. "From human factors to human actors: The role of psychology and human-computer interaction studies in system design." In *Readings in human–computer interaction*, edited by Ronald M. Baecker, Jonathan Grudin, William A.S. Buxton and Saul Greenberg, pp. 205–214. San Francisco, CA: Morgan Kaufmann.

Bannon, Liam J., and Susanne Bødker. 1989. "Beyond the interface: Encountering artifacts in use." *DAIMI Report Series* 288.

Barbour, James Murray. 1953. *Tuning and Temperament: A Historical Survey*. East Lansing: Michigan State College Press.

Benson, Dave. 2006. *Music: A mathematical offering*. Cambridge, UK: Cambridge University Press.

Boulez, Pierre. 1954. "Le marteau sans maître" in *Music of the twentieth century: an anthology*, edited by Bryan Simms. New York: Schirmer, 1986.

Bridges, Brian. 2012. "Towards a Perceptually-grounded Theory of Microtonality: issues in sonority, scale construction and auditory perception and cognition." PhD Diss. National University of Ireland, Maynooth.

Bush, Vannevar. 1945. "As We May Think." *The Atlantic Monthly*. 176, no. 1: 101–108.

Card, Stuart K., Allen Newell and Thomas P. Moran. 1983. *The Psychology of Human-Computer Interaction*. Hillsdale, New Jersey: Lawrence Erlbaum Associates.

Carman, Taylor. 1999. "The Body in Husserl and Merleau-Ponty." *Philosophical topics* 27, no. 2: 205–226.

Chagas, Paulo C. 2008. "Composition in circular sound space: Migration 12-channel electronic music (1995–97)." *Organised Sound* 13, no. 3: 189–198.

Church, Alonzo. 1936. "A note on the Entscheidungsproblem." *J. Symb. Log.* 1, no. 1: 40–41.

Damasio, Antonio R. 1994. *Descartes' error: Emotion, reason, and the human brain.* New York: Putnam Publishing.

Darcy, Warren J. 1994. "The metaphysics of annihilation: Wagner, Schopenhauer, and the ending of the Ring." *Music Theory Spectrum* 16, no. 1: 1–40.

Descartes, René. (1641) 2013. *René Descartes: Meditations on first philosophy: With selections from the objections and replies*. Cambridge, UK: Cambridge University Press.

Doornbusch, Paul. 2017. "Early Computer Music Experiments in Australia and England." *Organised Sound* 22, no. 2: 297–307.

Dourish, Paul. 2004. *Where the action is: the foundations of embodied interaction*. Cambridge, MA: MIT Press.

Dreyfus, Hubert L. 1965. *Alchemy and artificial intelligence*. No. P-3244. The Rand Corporation, Santa Monica, CA.

Dunn, David. 1992. "A history of electronic music pioneers." *Eigenwelt der Apparatewelt: Pioneers of Electronic Art*. Linz, Austria: Ars Electronica.

Eimert, Herbert. 1957 "Musique électronique." *Lá Reveu Musicale,* no 236: 45–49.

Fechner, Gustav Theodor. 1860. *Elemente der psychophysik*. Vol. 2. Leipzig: Breitkopf u. Härtel.

Fodor, Jerry A. 1975. *The language of thought*. Vol. 5. Cambridge, Mass.: Harvard University Press.

Fyans, A. Cavan, and Michael Gurevich. 2011. "Perceptions of Skill in Performances with Acoustic and Electronic Instruments." In *NIME Proc.*, vol. 11: 495–498.

Gann, Kyle. 1998. "Breaking the Chain Letter: An Essay on Downtown Music." Accessed September 29, 2020. http://www. kylegann.com/downtown.html.

Gardner, Howard. 1985. *The mind's new science: A history of the cognitive revolution*. New York: Basic Books.

Gloag, Kenneth. 2012. *Postmodernism in music*. Cambridge, UK: Cambridge University Press,.

Godøy, Rolf Inge. 2006. "Gestural-Sonorous Objects: Embodied Extensions of Schaeffer's Conceptual Apparatus." *Organised sound* 11, no. 2: 149–157.

Grant, Morag Josephine. 2005. *Serial music, serial aesthetics: compositional theory in post-war Europe*. Vol. 16. Cambridge, UK: Cambridge University Press.

Harnad, Stevan. 1990. "The symbol grounding problem." *Physica D: Nonlinear Phenomena* 42, no. 1–3: 335–346.

Heidegger, Martin. 1927. *The Basic Problems of Phenomenology*, translated by Albert Hofstadter. Bloomington: Indiana University Press.

Heinaman, Robert. 1990. "Aristotle and the mind-body problem." *Phronesis* 35, no. 1–3: 83–102.

Hermann, T., A. Hunt and J. G. Neuhoff. 2011. "Chapter 1 Introduction." In *The Sonification Handbook*, edited by Thomas Hermann, Andy Hunt and John G. Neuhoff, 9–39. Berlin: Logos Verlag.

Hurtienne, Jörn. 2009. "Cognition in HCI: An ongoing story." *Human Technology: An Interdisciplinary Journal on Humans in ICT Environments*. Volume 5 (1): 12–28. https://humantechnology.jyu.fi/archive/vol-5/issue-1/hurtienne5_12-28.

Hurtienne, Jörn, and Luciënne Blessing. 2007. "Design for Intuitive Use-Testing image schema theory for user interface design." In *DS 42: Proceedings of ICED 2007, the 16th International Conference on Engineering Design, Paris, France, 28–31.07.2007*: 829–830.

Imaz, Manuel, and David Benyon. 2007. *Designing with blends: Conceptual foundations of human-computer interaction and software engineering.* Cambridge, MA: MIT Press.

Jorgensen, Larry M. 2012. "Descartes on music: Between the Ancients and the Aestheticians." *The British Journal of Aesthetics* 52, no. 4: 407–424.

Kane, Brian. 2007. "L'Objet Sonore Maintenant: Pierre Schaeffer, Sound Objects and the Phenomenological Reduction." *Organised Sound* 12, no. 1: 15–24.

Keislar, Douglas. 1987. "History and principles of microtonal keyboards." *Computer Music Journal* 11, no. 1: 18–28.

Kurth, Ernst. 1920. *Romantische Harmonik und ihre Krise in Wagner's "Tristan".* Berne: Haupt.

Lakoff, George, and Mark Johnson. 1980. *Metaphors we live by.* Chicago: University of Chicago Press.

Lakoff, George, and Mark Johnson. 1999. *Philosophy in the flesh: The embodied mind and its challenge to western thought.* New York: Basic Books.

Landy, Leigh. 2007. *Understanding the art of sound organization.* Cambridge, Mass.: MIT Press.

Lerdahl, Fred. 1992. "Cognitive constraints on compositional systems." *Contemporary Music Review* 6, no. 2.

Lukasik, Stephen. 2010. "Why the ARPANET was built." *IEEE Annals of the History of Computing* 33, no. 3: 4–21.

Mathews, Max V., and F. Richard Moore. 1970. "GROOVE—a program to compose, store, and edit functions of time." *Communications of the ACM* 13, no. 12: 715721.

McAdams, Stephen. 1989. "Psychological constraints on form-bearing dimensions in music." *Contemporary Music Review 4*, no. 1: 181–198.

McCulloch, Warren S., and Walter Pitts. 1943. "A logical calculus of the ideas immanent in nervous activity." *The bulletin of mathematical biophysics* 5, no. 4: 115–133.

Meyer, Leonard B. 1956. *Emotion and meaning in music.* Chicago: University of Chicago Press.

Meyer-Eppler, Werner. 1949. *Elektrische Klangerzeugung: Elektronische Musik und synthetische Sprache: mit 122 Abbildungen.* Bonn: Ferd. Dümmler.

Middlebrooks, John C., Ewan A. Macpherson and Zekiye A. Onsan. 2000. "Psychophysical customization of directional transfer functions for virtual sound localization." *The Journal of the Acoustical Society of America* 108, no. 6: 3088–3091.

Morabito, Carmela, and Mattia Della Rocca. 2010. "Epistemological models in psychoacoustics: a historical overview." *Proceedings of Fechner Day* 26: 327–332.

Neuhoff, John G. 2004. "Ecological psychoacoustics: Introduction and history." *Ecological psychoacoustics* 1: 13.

Newell, Allen, and Herbert A. Simon. (1975) 2007. "Computer science as empirical inquiry: Symbols and search." In *ACM Turing award lectures.* New York: Association for Computer Machinery. https://doi.org/10.1145/1283920.

Pace, Ian. 2009. "Notation, Time and the Performer's Relationship to the Score in Contemporary Music." In: Crispin, D. (ed.), *Unfolding Time.* (pp. 151–192). Leuven: Leuven University Press.

Paine, Garth. 2009. "Towards unified design guidelines for new interfaces for musical expression." *Organised Sound* 14, no. 2: 142–155.

Parncutt, Richard, and Graham Hair. 2018 "A psychocultural theory of musical interval: Bye bye Pythagoras." *Music Perception: An Interdisciplinary Journal*, 35, no. 4: 475–501.

Puckette, Miller. 1996. "Pure Data: another integrated computer music environment." *Proceedings of the second intercollege computer music concerts*: 37–41.

Putnam, Hilary. 1967. "Psychological predicates." *Art, mind, and religion 1*: 37–48.

Schaeffer, Pierre. 1966. *Traité des objets musicaux [Treatise on musical objects].* Paris: Editions du Seuil.

Schafer, R. Murray. 1993. *The soundscape: Our sonic environment and the tuning of the world.* New York: Simon and Schuster.

Schoenberg, Arnold. 1941. *Style and idea: Selected writings.* Berkeley: University of California Press.

Schopenhauer, Arthur. 1883. *The World as Will and Idea*, translated by R.B. Haldane and J. Kemp. London: Keegan Paul, Trench, Trübner & Co.

Searle, John R. 1980. "Minds, brains, and programs." *The Behavioural Sciences 3*: 417–457.

Sethares, William A. 2005. *Tuning, timbre, spectrum, scale*. London: Springer.

Shannon, Claude E. 1938. "A symbolic analysis of relay and switching circuits." *Electrical Engineering* 57, no. 12: 713–723.

Steels, Luc, and Rodney Brooks. 1995. *The artificial life route to artificial intelligence: Building embodied, situated agents*. London: Routledge.

Stein, Jack Madison. 1960. *Richard Wagner & the synthesis of the arts*. Detroit: Wayne State University Press.

Stevens, Stanley S. 1957. "On the psychophysical law." *Psychological review* 64, no. 3: 153.

Stevens, Stanley S. 1975. *Psychophysics: Introduction to its perceptual, neural and social prospects*. Abingdon, Oxon: Routledge.

Stockhausen, Karlheinz. 1956. "Gesang der Jünglinge". *Gesang der Jünglinge/Kontakte*. Deutsche Grammophon, 138 811 SLPM, 1962, LP.

Sutherland, Ivan E. 1964. "Sketchpad a man-machine graphical communication system." *Simulation* 2, no. 5: R-3.

Taruskin, R. 2008. *The Dangers of Music and Other Anti-Utopian Essays*. Berkeley: University of California Press.

Teruggi, Daniel. 2007. "Technology and musique concrète: the technical developments of the Groupe de Recherches Musicales and their implication in musical composition." *Organised Sound* 12, no. 3.

Todes, Samuel. 2001. *Body and world*. Cambridge, MA: MIT Press.

Turing, Alan Mathison. 1936. "On computable numbers, with an application to the Entscheidungsproblem." *J. of Math* 58, no. 345–363: 5.

Turing, Alan Mathison. 1950. "Computing Machinery and Intelligence." (O. Academic, Ed.) *Mind*, 59 (236), 433–460. https://doi. org/10.1093/mind.

Uttal, William R. 2004. *Dualism: The original sin of cognitivism*. London: Routledge.

Varela, Francisco J., Evan Thompson and Eleanor Rosch. 1991. *The embodied mind: Cognitive science and human experience*. Cambridge, MA: MIT press.

Vercoe, Barry. 1986. *The CSound Manual Version* 3. Cambridge, MA: MIT Media Lab.

Von Neumann, John. (1945) 1993. "First Draft of a Report on the EDVAC." *IEEE Annals of the History of Computing* 15, no. 4: 27–75.

Wagner, Richard. (1875) 1983. *My life*. Translated by Andrew Gray. Edited by Mary Whittall. Cambridge, UK: Cambridge University Press.

Wagner, Richard. 1879. "The Work and Mission of My Life. Part I." *The North American Review* 129, no. 273: 107–124.

Walker, Bruce N., and Gregory Kramer. 2004. "Ecological psychoacoustics and auditory displays: Hearing, grouping, and meaning making." In *Ecological psychoacoustics*, edited by John G. Neuhoff, 150–175. Cambridge MA: Elsevier Academic Press.

Walker, Bruce N., Michael A. Nees. 2011. "Theory of Sonification." In *The Sonification Handbook*, edited by Thomas Hermann, Andy Hunt and John G. Neuhoff, 9–39. Berlin: Logos Verlag.

Weber, Ernst Heinrich. 1846. *Zusätze zur Lehre vom Baue und den Verrichtungen der Geschlechtsorgane*. Vol. 8. Leipzig: Weidmann.

Wiener, Norbert. 1948. *Cybernetics or control and communication in the animal and the machine*. Cambridge, Mass.: MIT Press.

Wishart, Trevor. 1996. *On Sonic Art*, edited by Simon Emmerson. Netherlands: Harwood Academic Publishers.

Worrall, David. 2010. "Parameter mapping sonic articulation and the perceiving body." *Proceedings 16th International Conference on Auditory Display* (ICAD), Georgia Institute of Technology.

Worrall, David. 2013. "Understanding the need for micro-gestural inflections in parameter-mapping sonification." *Proceedings 19th International Conference on Auditory Display* (ICAD), Łódź University of Technology *6–10.07.2013*: 197–204.Zachary, G. P. 1999. *Endless frontier: Vannevar Bush, engineer of the American century*. Cambridge, MA: MIT Press.

3

Mapping Space and Time in the Soundtrack
Embodied Cognition and the Soundtrack's Spatiotemporal Contract

Adam Melvin and Brian Bridges

1 Mapping Space Within the Soundtrack

> "Movement is the unfolding relationship between space and time as *event.*"
>
> (Ward 2015, 159)

Cinema is often regarded as a medium whose structural language is, at its most fundamental, concerned with the relationship between space and time. As Ward argues, "what makes cinema *cinematic* is the expressive manipulation of perceived space and time" ((ibid., 159), 166). Where audio is concerned, it is the relationship with space, rather than the more straightforward parameter of time, that has recently attracted scrutiny from practitioners and scholars alike. Developing practices prompted by advances in multi-channel sound technology have provoked several writers to reconsider how we interrogate the role of sound in articulating and shaping *space* (in its various cinematic connotations), resulting in a number of important points of reference, e.g., Chion's (1994, 150–152) concept of the *superfield* (Dolby Stereo), its successor, Kerins' (2011) *ultrafield* (encompassing higher channel count multichannel audio) and broader theoretical frameworks such as Brophy's concept of *architecsonics* (1990). In this context it is important to remember that cinema, particularly with regard to contemporary practices, is not simply the relationship between image and sound, but between image and the soundtrack as an integrated material entity (see Greene and Kulezic-Wilson 2016; Mera 2016b).

DOI: 10.4324/9780429356360-4

1.1 Space and Spatial Relationships in the Soundtrack

Despite the theoretical viewpoints cited above, the parameter of space can prove somewhat problematic for those seeking to establish analytical frameworks addressing sound. One particular problem highlighted by Kerins (2011), in hypothesizing his *ultrafield* concept, is that many of the more established analytical models upon which our understanding of the spatiotemporal are based do not align particularly well with the *surround sound* era, on account of their tendency to assume sound-image interactions in terms of single specific relationships. As Kerins (ibid., 238) notes: "In reality, the soundtrack is plural, including multiple distinct channels ... which may include several sounds, each relating to the image in a different way." As relevant as Kerins' (ibid.) argument is, particularly if one considers the evolution of object-based approaches with high track/channel counts, such as Dolby Atmos, it is, nonetheless, indicative of a broader, perhaps inevitable, tendency amongst writers to place the location of sound – essentially Chion's *point of audition* (1994, 90–92) and the resultant three-dimensional sonic environment – at the forefront of discussions concerning spatialization, as opposed to the treatment of space as more broadly relational, and evoked through a variety of means. The potential for gesture, and in particular timbre, to evoke and shape cinematic space is often overlooked. As Smith (2013, 336) notes, instances of the type of rich spatial experience that Kerins' approach addresses occur relatively infrequently in contemporary cinema and are often confined to set pieces (a point that gains further traction when one considers that today, we often experience cinema as much out of the movie theatre as we do within it). What is proposed in this chapter, therefore, is a broader consideration of the soundtrack's plurality to include the means by which gesture, and in particular timbre, can articulate and shape cinematic space in terms of causal and associative relationships. The model we will propose considers the blending of multimodal materials as the key to integrating metaphors, based on the dimensions and features of the contemporary soundtrack's functionality, which Ward (2015, 161) terms "a playful recombination of auditory and visual fragments and a heightened manipulation of auditory spatialization, temporal resolution, and timbre."

1.2 Spatial Models of Music and Timbre

A number of established models of musical structure are notable for their spatial implications. Brower (2000) sees Western music's tonal structures as based upon embodied *image schemata* (Johnson 1987; Lakoff and Johnson 1999), with tonality becoming a structure based on *center –periphery*, *verticality* and goal-directed movement. Indeed, even more traditionally formalist theories (Lerdahl 2001) also embody spatial associations and dynamics (e.g., Lerdahl's modelling of tonal attraction based on a gravitation-style inverse-square law). This chimes with the theories of proponents of embodied models of musical structure, such as Johnson (2008), who proposes that our conceptualization of music is based on metaphors such as a *music-as-moving-force* equivalence (see Melvin and Bridges 2020). More broadly, theories such as the above can be seen as being encompassed by the interdisciplinary field of *embodied cognition*, whereby conceptual structures are modelled using aspects of the body and/or the environment (Varela et al. 1991; Lakoff and Johnson 1999; Shapiro 2011).

Indeed, when discussed in the context of the soundtrack, the spatial associations of musical structures are often invoked through discussions of contour, register and more specific tonal relationships; it could even be stated that descriptions implicitly indebted to embodied discourses are fairly common in film scoring (see Chattah 2015). However, timbre's spatial implications are another matter. Whilst some phenomenologically-based descriptions (Schaeffer 1966; Erickson 1975; Smalley 1997; Demers 2010; Godøy 2006) frequently highlight how certain broad types of timbre help to articulate or delineate a sense of spatial perspective, within cinema the idea that the percept of timbre may itself constitute a space which may act as a counterpart to visual or conceptual spaces has tended to be more implicit than explicit. Perhaps this failure to account for timbre as a spatial, as opposed to merely textural, entity can be related to a consideration of the multidimensional complexity of timbre as either (a) difficult to resolve or reduce, and/or (b) quite distinct from other aspects of music's spatiality. There is, perhaps, an irony that musical timbre, which is expressly physical and spatial in its origin (through the sizes and shapes of vibrating bodies) may be less frequently discussed in spatial terms than the more abstract world of pitch relationships.

Work which addresses timbre through models which are not only spatial, but also based on embodied relationships, has the potential to influence new, more holistic, theories of cinematic space. Where Grey and Gordon (Grey 1977; Grey and Gordon 1978) proposed a three-dimensional timbral space – with dimensions correlating with acoustic features including the presence or absence of *attack transients*, synchronicity or divergence of partials' amplitude envelopes, and spectral 'center' (*spectral centroid*) as a vertical axis – these early studies did not have a particular focus on the physicality (or its implications) of the sounding gesture. It could even be argued that such early studies imply a somewhat decontextualized, formalist stance; they addressed the cognition of timbre much as they would any other information processing task, treating the particular domain as unimportant with regard to the structure which would be uncovered by statistical approaches such as multidimensional scaling, whereby a complex multi-parametric data set would be collated as points within an abstract, Cartesian space. However, this would be to miss the embodied implications of these findings. The space which Grey and Gordon (1978) obtained for their timbral studies correlates with acoustic dimensions which we frequently recognize as perceptually important. We argue here that later studies (McAdams, Chaigne and Roussarie 2004; Giordano and McAdams 2006) reinforce the *embodied and physical implications* of the original Grey and Gordon timbre–space models, particularly regarding their focus on the attack transient and, hence, the distinctive materiality of acoustic systems being driven into action.

Theories of timbre and timbral relationships derived from electroacoustic music also focus on embodied perspectives, from Godøy's (2006) reading of Schaeffer's (1966) *sonorous objects* as gestural and embodied, to Smalley's (1997) *spectromorphology*, whose timbral-gestural relationships are advanced via moving force metaphors which are compatible with Johnson's (2008) embodied *qualitative dimensions of movement* (Graham and Bridges 2014). Indeed, the connection between Smalley (1997) and Johnson's (2008) conceptions of movement, and with Lakoff and Johnson's embodied image schema theory, allows for an embodied perspective on Grey and Gordon's (1978) timbre-space (Graham et al. 2017; Roddy and Bridges 2018), with dimensions of *verticality* (spectral centroid), *linearity* (Johnson 2008) or *dynamism versus inertia* (synchrony of profile of partials) across the X axis, and *presence/diffusion* (presence or absence of attack transients) across the Z axis; see figure 1, below. A conception of timbre on the basis of height, dynamism and presence is clearly one which has potential for integrating with narrative and affective design within the soundtrack, especially when paired with sound choices which are evocative in their materiality.

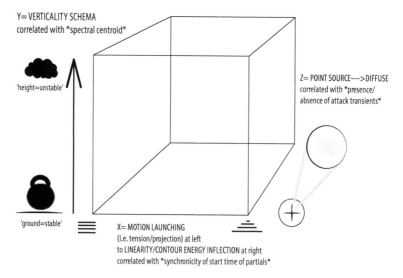

Figure 3.1 Applying embodied-cognitive rubrics to the classical three-dimensional timbre-space model of Grey (1977) and Grey and Gordon (1978).

Graham et al. (2017); Roddy and Bridges (2018); image © B. Bridges 2017.

1.3 Acoustic and Perceptual Space; Presence and Embodiment

Moving on from the cognitive, conceptual and figurative spaces of sound texture, the acoustic spaces of the soundtrack also take into account an even more direct spatiality: that of rendering or suggesting the acoustic space of the sound field. The stereo/early surround *superfield* (Chion 1994, 150–151) and more extensively multichannel *ultrafield* (Kerins 2011) allow us to situate cinematic action and affective, figurative presentation within a field significantly more expansive than the (generally) two-dimensional screen and its comparatively constrained point of view. Here, the performative, affective, aesthetic and environmental dimensions of the soundtrack may be integrated and diffused across a wider field of acoustic territories, both acoustically 'real' (in this context, obtained from location recording, Foley, and soundstage dialogue recording) and imagined (through sound design, mixing, editing and composition). Even as a wider range of spatial audio technologies have become more commonplace within cinema, electroacoustic music provides an example of a field in which the aesthetic exploration of sonic space is already well developed. This has led to the development of significant commentary on the perceptual framing of acoustic territories, with two prominent models being Emmerson's (2007) *local/field space frame* typology and Smalley's (2007) *space form*. The former is a descriptor of sonic focus and salience via concentric circles, differentiating *local* (localized, foregrounded) events from an accompanying *field* (ambient backdrop). Smalley's *space form* has a similar concern for the shapes and implications of acoustic territories and their perceptual and aesthetic implications, treating the grouping of sounding gestures, auditory ecologies, mediatized sound and spectral space as integral to the structuring of a listener's unifying perceptual perspective.

A comparable cinematic model for considering sonic space is found in Brophy's (1990) dual concepts of *architecsonics* and *psycho-diegetics*. Primarily concerned with causality, *psycho-diegetics* addresses "the psychological comprehension of dimensionality in film" via "the reconciliation of perceptual modes with representational modes … based on the formal relationship between established and accepted time/space factors" (ibid., 95). Essentially then, it is how the depiction of the fictional realms onscreen relate to our own physical (we might add 'embodied') experience both of the real world as well as that of the audience/audiovisual experience within the film theatre itself. We detect similarities between this idea and Smalley's (1997) concept of "gestural surrogacy" relating and integrating sound structures, and a broader dynamic of "apparent causality" (ibid.) within their relationships. *Architecsonics*, meanwhile, provides a more holistic framework for considering "the spaces constructed for sounds and musics in the cinematic text, and the consequent ways in which sounds and musics are employed in the construction of narrative" (ibid., 97), relevant to the *ultrafield* era. It can be understood in terms of "production of space" (Lefebvre 1991) as a framework for scrutinizing the creation and crossing of spaces, similar to the broadly ecological model of spatial territories as structural entities in Smalley's (2007) *space form*, in this instance via the cinematic narrative made temporal. In addition, Mark Ward's explanation of his own approach to the sound design of *In the Cut* (dir. Campion 2003) provides a useful practice-based point of reference which, in many ways, reflects Brophy's concept. Ward's partitioning of diegetic space into audio/spatial zones both vertically (from sub-terranean to high above street-level) and horizontally (according to zones of proximity to the lead protagonist, including her own thought space) results in a framework of "acoustic container(s) designed to induct a specific feeling-state in the body of the audience" (Ward 2015, 179) that resonates with the concepts regarding space form outlined above (particularly, the *local/field* structure Emmerson (2007) outlines). Taken together with the tonal and timbral models discussed above, such accounts of spatiality of sound within an aesthetic domain may provide means by which the soundtrack constructs not just an audiovisual, but a spatiotemporal, and embodied, contract within its varied relationships. Elaborating upon the embodied timbre-space model above and relating it to Ward's audio/spatial zones (swapping his horizontal for our depth dimension), we obtain the following points of comparison (Table 3.1):

Table 3.1 Comparing timbre-space (Grey 1977; Grey and Gordon, 1978), embodied timbre-space (Graham et al. 2017; Roddy and Bridges 2018) and Ward's embodied-spatial model of sound design (2015).

Dimension	Acoustic timbre-space; after Grey and Gordon (1978)	Embodied timbre-space (Graham et al. 2017; Roddy and Bridges 2018)	Audio/spatial zones; after (Ward 2015)	Commentary
X/horizontal	'Envelope': synchronicity of partials	Tension (fast attack/ synchronous partial entry) versus inertia	Added temporal dimension within *or between* timbral events: synchronous or near-synchronous events with little change in properties to slowly-changing or slowly-following events.	See also Bregman's (1993) auditory scene analysis and environmental regularities; 'streams of sound [i.e., sounds from related/relatable sources] change their properties relatively slowly'

Y/vertical	'Brightness': frequency height/ spectral centroid	Pitch/frequency height (verticality schema); ground/ weight=stable versus height/ lightness=unstable	Physical or auditory verticality (vertical position of source or vertical metaphor related to source)	A straightforward mapping 'tracking' perspectival position with pitch/frequency height to reinforce vertical metaphor or position
Z/depth	Presence or absence of attack transients	Presence (salience) to diffusion	Proximity to (or distance from) protagonist	'Focusing': closer microphone technique, on-axis microphone technique or drier acoustic; more audible transient detail 'Depth/diffusion/ immersion': distant microphone placement, reverberation, mixing/blending other sounds, less transient detail

In such a combined model, the verticality (pitch/frequency height), synchronicity and presence/proximity of cinematic sound come into focus as key elements of (to borrow Brophy's term) cinema's *dimensionality* (1990, 95). We will now consider how these relationships, particularly vertical and depth/immersion dimensions, articulate cinematic space.

1.4 Case Study 1: Mapping Space with Timbre and Pitch in Chernobyl (2019), Blade Runner 2049 (2017) and Arrival (2016)

The HBO miniseries, *Chernobyl* (dir. Renck 2019) provides a useful example with which to consolidate our discussions of spatial models of timbre and music within the soundtrack. The single most prominent feature of the series' soundtrack is a distinctive timbral gesture, a steady, pulsing, low-register, non-diegetic throb, heard just a few moments into the first episode where we witness Legasov, Director of the Institute of Atomic Energy, recording and hiding cassette tapes of his testimony concerning the Chernobyl disaster. The embodied associations of pulse, energy and even pain, mean that we immediately connect the presence of the throb with the negative impact of the disaster and perhaps even the radiation emitted from the reactor explosion itself. In contrast to more common sonic tropes associated with radioactivity – the most obvious perhaps being the clicking of a Geiger counter – the throb's grinding timbre evokes a sense of weight/density, force/resistance that affects the presence of a substantial, quasi-physical entity more representative of volatile, once inert elemental material, made radioactive (perhaps even the material structure of the reactor) rather than the radiation itself. This is heightened by the fact that the material of the sonic gesture is somewhat unstable; although remaining temporally steady and generally low in register, it changes pitch, length and trajectory (sliding up and down), as well as timbre, each time it is sounded.

The presence of the throb in *Chernobyl* resonates with Mera's (2016b, 93–95) definition of immersion or spatial presence: "a cognitive and experiential process … that can be enriched but does not entirely depend on external mediatized information" (ibid., 94). Similarly, the throb constitutes a largely timbral gesture, rather than one necessarily governed by spatial diffusion (*Chernobyl* is a TV production, and so utilizes stereo) the presence of which reinforces both an important narrative theme of the series as well as its overall mood primarily through its haptic quality. Ward (2015, 164) terms this manner of approach, perceptual design: "(the induction of) core affect, mood and feeling-states within the body of the audience through… pre-attentional, nonconscious mechanisms…" as opposed to "attentional, conscious mechanisms" (narrative design). In the absence of any visual confirmation of the radioactive fallout itself – we only witness its effects – the throb functions rather like the spectral (in the ghostly sense) presence that Donnelly (2005) attributes to film scores as a whole, utilizing embodied affect to ominously intrude on the diegetic world of the characters at various junctures. Furthermore, its timbre reconciles the representational and perceptual (Brophy 1990, 95) via the type of time-space models outlined above. The muffled, cavernous quality of the throb's first occurrence evokes a sense of the subterranean, mapping both a physical (perceived) and conceptual zone of unsettling volatility (the 'ground' made unstable) that expands the diegetic space onscreen (several of the series' key scenes take place underground beneath the reactor space). It is worth noting, too, that as the throb reoccurs, its timbre evolves. For example, in a later episode, the throb is heard as we see slowly panned, ground-level shots of the empty Chernobyl landscape (with abandoned vehicles, pylons, etc.). Here, its timbre is altogether more 'physically present', featuring metallic, brushed sounds and higher-frequency sonorities (fluttering gestures, bowed cymbals) mixed within a more 'naturally' reverberant space that seems more closely aligned to the diegetic environment depicted on screen. Spatially, image and sound combine to forge a similar dimensional foundation that articulates a more panoramic, horizontal dimensional space than the arguably more vertical spatial construct of the opening scenes.

Further instances of implied or *mapped* spatiality can be found in a number of recent cinematic examples. Denis Villeneuve's *Arrival* (2016) and *Blade Runner 2049*

(2017), for example, both contain sequences where timbre plays an important role in evoking architectural or physical space and, in doing so, an embodied perspective within that (mapped) space that spotlights that of the lead character(s). For instance, in one pivotal scene in *Blade Runner 2049*, we see the character of K (Ryan Gosling) return to the spot where he remembers hiding a small toy horse as a child. As we observe K traversing the lower basement area of a tall, open industrial space, a descending Shepard-tone-like drone is heard, giving the effect of sonically converging on the point where the horse is hidden (a furnace low to the ground). As K steadily walks, the high pitch of the Shepard tones serves to initially retain the height of the architectural setting (off screen) before gradually compressing the apparent height of the audiovisual space until sustained whistling tones announce that we've arrived at our destination. By comparison, the scene in *Arrival* where Banks (Amy Adams) and Donnelly (Jeremy Renner) first enter the shell-like alien craft is accompanied by layers of sustained, drone-like, bowed, rumbling and even quasi-vocal (humming) sounds that sonically capture the immensity of the alien ship. These eventually give way to an isolated low-pitched klaxon gesture that ascends (sliding) up a tone before falling by step to its starting pitch. At first the gesture is almost chant-like in quality but then becomes more prominent, timbrally rich and chordal as the crew near the aliens themselves. As Kulezic-Wilson (2020, 42–43) notes, the soundtrack's emphasis on textural and frequential density here results in an overtly visceral experience; it "flood(s) the audiovisual space." Yet this has further spatial connotations. The sound's distinctive resonance (as if produced from a large object or a formant-shifting audio process) provides an embodied audio reference point for the ship's length; although not obviously diegetic, it is almost as if the sound is being physically sustained by the ship's long, tube-like interior, itself reminiscent of an aerophone such as an alpine horn. The more resounding chordal guise of the klaxon sound also affirms the craft's structure albeit in a different way. Here, the gesture's, density, richness and volume seem immediately at odds with the enclosed, low ceiling of the flat letterbox space within which the characters stand; it is a reminder of a bigger presence both in the physical and dramatic sense.

2 Mapping and Metaphors: Conceptual Blending, Spatiotemporal Relationships and Materiality

2.1 Conceptual Blending, Spatiotemporality and the Soundtrack

Aside from the application of embodied cognition's models of logical structures as spatial relationships (Lakoff and Johnson's (1999) image schema theory), musical and sonic relationships may also be treated via a broader model, *conceptual blending* (Fauconnier and Turner 1998). In this theory, metaphors are treated as spaces and spatial relationships (similar to that of image schema theory), however more consideration is taken of their combination; see figure 3.2, below. The intersection of different *cognitive spaces* is taken to describe more complex ideas with emergent properties derived from the congruence or divergence of *input* spaces. We argue here that the application of this theory may hold one of the keys to an explanation of the spatial soundtrack away from a simple discussion of the technologies of sound spatialization and individual elements within the soundtrack towards an integrated account addressing the combination of timbre, space and more traditional structural elements such as pitch contour, rhythm and even harmony.

The combination of different elements across the musical (timbre, pitch, rhythm, dynamic), and the visual (cinematography, set/landscape, performance) may contribute to conceptual blending. Indeed, even within the solely auditory domain, we have an extensive range of elements which are either directly spatial (via spatial/surround mixing) or can be treated as spatial (as discussed earlier in relation to timbre, and as covered in embodied models of musical pitch, rhythm and harmony). Moreover, the spatial metaphors associated with these sonic attributes also embody an associated temporality. The integration of these two domains via embodied metaphors of physicality can be found in Smalley (1997), Brower (2000) and Johnson's (2008)

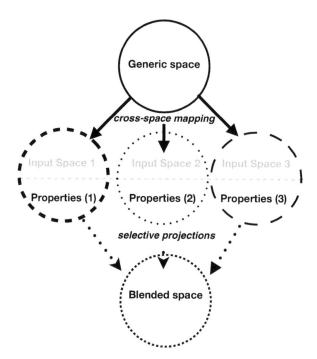

Figure 3.2 Conceptual blending as mapping ('selective projection') from various component spaces to composite ('blended') space, after Fauconnier and Turner (1998).

Table 3.2 Comparison of Johnson's (2008) *qualitative dimensions of movement* with Smalley's (1997) *energy–motion profiles* and embodied associations, after Graham and Bridges (2014).

Johnson	Embodied Association	Smalley
Tension	Rate–effort=>overcoming inertia	Motion rootedness
Projection	Sudden rate-change/transient movement	Motion launching
Linearity	Coherence of path	Contour energy/inflection

music-as-moving-force metaphor. Johnson's (ibid.) proposal for an embodied theory of aesthetics elaborates upon image schema theory via a series of conceptual blends, with musical motion as based on moving objects and musical syntaxes as based on *force metaphors* affecting these virtual objects. Furthermore, the aesthetic experience is seen as being strongly influenced by the dynamism and coherence of the path traced by these movements; see Johnson's "qualitative dimensions of movement" (ibid.), including *tension*, *projection* and *linearity*. Graham and Bridges (2014; 2015) and Roddy and Bridges (2016) have discussed parallels between this theory and that of Smalley (1997), which sees electroacoustic music as based upon a language of moving sound masses, whose position in frequency, amplitude and temporal space is influenced by similar ideas of dynamism and inertia (Table 3.2, below).

As may be apparent, these concepts, when applied to music, could be thought of as conceptual blends whose influence is exerted in multiple directions and domains (see Figure 3.3).

Sounds (in Smalley's theory) and musical pitches and rhythms (in Johnson's) are treated as moving objects, whose perceptual framing is at least partly derived from their apparent similarities and distinctions of their physical attributes (coherence of path, sudden instantiation of force and the degree of impeding force, i.e., inertia). A corollary of this embodied conceptualization of perceptual – and aesthetic – structure is that it is an integrating one, drawing potentially disparate elements into strong relationships.

2.2 Case Study 2: Mapping and Framing Space and Time in Blade Runner 2049 (dir. Villeneuve, 2017) and Arrival (dir. Villeneuve, 2016)

If this spatiotemporal conceptual blending can be seen as happening within the purely auditory domain, we argue that it also has potential utility within the broader cinematic text. As Ward notes, cinema can be defined as an "affective spatiotemporal system," the most fundamental component part of which is a "burst of presence" (2015, 166). When sequenced together these "bursts" form what Ward terms *proto-narrative*, "an affective linking of the body's experience of mediated space through

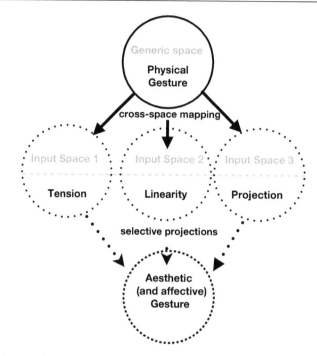

Figure 3.3 'Qualitative dimensions of movement' (Johnson, 2008), considered as conceptual blending (Fauconnier and Turner, 1998).

time" that underpins a film's narrative proper (ibid., 166–167). It is during particularly overt affectively-driven bursts where the manipulation of subjective time may occur, something Ward identifies within his own aforementioned sound design for *In the Cut* as moments of "cognitive overflow" or "leakage" from the otherwise contained mapped spatial fields noted above (ibid.), constituting a third schema of sorts. However, what is interesting about the scenes that Ward selects to illustrate these instances of "leakage" or "non-containment" (ibid., 181) is that it is not just time that is warped, but arguably, space as well. For example, an auditory and spatiotemporal "dematerialisation" (ibid.) of an apartment space occurs on two prominent occasions: the gentle removal of the muffled yet prominent music from a ground floor strip club when the protagonist, Frannie (Meg Ryan) discovers her half-sister murdered in the apartment above and the sudden cessation of a diegetic buzzing (presumably from some household appliance) when Frannie sense an intruder in her own home. Sonically, the established spatial frame of each apartment evaporates; the dilation of time coincides with the deconstruction of space. We might conclude therefore that if time is skewed, then so too is space, and vice versa.

A similar, more pronounced spatiotemporal integration occurs in the Villeneuve examples, in that sound functions not just to evoke and delineate space, as previously discussed, but also to shape movement *within* space and, by consequence, the temporal trajectory of each sequence (interestingly, both films explore concepts of non-linear time in their respective narratives). The continuous motion of the descending Shepard tones in *Blade Runner 2049* has the effect of smoothing the stops and starts that K makes as he walks, thereby steadying the temporal flow of the sequence and emphasizing the scene's trajectorial progression towards the diegetic 'goal' of the furnace. In *Arrival*, the floating glissando ascent and return to point-of-rest of the klaxon sound articulates a gestural shape that actually preempts and mimics the crew's physical movements during their gravity-defying change of orientation from vertical to horizontal while inside of the ship (they physically float upwards before gently coming to rest on the side-now-floor of the ship's interior), broadly reflecting the sudden plasticity and skewed sense of perception of the physical world within the diegesis. The placement of sonic entries meanwhile evenly punctuates the progression of the crew as they approach the screen, providing a temporal regularity that is at odds with the external diegetic world. Thus, to adapt Kerins' term, a *plurality* of sound-image and spatiotemporal interactions within the sequence's material form a *proto-narrative* (Ward) that serves to partition and isolate the dominant spatial structure of the sequence – the spacecraft – in multiple ways; Ward's (2015, 166) "bursts of presence" are reframed as "bursts" of agency and motion, integrated across perceptual, and conceptual, domains.

2.3 Material Metaphors and the Haptic

In the previous section, we discussed various ways in which perceptual dimensions of timbre could be related to embodied concepts articulated in space and developed over time. As we conceive it, different locations in timbre space may be joined by trajectories (and their attributes) which are at least partly defined by Johnson's (2008) dimensions. The embodied dimensions of the timbre-space model (*height* as perceptual *brightness*, *dynamism/inertia* and *presence/diffusion*) can be seen as connected via conditions based on the above (the tension of movement in this space generates a rapid, dynamic transition between ends of these continua, i.e., a projection, whether it is an abrupt transition between low/high frequency, a dynamic and rapid envelope articulation or a particularly salient attack transient). The degree of linearity (coherence) of a sonic object tracing various positions within this space over time may help determine the "apparent causality" (Smalley 1997) or relatedness of sonic objects, i.e. sounding events. These dimensions could be seen as embodied forces which contribute to our sense of relatedness of different timbral gestures, framing spatiotemporal relationships based on their predictability. They could be conceptualized in the manner of the focusing of a lens, providing a sense of containers and boundaries, as opposed to an undifferentiated continuum.

We have concerned ourselves thus far primarily with how relationships are conceptualized through motion, but there are other important aspects of embodied experience to consider. Adlington (2003) has explored music's relationship with time and materiality. Moving beyond the metaphor of musical motion, Adlington explores ideas of music's structural unfolding as material metaphors and, in particular, how these virtual materials change state in response to musical processes. He advances this view with particular reference to contemporary (post-tonal) musical examples, and, in a parallel with the embodied timbre-space model discussed above, asserts that in some (texturally-based) contemporary music, "a lack of sharp outlines works against a conceptualization in terms of discrete 'objects,' thus making metaphors concerning weight, solidity, or relative location, less appropriate" (Adlington 2003, 316). Adlington (ibid.) here proposes an alternative: that music may be understood "in terms of a variety of physical metaphors," and that motion should not be preeminent but "intermingles with metaphors of heat, light, weight, tension, and so on" (ibid., 318). Restating this proposition, musical structures may be understood in relation to physical metaphors inherited from our embodied exploration of the environment, including not just motion but also attributes of objects: weight, tension, heat and light. This implies that the interaction of these different metaphors is important; they frame our experience through their integration of information and ideas from a number of source domains.

Cases in which a wider range of source metaphors within the musical materials are to be found may be particularly productive when considering the haptic qualities of many contemporary soundtracks (Ward 2015, 167; Connor 2013; Coulthard 2013; Mera 2016a): the tendency for the score and/or sound to prompt the audience to "simultaneously feel as much as we hear" (Mera 2016a, 171). For example, Adlington discusses Ligeti's *Atmosphères* as a work in which directionality and goal-directed movement are superseded by an association with density and light. This piece, with its metaphor of materiality, evoking other-worldly environments, of course, famously features as a key component of a soundtrack, that of *2001: A Space Odyssey* (dir. Kubrick 1968). It is perhaps no accident that as the soundtrack has developed as an artform (in parallel with the development and accessibility of various sound design technologies), a wider range of sonic models and metaphors, ripe for a range of conceptual blending relationships, can be seen to be deployed.

2.4 Case Study 3 and Model: Material Metaphors and Conceptual Blending in Chernobyl (2019)

Returning to *Chernobyl* (2019), we find a number of instances where materiality, timbre, affect and space provide fruitful domains for conceptual blending, spearheaded by the reprising of the aforementioned throb and its embodied associations. In one of the series' most striking sequences, that of the initial aftermath of the explosion (episode 1), the throb is echoed in a number of diegetic sonic gestures that are heard in succession: the sirens of the fire engines that arrive at the site of the disaster have a similar pulsed trajectory, as do the alarms within the interior of the power station itself. This reaffirms the throb's status as a warning signal whilst helping to bind the sequence sonically. As we observe the firemen about to scale the rubble to access the roof of the reactor (an ultimately fatal endeavor) the throb re-emerges in its original guise, growing in volume and instability. Here, the combination of ground-level camera position, aligning with the perspective of the emergency workers, and the sound's aforementioned embodied associations, evoke a sense of *terra firma* that contrasts with the plume of glowing matter we see rising from the exploded reactor core (the radiation itself is of course airborne rather than at ground level). In addition, the throb prompts multiple associations with the radioactive fallout that is confirmed by other sensory indicators within the diegesis, e.g., the fragments of radioactive graphite that the firefighters' (and our) attention is drawn to and the taste of metal they mention. Meanwhile, high-frequency, sustained, fluttering timbres and reverb seem to attach themselves to the clouds of water vapor and dust on screen (they are not unlike whistling steam), evoking an enclosed audio space that is at odds with the open-air setting of the exterior of the power station. Although we do not see the roof, the images and higher sonic frequencies do articulate an embodied sense of space; the throb provides the foundation for a vertical schema (one that is larger than the space we witness on screen) that ultimately oppresses the audiovisual space and those within it. To summarize, the mapping of space and the causal relationships created by the explosion are achieved through a combination of more straightforward frequency/time gestures and *haptic* audio materials tracing timbral-spatial connections, enriching the resultant conceptual blending through the embodied associations of their (almost tangible) auditory materiality (see Figure 3.4 and Table 3.3).

In examples such as this, it is not just the fact that we experience a *material score* that is significant, but the material itself, its evolution and place within the broader spatiotemporal frame of the various instances it is heard within the series. In tracing the projections from the physical domain, via the perceptual, to the cognitive, we see how physical attributes are mapped to conceptual associations that have similar spatial structures, importing aspects of these structures into the new, metaphorical domain. Thus, size, weight and density are equated with significance. Diffusion and dispersion, or their opposite, a clear location, are obviously associated with presence and agency.

Such associations, when presented here in diagrammatic or tabulated from, appear obvious; indeed, they have the potential to be overly didactic. What is arguably crucial in a successful creative use of a conceptual blend is the emergent complexity of the combination, with the tension between the various spaces giving rise to interesting associations and features within their projection. On the input side,

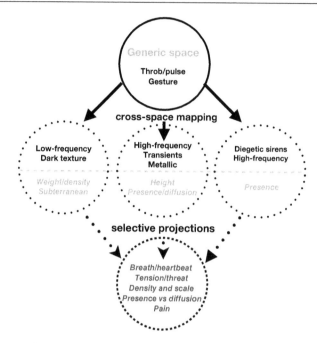

Figure 3.4 Conceptual blending of spatial and material metaphors and timbre in *Chernobyl* (dir. Renck 2019).

Table 3.3 Mapping spatiality to conceptual metaphors, blends and affective qualities in *Chernobyl* (dir. Renck 2019).

PHYSICAL SPACE	CONCEPTUAL AND AFFECTIVE 'SPACE'
Size and location	Perceived magnitude of object/process Subterranean or airborne location; significance
Materiality, surface texture, density, weight	Associations of tangible and observable properties; significance
Clear location versus diffusion/ dispersion	Presence and agency versus absence/dispersion/ inaction
Embodied presence/agency	Proximity and distance; stability versus instability; safety versus threat

a rich range of sound design processes and timbral gestures can be explored whilst still maintaining some clear relationships to an originating gesture, as can be seen with the frequently transmuting throb. Furthermore, the rich range of potential associations can be exploited through the combination of similar sound design figures with new materials in the visual domain. It follows from this that the cinematic space forms a spatiotemporal contract integrating the visual and auditory fields within a conceptual space which blends some of their component dimensions and associations in a way that serves narrative, aesthetic and, bringing the two together, affective expression. We believe that such an approach, modelling sound design relationships through timbral and diegetic dimensions and conceptual associations, may have broad utility in mapping and integrating the diverse materials and relationships within the contemporary soundtrack.

3 Conclusions: Transducing and Transmuting Cinematic Relationships

We began our discussion with the proposition that cinema is the structuring of movement (recorded, encoded and/or edited; auditory and/or visual) across space and time. At key moments, these domains align to produce some of the medium's most striking schemas; metaphor-laden aggregates of audio, music and image that embody concepts, define spaces and provide affective context. Taking as our points of departure the ideas that cinematic space may be evoked as much through timbre as through auditory spatialization techniques, we have explored relationships between acoustic, perceptual and conceptual spaces. In particular, we have sought to integrate more general concepts of sound's spatiality from electroacoustic music theory and music cognition with those from soundtrack studies, leading to the development of a framework which allows us to treat the production of cinematic space as operating across multiple domains whilst nonetheless cohering into broadly consistent metaphors.

We argue here that cinema's broad multi-modal embodiment helps us transduce and integrate a diversity of gestures and materials, with the material turn in many contemporary soundtracks immersing us within a network of plural relationships from which we draw our sense of the work's structure and conceptual metaphors. In taking this approach, we have sought to find ways in which this embodied definition of space intersects with temporal dimensions, giving rise to senses of presence, causality, agency and the reinforcement of viewers' involvement. Whilst spatial audio technologies may have prompted our field to consider the soundtrack's spatiality, in the era of multiple viewing platforms – from the 'ideal' experience of the cinema, with its expansive scale and immersivity, to the personal sphere of the smartphone – it is important that certain key aspects of cinematic relationships be comparable. Where spatiality is as much a part of the cinematic experience as temporality, it is important that it be reiterated across multiple modes.

The structures resulting from the haptic and spatial qualities of sound design help to connect and blend images with affective and conceptual relationships, mapping and focusing our attention from enveloped immersion through to the eruption of presence and erosion of distance. We contend that the integration of auditory, audiovisual and broader conceptual relationships through these unifying gestures is at the heart of cinema's audiovisual contract, now reframed as an embodied, spatiotemporal contract. We propose the model discussed above, adapting Fauconnier and Turner's (1998) conceptual blending, as a framework for investigating the various connections within and between the spatiality and haptic sense engendered by the soundtrack and its broader audiovisual context. We believe that such connections will become ever more important as cinematic media increasingly addresses the challenges of multiple platforms and spatialized and immersive tendencies.

References

Adlington, Robert. 2003. "Moving beyond motion: Metaphors for changing sound." *Journal of the Royal Musical Association* 128, no. 2: 297–318.

Bregman, Albert S. 1993. "Auditory scene analysis: Listening in Complex Environments." In *Thinking in Sound: the cognitive psychology of human audition*, edited by Stephen McAdams and Emmanuel Bigand, 10–36. Oxford University Press.

Brophy, Philip. 1990. "The Architecsonic Object: Stereo Sound, Cinema and *Colors*." In *Culture, Technology and Creativity in the Late 20th Century*, edited by Philip Hayward, 91–110. London: John Libby Press.

Brower, Candace. 2000. "A cognitive theory of musical meaning." *Journal of Music Theory* 44, no. 2: 323–379.

Campion, Jane, dir. *In the Cut*. 2003. Burbank, CA: Walt Disney Studios HE, 2003. DVD.

Chatta, Juan. 2015. "Film Music as Embodiment." In *Embodied Cognition and Cinema*, edited by Maarten Coëgenarts and Peter Kravanja, 81–112. Leuven: Leuven University Press.

Chion, Michel. 1994. *Audio-Vision: Sound on Screen*. Translated by Claudia Gorbman. New York: Columbia University Press.

Connor, Steven. 2013. "Sounding Out Film." In *The Oxford Handbook of New Audiovisual Aesthetics*, edited by John Richardson, Carol Vernallis and Claudia Gorbman, 107–120. Oxford and New York: Oxford University Press.

Coulthard, Lisa. 2013. "Dirty Sound: Haptic Noise in New Extremism." In *The Oxford Handbook of Sound and Image in Digital Media*, edited by Carol Vernallis, Amy Herzog and John Richardson, 115–26. Oxford and New York: Oxford University Press.

Demers, Joanna, 2010. *Listening through the noise: the aesthetics of experimental electronic music*. Oxford and New York: Oxford University Press.

Donnelly, Kevin. 2005. *The Spectre of Sound: Music in Film and Television*. London: British Film Institute.

Emmerson, Simon. 2007. *Living Electronic Music*. Aldershot: Ashgate.

Erickson, Robert, 1975. *Sound structure in music*. University of California Press.

Fauconnier, Gilles, and Mark Turner. 1998. "Conceptual integration networks." *Cognitive Science*, 22 (2): 133–187.

Giordano, Bruno L., and Stephen McAdams. 2006. "Material identification of real impact sounds: Effects of size variation in steel, glass, wood, and plexiglass plates." *The Journal of the Acoustical Society of America* 119, no. 2: 1171–1181.

Godøy, Rolf Inge. 2006. "Gestural-Sonorous Objects: embodied extensions of Schaeffer's conceptual apparatus." *Organised sound* 11, no. 2: 149–157.

Graham, Ricky, and Brian Bridges. 2014. "Gesture and Embodied Metaphor in Spatial Music Performance Systems Design." In *NIME Proceedings* (Proceedings of the International Conference on New Interfaces for Musical Expression), Goldsmiths, University of London: 581–584.

Graham, Ricky, and Brian Bridges. 2015. "Managing musical complexity with embodied metaphors." In *NIME Proceedings* (Proceedings of the International Conference on New Interfaces for Musical Expression), Louisiana State University: 103–106.

Graham, Richard, Brian Bridges, Christopher Manzione and William Brent. 2017. "Exploring pitch and timbre through 3d spaces: embodied models in virtual reality as a basis for performance systems design." In *NIME Proceedings* (Proceedings of the International Conference on New Interfaces for Musical Expression), Aalborg University, Denmark: 157–162.

Greene, Liz and Danijela Kulezic-Wilson. 2016. "Introduction." In *The Palgrave Handbook of Sound Design and Music in Screen Media: Integrated Soundtracks,* edited by Liz Greene and Danijela Kulezic-Wilson, 1–14. London: Palgrave Macmillan.

Grey, John M. 1977. "Multidimensional perceptual scaling of musical timbres." *Journal of the Acoustical Society of America.* 61 (5):1270–1277

Grey, John M., and John W. Gordon. 1978. "Perceptual effects of spectral modifications on musical timbres." *The Journal of the Acoustical Society of America* 63, no. 5: 1493–1500.

Johnson, Mark. 1987. T*he Body in the Mind: The Bodily Basis of Meaning, Imagination, and Reason.* Chicago, IL: University of Chicago Press.

Johnson, Mark. 2008. *The meaning of the body: Aesthetics of human understanding.* Chicago, IL: University of Chicago Press.

Kerins, Mark. 2011. *Beyond Dolby (Stereo): Cinema in the Digital Sound Age,* Bloomington and Indianapolis: Indiana University Press.

Kubrick, Stanley. 1968. dir. *2001: A Space Odyssey.* Burbank, CA: Warner Home Video. 2019. DVD.

Kulezic-Wilson, Danijela. 2020. *Sound Design is the New Score: Theory, Aesthetics, and Erotics of the Integrated Soundtrack.* Oxford: Oxford University Press.

Lakoff, G., and Johnson, M. 1999. *Philosophy in the Flesh.* New York: Basic Books.

Lefebvre, Henri. 1991. *The production of space.* Translated by Donald Nicholson-Smith. Oxford: Blackwell.

McAdams, Stephen, Antoine Chaigne and Vincent Roussarie. 2004. "The psychomechanics of simulated sound sources: Material properties of impacted bars." *The Journal of the Acoustical Society of America* 115, no. 3: 1306–1320.

Melvin, Adam, and Brian Bridges. 2020. "Music Theory for Sound Designers." In *Foundations in Sound Design for Linear Media: A Multidisciplinary Approach*, edited by Michael Filimowicz, 227–247. New York and London: Routledge.

Mera, Miguel. 2016a. "Materializing Film Music." In *The Cambridge Companion to Film Music,* edited by Mervyn Cooke and Fiona Ford, 157–172. Cambridge: Cambridge University Press.

Mera, Miguel. 2016b. "Towards 3-D Sound: Spatial Presence and the Space Vacuum." In *The Palgrave Handbook of Sound Design and Music in Screen Media: Integrated Soundtracks,* edited by Liz Greene and Danijela Kulezic-Wilson, 91–112. London: Palgrave Macmillan.

Renck, Johan. 2019. dir. *Chernobyl.* 2019. New York: HBO International and Acorn Media, 2019. DVD.

Roddy, Stephen, and Brian Bridges. 2016. "Sounding human with data: the role of embodied conceptual metaphors and aesthetics in representing and exploring data sets." In *Proceedings MusTWork16–Music Technology Workshop 2016*, 64–76. University College Dublin.

Roddy, Stephen, and Brian Bridges. 2018. "Sound, Ecological Affordances and Embodied Mappings in Auditory Display." In *New Directions in Third Wave Human–Computer Interaction*, edited by Michael Filimowicz and Veronika Tzankova, vol. 2, 231–258. Basel, Switzerland: Springer.

Schaeffer, Pierre. 1966. *Traité des objets musicaux [Treatise on musical objects].* Paris: Editions du Seuil.

Shapiro, Lawrence. 2010. *Embodied Cognition.* Abingdon, Oxon: Routledge

Smalley, Denis. 1997. "Spectromorphology: explaining sound-shapes." *Organised Sound* 2, no. 2: 107–126.

Smalley, Denis. 2007. "Space-form and the acousmatic image." *Organised Sound* 12, no. 1: 35–58.

Smith, Jeff. 2013. "The Sound of Intensified Continuity." In *The Oxford Handbook of New Audiovisual Aesthetics,* edited by John Richardson, Carol Vernallis and Claudia Gorbman, 331–356. Oxford and New York: Oxford University Press.

Varela, Fransisco J., Thompson, Evan and Rosch, Eleanor. 1991. *The Embodied Mind, Cognitive Science and Human Experience.* Cambridge, MA: MIT press.

Villeneuve, Denis. 2016. dir. *Arrival.* London: Universal Pictures UK, 2017. DVD.

Villeneuve, Denis. 2017. dir. *Blade Runner 2049.* Culver City, CA: Sony Pictures, 2018. DVD.

Ward, Mark. 2015. "Art in Noise: An Embodied Simulation Account of Cinematic Sound Design." In *Embodied Cognition and Cinema,* edited by Maarten Coëgenarts and Peter Kravanja, 155–186. Leuven: Leuven University Press.

Measuring People's Cognitive and Emotional Response

Psychophysiological Methods to Study Sound Design

Emma Rodero

1 Introduction

One of the most extended methods in the field of psychology is to measure people's physiological responses. If we want a clear picture of what happens when someone listens to audio, we can complement the self-report methods (e.g., questionnaires, surveys) with psychophysiological techniques. As sound is very connected to emotions, and research evidence indicates that hearing and emotions are related, measuring people's body response is an adequate method. Besides, psychophysiological measurements have some advantages. The main benefit when we analyze audio is that we can register people's response while they are listening. Therefore, we can detect body changes second by second and, in this manner, study the effect of the different sound elements (e.g., music, sound effects) at the specific moment in which they are used in audio production.

This chapter is divided into two parts. In the first part, we will explain what the psychophysiological methods are, their main advantages applied to audio and the techniques that we can use to study sound. In the second part, we will review some studies that have analyzed audio messages in communication using psychophysiological methods. We will explain some of the existing research about voice, music and sound effects.

2 Psychophysiological Methods: Advantages of analyzing audio design

Psychophysiology is a consolidated discipline in the branch of psychology that studies the physiological correlations of psychological processes. Psychophysiology aims to understand the relationship between what people think, feel and do with their physiological reaction. It is, therefore, based on understanding psychological processes by measuring the body's physiological response. As a formal discipline, psychophysiology was born with the formation of the Society for Psychophysiological

DOI: 10.4324/9780429356360-5

Research in 1960 and the publication of the scientific journal *Psychophysiology* in 1964 (Berntson & Cacioppo, 2002; Vila & Guerra, 2009). In the 2020s, it has become a discipline with significant development, growth and popularity in many fields, mainly as it focuses on the relationship between mind and body.

As psychophysiology studies people's cognitive, emotional and behavioral responses, it is an adequate discipline to analyze the communication phenomena. In communication, some researchers are using psychophysiological measures to analyze the audience's cognitive and emotional response to media messages, although there are not many studies yet. Annie Lang and colleagues at the Institute for Communication Research (Indiana University) have been pioneers in applying psychophysiological measures to communication and proposing a theory, the Limited Capacity Model of Motivated Mediated Message Processing (LC4MP), about how people process media messages (Lang, 2009). These techniques are very appropriate for studying audio, as the listening activity elicits a physiological response. Rodero (2018) showed that there were higher activation and physiological arousal when participants listened to stories versus reading them. Also, scientists now know the connection between hearing and touch. What we hear affects what we feel (Ro et al., 2009). Therefore, measuring the physiological and emotional responses adds critical information in the analysis of audio perception and processing.

The use of these techniques applied to communication and, therefore, to audio has some advantages:

1. registering the response while people are listening to the audio stimulus
2. measuring the individuals' unconscious response
3. obtaining an objective measure of the sound stimulus

The first and more important benefit is that it allows registering the individual's response in real-time during the exposure to the audio messages. If this aspect is essential in analyzing messages in general, it is even more relevant when we examine the audio design. Imagine that you have an audio message in which you put a fragment of music and a sound effect at a precise moment. Traditional methods, especially surveys (which are the most common), would ask people about music and sound effects after listening to the message. Yet a) individuals probably forgot what they thought or felt about these audio elements, or b) they are not good at making judgments about their cognitive states, such as estimating their attention level at the moment they listened to music or reporting how they felt about one specific sound effect. Many individuals have difficulty identifying their reactions, thoughts or emotions, let alone transmitting or communicating them. Some simply do not want to do so. Psychophysiological methods allow us to register the individuals' emotional response millisecond after millisecond with temporal precision. We can use time-series analysis techniques and register listeners' phasic changes *throughout* the message and at *specific* moments of the message. In this manner, we do not have to trust in what people remember *after* listening to an audio piece.

The second major advantage of these techniques is that they allow us to register people's unconscious response. The use of physiological methods provides additional information crucial to analyze the people's cognitive and emotional reaction to audio: the implicit or unconscious response. Further, people cannot manipulate their physiological responses. They do not have voluntary control about their heart rate or arousal, so they cannot lie about a particular physiological body reaction, which is something that they can do in a survey.

The third advantage is that while subjective or perception questionnaires or surveys are conditioned for its design, the way to ask the questions, or the people's comprehension and memory, the psychophysiological techniques have not these influences and can be applied equally to everyone. The measurement here does not depend on people's capacities, skills or perception. Therefore, objectivity increases reliability (Dirican & Göktürk, 2011).

But psychophysiological measures also have some disadvantages:

1. expensive equipment not easy to use
2. more invasive methods
3. multiple methods to identify a psychological construct

The most significant inconvenience is that the application of these techniques is expensive in equipment, lab costs and personnel training. The psychophysiological equipment is expensive and requires specialized personnel to install it. The lab must set in a Faraday cage to avoid interferences and noises from different electric devices. Also, lab maintenance is costly and requires laboratory supplies. Finally, specialized training is necessary to manage this equipment, to learn how to place the electrodes on the individual's body and to analyze the psychophysiological data.

The second disadvantage of these techniques is that they are more invasive than filling out a simple questionnaire. Psychophysiological measures require attaching electrodes to the people's bodies. As having electrodes in the body is not something habitual, the individual can show a slight feeling of uncertainty. Of course, we know from experience that participants soon forget the sensors. Hence, a system with the electrodes does not interfere with the task. This is why some authors as Dirican and Göktürk (2011) consider this technique unobtrusive. Still, it can be regarded as more invasive, at least initially, than traditional methods.

The third weakness is that the relationship between the physiological measures and the psychological correlates is not exact. To understand how the psychophysiological response is interpreted, finding the correlation or indicator of the psychological process to the different physiological changes is necessary. The problem is that we cannot establish a unique correspondence, as a physiological pattern can identify different mental constructs. Therefore, using multiple registers to clearly define a specific psychological construct is recommended (Cacioppo & Tassinary, 1990; Dirican & Göktürk, 2011).

Comparing the pros and cons, psychophysiological measures have more advantages than problems and provide essential information about the body's reaction that it is crucial to identify cognitive processes and especially emotional reactions. Also, if psychophysiological measures complement other techniques (e.g., auto perception, recall tests), we can obtain a very accurate radiography of people's cognitive and emotional processing using reliable data.

3 Psychophysiological measures to analyze the audio design

Psychophysiological techniques are focused on measuring the human nervous system. The nervous system has two different parts, the central nervous system (CNS) and the peripheral nervous system (PNS).

The CNS is formed by the brain and the spinal cord. The CNS branches off throughout the body via the PNS. The PNS is the set of nerves that run from the CNS to all the areas of the body and allows the communication between the brain and the spinal cord. The PNS performs sensory and motor functions.

The PNS is divided into the somatic and the autonomic nervous system. First, the somatic nervous system consists of fibers afferent and efferent of the sensory and motor organs. This structure is responsible for sending sensory information to the central system and controlling muscle activity. Secondly, the autonomic nervous system (ANS) is made up of afferent and efferent nerves of the internal organs. The system is independent of our voluntary control and can be divided into the sympathetic (SNS) and parasympathetic nervous system (PNS).

The SNS is formed by the nerves coming from the thoracic and lumbar spinal cord. It is the system that activates the body automatically, like in a situation of immediate threat, and predisposes it to action/activity. This is the system that activates the body and produces the fight-flight response. The parasympathetic nervous system is made up of the nerves that come from the cervical area. This system coordinates resting activity in processes such as digestion or sleep. This is the system that decelerates our bodies.

Depending on which system is analyzed, there are techniques of psychophysiological analysis that measure the CNS or the PNS (Vila & Guerra, 2009). We will now explain the main measures used in communication divided into these two systems.

3.1 Central Nervous System (CNS)

EEG Electroencephalography, or EEG, is a method to register the electrical activity of the brain by attaching electrodes to the participant's scalp. EEG records brain voltage fluctuations over a period of time. This technique can determine low or high perceptual and cognitive processes. It is frequently used to diagnose epilepsy or sleep disorders. EEG can measure attention and memory processes by detecting changes in the P3 components. The P3a component is associated with an orienting response produced by a novel stimulus and P3b with memory processing and decision making (Polich, 2007). One of the main advantages of this technique is that the participants can move, which is not possible with other CNS imaging methods such as fMRI. The main disadvantage is that it is susceptible to biological and electrostatic artifacts.

ERP Event-Related Brain Potentials, or ERP, is a measure to register the electro-physiological response or voltage changes to specific events or stimuli. It is measured using an EEG. This technique allows us to know what is happening in the brain in response to a specific stimulus. For example, after a stimulus presentation, the participants have to perform sensory, cognitive control, affective or memory tasks. A sensory-related operation could be the estimation of color. A cognitive control operation could be answering a question. An affective task could be to associate the stimulus (for example, a sound) with positive or negative emotions. Finally, a memory task is based on recalling an item (Kropotov, 2016). Therefore, it is an interesting technique for exploring audio.

fMRI This method measures brain activity by detecting small changes in the blood flow. As blood flow and neuronal activation are associated, this technique is based on registering these blood flow changes to identify the neuronal activation. fMRI is very precise way to examine the brain's functional anatomy. But the problem is that this method is very invasive, as the participant has to be inside a scanner, which can provoke claustrophobia problems. Also, it is not a natural environment to analyze communication messages and is very expensive. For these reasons, it is not a standard method used in communication studies.

3.2 Peripheral Nervous System (PNS)

HR-HRV The most common cardiovascular measures used in communication are Heart Rate (HR) and Heart Rate Variability (HRV). HR and HRV are measured through an electrocardiogram (ECG). Three electrodes are placed on the participant's body. The pulse or heart rate is the primary psychophysiological measure of cardiovascular activity and is registered in beats per minute (BPM). An HR deceleration in five to seven seconds is indicative of increased cognitive effort, or mental workload, and attention (Potter & Bolls, 2012). The deceleration of HR also is related to the orientation reflex, or orienting response; thus, an increase in attention. HVR represents the variation in the time interval between consecutive heartbeats. A high HVR indicates a low cardiac activity, and a low HVR reflexes a heart rate acceleration, which can reveal stress. Figure 4.1 shows the electrocardiogram and HR signal.

EDA Electrodermal Activity, or EDA, is also called Skin Conductance or Galvanic Skin Response (GSR). This method records changes in the electrical activity at the skin surface by measuring the activity of the eccrine glands. These glands are mainly found in the palm of the hands and the soles of the feet. So, in this case, the electrodes are placed on these parts. The eccrine glands not only have a thermoregulatory function but also respond strongly to emotional stimuli. Its psychological correlation is the level of activation or arousal represented by the intensity of emotional activation caused by a stimulus, as well as attention to messages with different levels of complexity. The greater the range of response, the greater degree of activation, emotional reaction and attention to the stimulus. Electrodermal activity is one of the most employed measures to analyze audio messages. Figure 4.2 shows the electrodes of EDA in the hand.

Facial EMG A facial electromyogram, or EMG, detects the facial muscle movements to identify emotional valence. Facial electromyography records the electrical

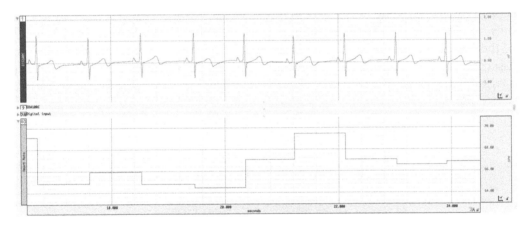

Figure 4.1 Electrocardiogram and HR signal

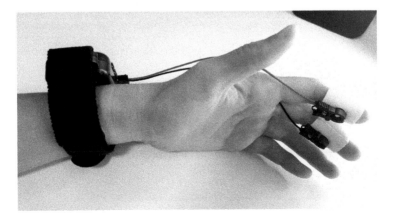

Figure 4.2 Electrodermal activity

activity from the contraction of muscle fibers (Vila & Guerra, 2009). The most common muscles to analyze are the zygomatic, the orbicularis and the corrugator.

The zygomatic is located at the top of the cheek. Its activation indicates a positive emotional response. The orbicular muscle surrounds the eye and also indicates a positive emotion. If someone smiles, these two muscles contract. In contrast, the corrugator, located on the nose between the eyebrows, indicates a negative emotional response, such as when someone frowns. Two electrodes are placed on each one of these muscles to measure their activity and the emotional dimension of valence. Therefore, we can detect if listeners are feeling a positive (zygomatic or orbicular activation) or negative emotion (corrugator). Figure 4.3 shows the electrodes in the orbicular and corrugator muscles along with EEG.

Pupillometry Eye movements such as fixations, saccades, gaze or blinks can provide valuable information about people's psychological states. The pupils have been called by some writers as "the windows to the soul" (Andreassi, 2000). In fact, the eyes are the only visible part of the brain. Pupillometry is the study of variations in the diameter of the pupillary aperture of the eye (Andreassi, 2000). Eyeblink rates and duration are associated with changes in the mood state, task demands and level of fatigue. When a task requires attention, the eyeblink rate decreases. Stress causes the contrary effect,

Figure 4.3   Facial Electromyography and EEG

and the rate increases. Pupil diameter and dilation changes are indicatives of arousal, cognitive workload and attention. You can easily note how the pupils dilate when someone is paying attention. Therefore, eye movements index changes in information processing and also respond to the emotional valence. There are some methods to measure eye movements and positions, such as electrooculography (EOG), but the most common in media is an optical eye tracker. This a standard measure in communication and marketing, but it has problems with listening to the audio, as many people close their eyes when they are very concentrated listening to a piece.

Respiration A substantial number of studies have concluded that a mental task requiring effort is associated with an increase in respiration rate and a decrease in the variability of respiration. Stressful tasks can lead to the same result. The respiratory pattern also can vary depending on different states, from excited and high arousal with an increase of respiration rate, to low arousal with a decrease in the rate. When exposed to a novel stimulus, respiration frequency decreases while amplitude increases. Therefore, respiration is a measure to analyze task demands, especially negative valance, and arousal. Respiration variations mainly affect HR and EDA. Respiration is usually measured by employing a band placed on the individual's chest.

Temperature Although this measure is not very used in media, variations in skin temperature can detect emotional activation. Temperature is associated with blood flow in skin tissue. Therefore, temperature variations are related to vasodilation and vasoconstriction of blood vessels (Andreassi, 2000). When the SNS is activated this may produce constrictions of blood vessels. This peripheral vasoconstriction provokes a decrease in the temperature of extremities, such as the nose (Gade & Moeslund, 2014). A cognitive effort causes changes in temperature variations in the forehead, nose, cheeks and chin regions (Marinescu et al., 2018). These temperature variations are registered by using a thermal camera.

Once we know what the psychophysiological methods are, their advantages, and the techniques that we can use analyzing audio, in the next section, we will review some audio studies that have used these methods. We will look at studies about voice, music and sound effects.

4 Studies about Sound Design using Psychophysiological Methods

When we produce an audio message in media and communication, we always have one goal in mind: to get the listener's attention. If we cannot catch listener's attention, they will not remember the message. Psychophysiological measures are techniques that can register claims of attention, as an orienting response, and attention changes while listening to a specific stimulus, such as a song or sound effect. Therefore, these measures allow us to have an objective way of analyzing if our sound design in a piece is compelling and engaging.

When we hear a novel stimulus as a new sound, for example, a bell sounding, we automatically direct the attention toward this object. This phenomenon is called an orienting response or an orientating reflex. An orienting response (OR) can be defined as an automatic and immediate reaction to a novel and significant stimulus (Graham, 1992; Sokolov et al., 2002). OR facilitates stimulus processing, as it causes a temporary increase in cognitive resources allocated to encoding this novel stimulus.

An OR produces distinct, measurable physiological changes. In this automatic response, we can detect an increase in arousal (measured by EDA), pupil dilation (measured by eye-tracking), significant cardiac deceleration cardiac (measured by ECG), EEG activation (P3), constriction of the blood vessels in the skin (usually measured in the fingertips) and dilation of the blood vessels of the head, among them. Therefore, we can objectively measure an OR and shifts in attention.

One of the main elements that elicit an orienting response is audio. If we are at home and, suddenly, we hear an intense noise, we automatically draw the attention to this point, as the physical features of the stimulus determine this reaction. Thereby, we can analyze how sound resources work in terms of attention and arousal to build an effective audio design. The first step is to identify the structural and formal features of the message, which can automatically grab the listener's attention to improve our audio design.

There is a substantial literature that measures the physiological response to determine the power of audio structural resources to increase the listener's automatic attention (Potter, 2000; Potter, Lang & Bolls, 2008). Some audio resources have been found to evoke orientating responses, among them: voice changes, pitch variations, music or sound effects. In this part of the chapter, we will review some studies that have analyzed sound resources by using psychophysiological measures. The review will not be exhaustive, as our goal is only to put forward some examples of this kind of research.

4.1 Research About Voice

Regarding this feature, we conducted one study to know what type of announcer's voice was most appropriate in commercials. Regarding psychophysiological measures, we registered electrodermal activity to index arousal. Arousal is one of the variables to measure emotion (Bradley & Lang, 2000), and, as explained before, it represents the level of activation provoked by a stimulus as a response of the sympathetic nervous system (Dawson, Schell & Filion, 2007). An increase in arousal signifies a task with high cognitive demand, attention and emotional activation. To measure the electrodermal activity, we placed two electrodes with a constant voltage (.5V) on the fingers of one of the participant's hands (Potter & Bolls, 2012). Low-pitched voices obtained a higher level of arousal than medium and high-pitched voices. In Figure 4.4, we can see the evolution of the level of participants' arousal, measured in microsiemens, while they were listening to the commercial.

As we can observe, low-pitched voices attained the best level throughout the commercial, indicating the highest level of activation or arousal. On the contrary, we can conclude that high-pitched voices obtained less degree of emotional activation, and we can also notice that the level was decreasing throughout the commercial. Therefore, this level of arousal not only is informing us that the low-pitched voices were more efficient in engaging the participants but also that they better maintain the level of interest throughout the ad.

Also, there are two more studies related to voice that we can highlight. Both of these studies are related to the attention changes provoked by replacing one speaker's voice by another when listening to an audio stimulus, in this case, radio commercials. Potter's work (Potter, 2000; Potter, Lang & Bolls, 2008) found that voice changes are auditory structural features capable of eliciting orienting responses. This author (2000) showed that voice changes elicited cubic cardiac deceleration in a six-to-ten second window following onset. Figure 4.5 shows the orienting cardiac response.

Therefore, when we change the voice in an audio message, for example, a male announcer for a female announcer, a cardiac deceleration is produced, indicating an orienting response or a claim of attention.

Another study that we conducted analyzed voice pitch in commercials (Rodero, Potter & Prieto, 2017). We explain this study here to show how we applied three of the main techniques (HR, EDA and EMG) to analyze audio. The question to answer in that study was whether different prosodic intonation strategies applied to

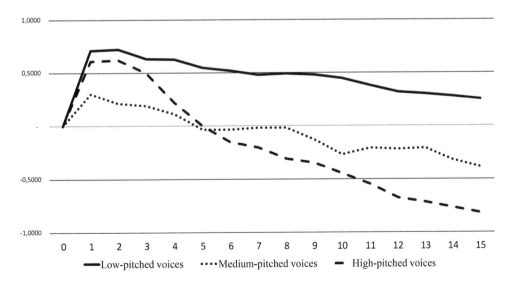

Figure 4.4 EDA register (microsiemens)

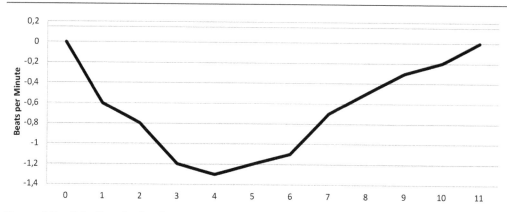

Figure 4.5 Orienting Cardiac Response

commercials could improve listeners' attention and memory. We tested three models of intonation. As psychophysiological techniques, we registered the participants' heart rate extracted from an electrocardiogram (ECG), electrodermal activity or skin conductance (EDA) and facial electromyography (EMG). First, we measured participants' attention and cognitive effort extracting heart rate in beats per minute from the electrocardiogram. The second measure was electrodermal activity to index arousal. Finally, we used facial EMG to measure the orbicularis and the corrugator activity to register a positive and negative emotional response.

These three measures allowed us to analyze the participants' attention, emotional activation and valence. The results showed that the models with pitch variations –

especially the High-Low pitch (HL) – attained the best levels in the analyzed variables. Acoustic changes provoked by pitch variations contributed to attracting the listener's attention, as shown in the HR levels. Although there were no significant differences, we can observe in Figure 4.6 that the HL intonation had more variations in the curve compared to high intonation and low intonation.

This meant that participants recovered their attention throughout the ad thanks to the high-pitched acoustic signals that this intonation model produced at certain moments. The previous high pitch acted as an orienting element of attention (increasing the level of attention and arousal) and then prepared the listener to hear the critical part. Therefore, we can conclude that this style was the catchiest.

Along with the level of attention, we measured the arousal by using electrodermal activity. These data were the clearest. The intonation achieving the highest level of arousal was the HL model, according to the HR results. In Figure 4.7, we can notice that this intonation model was not only the highest but also the one that better-maintained participants' engagement throughout the commercial.

These results were reinforced with the emotional valence measured using facial EMG. As an average, the HL model of intonation achieved a more positive valence and

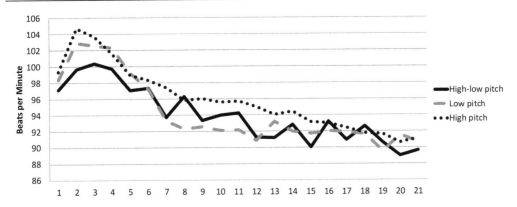

Figure 4.6   Heart Rate of Intonation Models

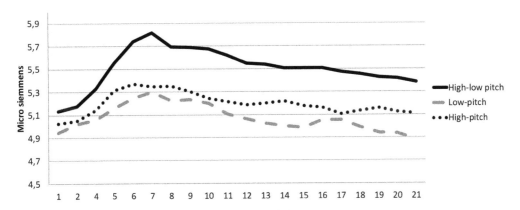

Figure 4.7   Electrodermal Activity of Intonation Models

especially with a less negative emotion, as we can see in Figures 4.8 and 4.9 showing the orbicularis muscle (positive emotion) and the corrugator muscle (negative emotion).

Bolls, Lang and Potter (2001) also employed Facial EMG to show that radio messages pretested with negative valence provoked a higher activity in the corrugator muscle than messages pretested as being more positive in their emotional tone.

4.2 Research About Sound Features

Along with the voice, effective audio messages should be designed according to an appropriate combination of sound resources, especially sound effects and music.

Sound effects represent the sounds of reality. Therefore, they are the sound elements that give verisimilitude to stories. Sound effects have proven to have a

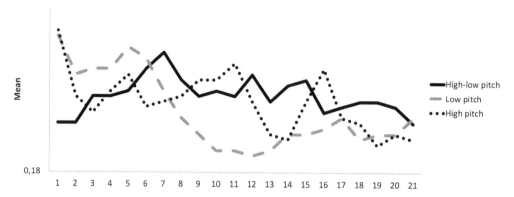

Figure 4.8   Electromyography: orbicularis muscle. Positive emotion

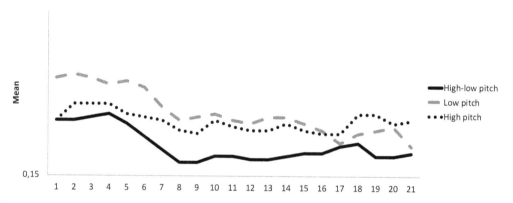

Figure 4.9   Electromyography: corrugator muscle. Negative emotion

significant capacity to increase the number of mental images that are created in the listener's mind, as they help to restore reality and create a realism sense (Miller & Marks, 1992; Bolls & Lang, 2003; Rodero, 2012 and 2019). In addition, we know that the use of sound effects in audio design improves listeners' attention (Potter & Choi, 2006; Rodero, 2015). This is why a study by Shinskey (2017) found that students had a more profound knowledge of words that were taught with an associated sound effect compared with words explained with no sound effect.

Some studies have analyzed the use of sound effects in radio commercials and audio fiction stories. Potter and colleagues (Potter, 2006; Potter, Lang & Bolls, 2008) showed an orientating response in messages with several auditory structural features, among them sound effects and production effects. A study by Potter and Choi (2006), in which participants listened to structurally complex messages with multiple voice changes, sound effects, music onsets and/or production effects, found that the levels of arousal were higher than in simple messages containing only a few such features. Also, Potter (2006) analyzed production effects defined as synthesized auditory effects. This research showed that the participants allocate attention to process short radio promos immediately following the onset of production effects. This attention was measured using heart rate. This author demonstrated that there was a strong cardiac deceleration in response to the presence of both laser and echo effects. Therefore, there was an orienting response to these effects.

In another study, Potter, Lang and Bolls (2008) analyzed the effect of different sound resources, among them production effects, in radio production. In this research, production effects also produced an orienting response measured with a deceleration in the heart rate after the onset. Along with attention, the OR increased recognition memory for the information after the structural features.

We also studied the psychophysiological response to descriptive sound effects in audio fiction stories (Rodero, 2020). Descriptive sound effects are the sounds that restore reality in audio fiction, for example, a door closing or a car accident. The results of this analysis confirmed that there was a psychophysiological response when, in an audio message, participants listened to significant sound effects, such as a car accident. Immediately after these sound effects, heart rate showed the pattern of an OR, respiration was faster, electrodermal activity was higher and emotional valence was more intense and negative than in the story without these effects.

Secondly, if there is one element of audio language that has a special relationship with emotions, it is music. Music is a powerful feature to elicit emotions and to modify mood states. Listening to music provokes a physiological reaction. When people listen to music, they produce chemical changes in their bodies that affect emotions and moods. That is why we can feel that our mood changes listening to music. Rodero (2019), in a study comparing different sound resources, concluded that when we process music we do so emotionally, with greater emotional activation. In another study, using fiction stories (Rodero, 2020), those with music attained higher arousal, emotional activation, attention, respiration was faster and positive emotional valence was higher than in the stories with no music.

There are also many studies analyzing the healing power of music and its ability to relax. Bartlett (1996) reviewed some studies related to music using physiological measures. Most of these research studies measured muscle tension and concluded that relaxing music could reduce it. This author also reported that relaxing music increased skin temperature in half of the revised studies. Pelletier (2004) also reviewed some experiments about the effects of relaxing music. These studies concluded that music could reduce significantly perceived arousal and physiological activation. Alves et al. (2019) showed that listening to music while driving in urban traffic can reduce stress. These authors registered nonlinear HRV changes indicating a reduction in the cardiac autonomic overload. Burns et al. (1999) found that finger temperature and skin conductance decreased when participants were listening to music. Rickard (2004) also showed that participants exhibited a higher skin conductance and chills when they were listening to music considered by them as emotionally powerful. The study by Krumhansl (1997) showed a decreased heart rate and increased breathing rate and blood pressure in college students when listening to music.

Along with these, some studies have analyzed the arousal of music tempo and mode. These studies have concluded that perceived arousal is higher with a faster tempo. Music mode (major or minor key) affects people's moods (Lundin, 1985). In this line, Kellaris and Kent (1993) showed that faster tempos led to more reported arousal and pleasure. The study by Dillman and Potter (2007) revealed a higher skin conductance response (SCR) while participants listened to music compared to silence. This study also examined two types of music: fast-paced and slow-paced music. According to the other studies, the skin conductance level (SCL) showed that fast-paced music elicited greater activation than slow-paced music.

Finally, in clinic studies, music has shown to be valid to reduce respiratory rate and anxiety in patients (Evans, 2002; Nilsson, 2008). Other studies have demonstrated a significant decrease in cortisol (Finn & Fancourt, 2018).

Consequently, all these positive benefits of music finally provoke an increase in the levels of attention and memory of the message (Alpert et al., 2005; Chebat, Gelinas & Vaillant, 2001)). These are enough reasons to use music in audio design.

All in all, the novelty introduced by an acoustic change provoked by using sound effects and music in audio design may cause an OR that attracts listeners' attention, increases arousal and favors recall to the information a few seconds after this moment. Therefore, the use of these acoustic features provokes a physiological, emotional response beneficial to engage our listeners.

5 Conclusions

Psychophysiological techniques are a very useful method for analyzing the audio design, as the sound is very connected to emotions, and there is scientific evidence that audio provokes a body response. The main advantage is that we can register the listeners' physiological reactions while they listen to the audio piece, millisecond by millisecond. This is an optimal manner of studying the impact of the different sound resources (voice, music, sound effects, etc.) at the specific moment in which they are used in audio design.

We can use different techniques to analyze the audio design. The most common are electroencephalography (EEG), electrocardiogram (ECG), electrodermal activity or skin conductance (EDA), Facial EMG, pupillometry, respiration and temperature. With these measures, we can know the physiological reaction to a specific audio input and determine the level or arousal or emotional activation, attention,

OR, mental workload and valence, among other aspects. Some studies using psychophysiological measures have shown that voice changes, pitch variations, sound effects and music can elicit an OR and provoke emotional activation. Consequently, these resources make messages easier to remember. The use of these resources in audio design is, therefore, recommendable to engage listeners.

In conclusion, if we complemented psychophysiological measures with other techniques (questionnaires, surveys and such) when studying audio design, we will attain a more accurate radiography of how people process, feel and respond to audio. Then this precise knowledge will allow us to design audio pieces more appealing, effective and memorable.

References

Alpert, M.I.; Alpert, J.I. & Maltz, L.N. (2005). Purchase occasion influence on the role of music in advertising. *Journal of Business Research*, 58, 369–376.

Alves, M.A., Garner, D.M., do Amaral, J.A., Oliveira, F.R., & Valenti, V.E. (2019). The effects of musical auditory stimulation on heart rate autonomic responses to driving: A prospective randomized case-control pilot study. *Complementary therapies in medicine*, 46, 158–164.

Andreassi, J.L. (2000). *Psychophysiology: Human behavior and physiological response* (4th ed.). Mahwah, NJ: Lawrence Erlbaum Associates Publishers.

Bartlett, D. (1996). Neuromusical research: A review of the literature. In D.A. Hodges (ed.), 343–385, *Handbook of music psychology*. San Antonio, TX: IMR Press.

Blascovich, J. (2004). Psychophysiological measures. *The SAGE Encyclopedia of Social Science Research Methods*. Thousand Oaks, CA: SAGE, 881–883.

Berntson, G. & Cacioppo, J. (2002). Psychophysiology, in *Biological Psychiatry*, 9, H.D. Haenen, J.A. Ben Boer & P. Willner (eds.), Hoboken, NJ: John Wiley & Sons, 123–138.

Bolls, P.D., & Lang, A. (2003). I saw it on the radio: The allocation of attention to high-Imagery radio advertisements. *Media Psychology*, 5 (1), 33–55.

Bolls, P.D., Lang, A. & Potter, R.F. (2001). The effects of message valence and listener arousal on attention, memory, and facial muscular responses to radio advertisements. *Communication Research*, 28, 627–651.

Bradley, M.M., & Lang, P.J. (2000). Affective reactions to acoustic stimuli. *Psychophysiology*, 37, 2, 204–215.

Burns, J., Labbe, E., Williams, K., & McCall, J. (1999). Perceived and physiological indicators of relaxation: As different as Mozart and Alice in Chains. *Applied Psychophysiology and Biofeedback*, 24, 197–202.

Cacioppo, J.T., & Tassinary, L.G. (1990). *Principles of psychophysiology: Physical, social, and inferential elements*. Cambridge: Cambridge University Press.

Chebat, J.C.; Gelinas, C. & Vaillant, D. (2001). Environmental background music and in-store selling. *Journal of Business Research*, 54, 115–123.

Dawson, M.E., Schell, A.M., & Filion, D.L. (2007). The electrodermal system. In J.T. Cacioppo, L.G. Tassinary & G.G. Berntson (eds.), 159–181, *Handbook of psychophysiology*. Cambridge University Press.

Dillman Carpentier, F.R., & Potter, R.F. (2007). Effects of music on physiological arousal: Explorations into tempo and genre. *Media Psychology*, 10 (3), 339–363.

Dirican, A.C., & Göktürk, M. (2011). Psychophysiological measures of human cognitive states applied in human computer interaction. *Procedia Computer Science*, 3, 1361–1367.

Evans, D., (2002). The effectiveness of music as an intervention for hospital patients: a systematic review. *Journal of Advanced Nursing*, 37, 8–18.

Finn, S., & Fancourt, D. (2018). The biological impact of listening to music in clinical and nonclinical settings: A systematic review. In *Progress in brain research*, vol. 237, 173–200.

Gade, R., & Moeslund, T.B. (2014). Thermal cameras and applications: a survey. *Machine Vision and Applications*, 25, 245–262.

Graham, F.K. (1992). Attention: The heartbeat, the blink, and the brain. In Campbell, B.A., Hayne, H., & Richardson, R. (eds.), 3–29, *Attention and Information Processing in Infants and Adults: Perspectives from Human and Animal Research*. London: Psychology Press.

Kellaris, J.J., & Kent, R.J. (1993). An exploratory investigation of responses elicited by music varying in tempo, tonality, and texture. *Journal of Consumer Psychology*, 2, 381–401.

Kropotov, J.D. (2016). *Functional neuromarkers for psychiatry: Applications for diagnosis and treatment*. Cambridge: Academic Press.

Krumhansl, C.L. (1997). An exploratory study of musical emotions and psychophysiology. *Canadian Journal of Experimental Psychology*, 51, 336–352.

Lang, A. (2009). The limited capacity model of motivated mediated message processing. In R. Nabi and M.B. Oliver (eds), 193–204, *The SAGE handbook of mass media effects*. Thousand Oaks, CA: SAGE.

Lundin, R.W. (1985). *An objective psychology of music*. New York: Ronald Press Company.

Marinescu, A.C., Sharples, S., Ritchie, A.C., Sánchez López, T., McDowell, M. & Morvan, H.P. (2018). Physiological parameter response to variation of mental workload. *Human Factors* 60, 31–56.

Miller, D.W., & Marks, L.J. (1992). Mental imagery and sound effects in radio commercials. *Journal of Advertising*, 21 (4), 83–93.

Nilsson, U. (2008). The anxiety- and pain-reducing effects of music interventions: a systematic review. *AORN Journal*, 87, 780–807.

Pelletier, C.L. (2004). The effect of music on decreasing arousal due to stress: A meta-analysis reference. *Journal of Music Therapy*, 41, 192–214.

Polich, J. (2007). Updating P300: an integrative theory of P3a and P3b. *Clinical Neurophysiology,* 118, 2128–2148.

Potter, R.F. (2000). The effects of voice changes on orienting and immediate cognitive overload in radio listeners. *Media Psychology*, 2 (2), 147–177.

Potter, R.F. (2006). Made you listen: The effects of production effects on attention to short radio promotional announcements. *Journal of Promotion Management*, 12 (2), 35–48.

Potter, R.F., & Bolls, P. (2012). *Psychophysiological measurement and meaning: Cognitive and emotional processing of media*. London: Routledge.

Potter, R.F., & Choi, J. (2006). The effects of auditory structural complexity on attitudes, attention, arousal, and memory. *Media psychology*, 8 (4), 395–419.

Potter, R.F., Lang, A. & Bolls, P.D. (1998). Orienting to structural features in auditory media messages. *Psychophysiology*, S66.

Potter, R.F., Lang, A., & Bolls, P.D. (2008). Identifying structural features of audio: Orienting responses during radio messages and their impact on recognition. *Journal of Media Psychology: Theories, Methods, and Applications*, 20 (4), 168–177.

Potter, R.F., Lynch, T. & Kraus, A. (2015). I've Heard That Before: Habituation of the Orienting Response Follows Repeated Presentation of Auditory Structural Features in Radio. *Communication Monographs*, 82 (3), 359–378.

Rickard, N.S. (2004). Intense emotional responses to music: A test of the physiological arousal hypothesis. *Psychology of Music*, 32, 371–388.

Ro, T., Hsu, J., Yasar, N.E., Elmore, L.C. & Beauchamp, M.S. (2009). Sound enhances touch perception. *Experimental Brain Research*, 195 (1), 135–143.

Rodero, E. (2012). See it in a Radio Story. Sound Effects and Shots to Evoked Imagery and Attention on Audio Fiction. *Communication Research*, 39, 458–479.

Rodero, E. (2018). The growing importance of voice and sound in communication in the digital age: The leading role of orality. *Anuario AC/E de Cultura Digital*, 80–94.

Rodero, E. (2019). The Spark Orientation Effect for improving attention and recall. *Communication Research*, 46 (7), 965–985.

Rodero, E. (2020). Audiobooks. What is the best format to tell a story? [Research project.] Leonardo Research Grants, Fundación BBVA.

Rodero, E., Potter, R.F. & Prieto P. (2017). Pitch range variations improve cognitive processing of audio messages. *Human Communication Research,* 43, 397–413.

Shinskey, J.L. (2017). Sound effects: Multimodal input helps infants find displaced objects. *British Journal of Developmental Psychology*, 35 (3), 317–333.

Sokolov, E.N., Spinks, J.A., Näätänen, R. & Lyytinen, H. (2002). *The orienting response in information processing*. Mahwah, NJ: Lawrence Erlbaum Associates Publishers.

Vila, J. & Guerra, P. (2009). *Una introducción a la psicofisiología clínica. [An introduction to clinic psycho-physiology]*. Madrid: Ediciones Pirámide.

5

Watching and Evaluating Simultaneously

An Approach to Measure the Expressive Power of Sound Design in a Moving Picture

Maximilian Kock

1 Introduction

Measuring and classifying emotional responses of recipients while they are experiencing (and also *after* they have watched) audiovisual media is an objective of many research studies. With these results it is possible to design audiovisual media more efficiently by creating bigger emotional and more impressive impacts on the viewers and, furthermore, one might presumably predict better the viewers' emotional reactions.

However, evaluating sound design in a moving picture generates a fundamental problem: A valuable assessment is only possible if at least two audio tracks with different sound design are presented successively to the viewer. However, if you present first audio track A and subsequently audio track B together with the identical moving picture, this process of "comparing" is not very viable. Because of habituation and other external leverages, there might be shifts in the resulting data of felt emotions and perception. For example, a data shift could be registered, when the test videos are presented in a different sequence (see the paragraph below). Usually, after each single test run, the participants fill out questionnaires about their felt emotions.

In the past, much research was done using this fundamental structural design of a study[1] in order to obtain meaningful emotional reactions and collect clusters of quantitative data[2] to analyze the viewer's responses. Using statistics, researchers evaluate the data, hoping the numbers will lead to results that can be applied to other moving pictures.[3]

This typical structural design of the test procedure mentioned above nevertheless brings at least two important aspects into focus:

First: The participant's emotional reaction is retrieved *after* they have seen the film or video. Therefore, the amount of the noted emotional intensity might have conceivably changed compared to an instantaneously measured affect.

DOI: 10.4324/9780429356360-6

Second: Participants might change their felt emotion when they are viewing the same video a second, third or fourth time, especially with different audio tracks. It could be incongruous to compare the reaction data of test participant A watching a video the first time with the reaction data of test participant B watching the identical video a second or even third time. The possible differences in the resulting data might be caused by the potential habituation of participant B to the film plot.

In consequence, a test procedure had to be established which rested only upon one 'untouched' test run for each participant with immediate measuring of his or her emotional reaction. This new approach led to the key questions of this study (see section 2.4, Experimental Questions) and thus might give us a new insight into the use of music and how sound effects might influence the viewer's perceived emotions.[4]

Furthermore, this new study design considered the viewer's mundane and natural way of behaving. Therefore, stress-inducing measuring devices like tracking of eye-movements,[5] measurement of heart rates/ skin conductance or fMRI-methods (functional magnetic resonance imaging) imaging the participant's brain activity should be avoided. Consequently, the subjects watched the test video on a tablet computer's touch screen with headphones in a "homey" surrounding providing a relaxed atmosphere, tagging their reaction in a two-dimensional x/y rating scale with a movable indicator directly on the tablet's screen. Bullerjahn argued that this type of experimental design could result in more sensible outcomes[6] than former test designs. Hence, the emoTouch application,[7] an open source tool for research operating with Apple's iOS on an iPad tablet computer, was used as a mapping tool and therefore complied with Bullerjahn's guidelines (see section 2, Method).

Additionally, this study complies with the pragmatic research methods of Creswell ("Truth is what works at the time")[8] in order to clarify the following questions: Which audio engineering and post production techniques work generally better in sound design? Which audio solutions are immersing the listeners and viewers more successfully into the story?

But how can one "measure" emotions reliably? In the past, a lot of research was done to describe and to classify emotions:[9] What are the decisive affects responsible for the emotional reactions when someone is watching audiovisual media?

In this study, two key audiovisual affects were defined to analyze the emotional reaction of the participants watching the test videos: 'suspense' and 'immersion.' Although these two affects cannot be exactly defined in scientific terms, 'suspense' and 'immersion' allow a pragmatic approach according to the work experience of sound designers in audio production.[10] The participants' aural envelopment can be described sufficiently by these two key media affects to depict their personal emotional situation. Further research might establish wider concepts as to how far these two affects might be intertwined: perceived suspense, originated by the film plot, might be required to feel immersion; if there is no suspense there will be perhaps no perceived immersion. Indeed, this principle works also in the other direction: efficient immersion might be the basis for suspense. Cohen employs in her own studies the notions of 'absorption' or 'involvement.' [11] The latter term is very close in meaning to 'perceived immersion.' Nevertheless, 'immersion' comprises both the technical-acoustic envelopment (e.g., by surround sound or using headphones) and the emotional involvement (e.g., by the film plot). The dependent variable 'suspense' is similar to the term 'arousal'[12] (see also the next section). 'Perceived suspense' nevertheless is more a media affect, presumably describing the possible impact of the variables music, sound effects and film type in concurrence with 'perceived immersion.' The varying combination of these three variables depicted the bedrock of this research study: how did these three variables affect the perception of the viewers regarding their perceived suspense and immersion?

It has to be mentioned that the author deliberately disregarded the audio type 'dialogue' (speech) because this audio element is special in its functional purpose: monophonic dialogue has to be understood, meaning that the immersive impact of dialogue is in principle lower compared to the usually stereophonic variables music and sound effects. Dialogue is typically the loudest element in an audio mix and therefore pushes the two other variables into the aural background. The main purpose of dialogue is mostly to transfer information, less, to be immersive.

2 Method

2.1 A Mapping Tool to Measure Emotional Responses

The emoTouch application[13] allows participants to continuously tag their reaction in a two-dimensional x/y rating scale with a movable indicator (smiley marker) directly on the tablet's screen. Originally, the emoTouch application was developed as a tool for research on musical emotions, referring to Russell's original two-dimensional emotion space model with its valence (negative – positive) and arousal (active – passive) dimensions.[14]

The emoTouch application was the ideal tool for this study, because it added a scaling background to the videos and thus allowed the viewers to give instant feedback. They marked their felt emotions on a tablet computer's touchscreen while looking at the video on the *same* screen and listening simultaneously to the audio variables (music and sound effects) via headphones (see Figure 5.1). After each test run, the resulting x/y data stream could be easily exported and subsequently analyzed with statistical software.

Figure 5.1   A participant during a test run.

2.2 Stimuli

Assuming that viewers perceive the sound design in animated movies and live action movies[15] differently, it was decided to produce two films. The first one was a computer-generated stop motion film, and the second one was a photographed movie, which was staged with (silent) actors and which was shot with a camera. Afterwards, the sound tracks were added to both films during the audio postproduction.

It has to be noted that the two test videos are only two samples in audiovisual design and certainly do not represent the vast design options in movie production. Therefore, the results of this study can only be related primarily to these two film stimuli. Further research is necessary to estimate the extent to which these results could apply to other videos and film types.

The plot of both films represents the inherent story of the original music that was chosen for the soundtracks. This music was taken from two excerpts of Modest Mussorgsky's *Pictures at an Exhibition* in the original piano version: "Samuel Goldenberg and Schmuyle" (Length 1:47, used with the animated film, see Figure 5.2) and "The Catacombs" (Length 1:12, used with the live-action movie, see Figure 5.3). These two excerpts proved to be perfect for the strategic objective of this study. First, the visual plot of the films could be easily derived because Mussorgsky's compositions tell a story of their own (like a typical movie sound track). Second, the piano music provided enough auditory space to add the sound effects afterwards. The special acoustic-envelope structure of the piano tones enabled the researcher to easily blend the music with the sound effects.

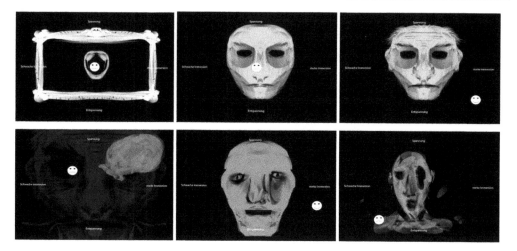

Figure 5.2 Animated film *Samuel Goldenberg and Schmuyle* within the emoTouch application with the smiley marker in different positions, moved by the participant (original video by Sebastian Bockisch © 2013).

The animated film *Samuel Goldenberg and Schmuyle* describes the struggle of a rich (yellow) and a poor (blue) Jew, fighting over money and alcohol in a pub. This fight was simultaneously staged, visually as a stop motion film and aurally as a radio play.

However, the viewer or listener perceives the fight more intensely by ear than by eye. You hear the splintering of glass, crashing of chairs, the shouting during the fight and finally the weeping of the loser: the poor Jew. The piano composition by Mussorgsky symbolizes each Jew with a leitmotiv: the rich one is represented by a droning bass melody, the poor one by a repetitive descant leitmotiv that sounds diffident and miserable. First, both guiding themes are presented consecutively. Later, Mussorgsky lets the two leitmotivs become 'controversial': they are played simultaneously, sounding dissonant.

The live-action movie *The Catacombs* was staged with actors in subterranean hallways, these represent the catacombs of Paris. In Mussorgsky's programmatic piano composition, the catacombs are musically figured by droning and gloomy bass chords; they are played very slowly. The simultaneous visual stimulus enacts a new film plot, taking place in these Parisian hallways: a man accidentally observes a murder.

In the closing scene, he is detected by the murderer and tries to escape. Eventually, the video comes to an end with a cliffhanger: it is not obvious for the viewer or listener whether the man's escape is successful or not. The (diegetic) sound effects were added and arranged in order to support acoustically this story visualized in the picture: You hear the steps of all actors and the moaning of the strangled victim. Heavy use of reverb illustrates sonically the hugeness of the catacombs.

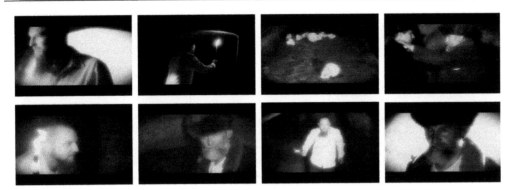

Figure 5.3   Live-action movie *The Catacombs* (original video by Robert Neuber © 2013).

Two students studying media production and media technology at Technical University Amberg-Weiden (Ostbayerische Technische Hochschule) in Germany produced both films as part of a student research project. Previously, the two pieces for piano were performed by the author and were recorded in the university's own music studio. The author subsequently designed the sound tracks for the two stimuli using sound libraries and additionally creating and arranging new specially designed sound effects.

The two newly produced film stimuli ensured that the participants had not seen them previously, meaning that each participant started the test runs at the same level of knowledge. Additionally, with this special test design the researcher got more control over the video content and the interplay of music and sound effects.

It has to be emphasized that both films were audiovisually conceived, designed and manufactured in a high production quality in order to prevent distraction by potential amateurish elements. Furthermore, the live action movie was staged with convincing actors.

2.3 Procedure

The two test videos were shown to the viewers in combination with one of four different audio tracks: without audio (silent), with music, with sound effects, and with complete sound design (music and sound effects). Each participant was exposed only to *one* video with a distinctive combination of film type and audio track in order to prevent any learning or habituation effects and to ensure a "fresh" emotional reaction of the viewer.

In a pretest, the participants could familiarize themselves with the setup. They practiced the handling of the device and the application with the first (animated or live action) video. The main experiment, which immediately followed, generated the test data using the second video. For example, if a participant had to rate the animated film in the main test run, the live action film was used for training and vice versa.

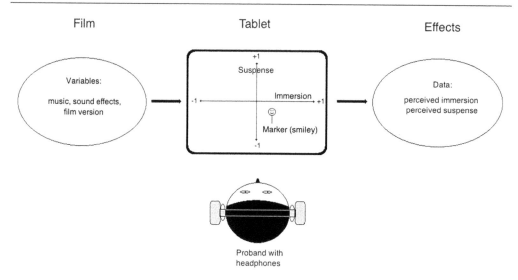

Figure 5.4 Scheme of the test procedure.

Watching the videos on the tablet computer 's screen in an office space of the university, the participants listened to the soundtrack via headphones individually (always adjusted with the same sound pressure level). The participants were asked to continuously mark their current emotional state on the tablet's touchscreen in two dimensions. The application's coordinate system was adjusted by the author: in the x-scale from -1 to +1, indicating the amount of the perceived immersion; in the y-scale from -1 to +1 describing the amount of perceived suspense. Without units, both scales represented thus only relative values (see Figures 5.2 and 5.4). The study design was based on three independent, dichotomous variables: music (yes/no), sound effects (yes/no) and film version (animated/live action). The emoTouch application continuously recorded the values of the dependent variables 'perceived immersion' and 'perceived suspense' (Figure 4), generating a continuous x/y data stream in steps of seconds. Using this data stream, the means, the maxima, the medians, the bandwidth and the evaluating graphics were calculated using the freely available statistics software R.

Immediately after the experiment the participants answered a short questionnaire (see section 5) consisting of eight questions about their former usage of recorded music and motion picture, about their musical experience (e.g., playing an instrument) and about their opinions and experience of the impact of audio on audiovisual perception. With these questions, the examiner intended to acquire an insight into each participant's aural history and aural bias. The participants answers could potentially interplay with the results of the recorded two-dimensional emotional data of the experiments.

2.4 Experimental Questions

How much did the stimuli with sound design increase (or decrease) the participant's perceived immersion and suspense compared to the silent stimulus?

To what extent was this change due to the music and/or the sound effects?

Did a certain combination of a film type (animated or live action movie) and audio type have a special impact on the viewers?

2.5 Participants

In total, at least 240 participants were necessary for this experiment. This number resulted from a minimum of 30 required cases per each condition consisting of two videos, each with four audio tracks. This resulted in a total of eight different conditions.

The average age was 25.74 years ($SD = 7.58$). One-third of the test participants were female. Of the participants, 82.5 percent were students enrolled in the study course Media Production and Media Technology at Technical University Amberg-Weiden in Germany. Although this signified a high affinity to audiovisual media as a producer and a consumer, only very few test participants were acquainted with the music of Mussorgsky beforehand. The examiner randomly inquired about this after the participant's test run. There was no compensation for participation.

3 Results

The collected data was prepared with a three-factor ANOVA using the factors 'film version', 'music' and 'sound effects'. The significance level was set to .05. This led to the following results, as can be seen in the tables 5.1 and 5.2 below:

The two factors 'music' and 'sound effects' significantly influenced the median values of the 'perceived immersion.' The maxima values of perceived immersion were only significantly influenced by the factor 'sound effects.' There was no significant effect found for the film type.

Table 5.1 Results of the significance test 'immersion'

Immersion	Medians (p)	Maxima (p)	Bandwidth (p) (Δ Max–Min)	Highest Median in
Music	*	.	-	Animated movie
Sound effects (FX)	**	***	.	Live-action movie
Film version	-	**	**	
Interaction	No interaction	No interaction	Film version: FX	

*** = p <.001, ** = p <.01, * = p <.05, . = p <.1, - = p >.1

Table 5.2 Results of the significance test 'suspense'

Suspense	Medians (p)	Maxima (p)	Bandwidth (p) (Δ Max- Min)	Highest Maxima in
Music	-	.	.	Live action movie
Sound effects (FX)	-	*	-	Animated movie
Film version	***	-	***	
Interaction	No interaction	No interaction	Film version: FX	

*** = p <.001, ** = p <.01, * = p <.05, . = p <.1, - = p >.1

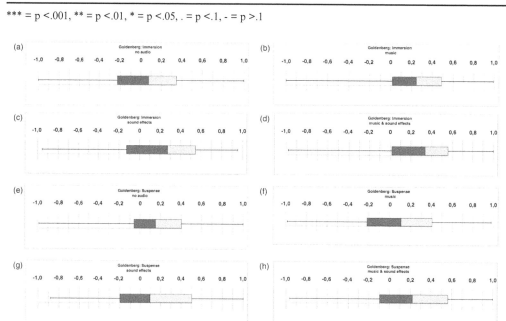

Figure 5.5 Boxplots of the participant ratings for the animated movie *Goldenberg*.

The film type significantly influenced the median values of 'perceived suspense.' Merely the maxima of the sound effects happened to have a significant difference regarding the emotional responses in suspense.

The Figures 5.5 and 5.6 provide an overview of the findings: the vertical black line between the black and gray boxes represents the median. On the left side of the black box is the 25-percent quartile, the right side of the gray box depicts the 75-percent quartile. The extrema of the participant ratings are indicated by the whiskers on the left and the right.

A preliminary note to the following paragraph: the values of the participants' emotional responses mentioned below are always juxtaposed to the values of the 30 null-version cases with no audio.

Watching both film types merely with sound effects, the participants experienced more than threefold higher values of perceived immersion (see 5.5 c versus a and 5.6 c versus a). However, showing the animated film with sound effects the perceived

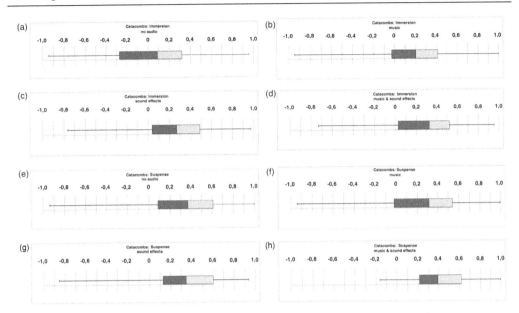

Figure 5.6 Boxplots of the participants ratings for the live action movie *Catacombs*.

suspense dropped by 40 percent (see 5.5 g versus e). Featuring the live action video with sound effects the perceived suspense declined by 10 percent (see 5.6 g versus e).

Median values of the perceived immersion augmented threefold for the animated film when only music was added (see 5.5 b versus a), and more than twofold for the live action movie when only music was added (see 5.6 b versus a). Suspense values were lower (see Figures 5.5 f and 5.6 f), but these changes were not significant.

Showing the live-action movie with music and sound effects (full audio mix) boosted the median values of perceived immersion 3.7 times (see 5.6 d versus a); screening the animated film with the full audio mix led to values of perceived immersion 4.4 times higher (see 5.5 d versus a). The values of perceived suspense rose in general, but these differences were only significant for the maxima of the sound effects (see Figures 5.5 h and 5.6 h as well as Tables 5.1. and 5.2).

A film with merely sound effects displayed a stronger impact on the participants' perceived immersion in contrast to a film with music (no matter which film type was shown). A track with a full audio mix, i.e., sound effects and music, indicated an even stronger impact on the participant's perceived immersion and perceived suspense compared to the videos provided just with music (no matter which film type was shown).

Using the free statistical computing software R, the variables of the experiments were checked on correlation with the answers of the post-test questionnaire. The test variables did not correlate with the participants' age, gender, personal aural history or their audiovisual habits (see section 5, Post-Test Questionnaire).

On the following pages the Figures 5.7 to 5.10 picture the mean immersion and suspense values of both film types on the y-axis. This y-axis is, as mentioned before, marked by relative units. The scale is defined from -1 to +1 in all the four figures. Each of the figures displays the audio conditions: no audio (dark gray graph), music (light gray graph), sound effects (gray graph) and the full audio mix (black graph). It has to be noted that at the beginning and end of the video timeline (x-axis), the four graphics seldom depict extreme high or low levels of perceived immersion or suspense. Presumably, the low number of active participants in the first or last two seconds of the videos might have caused these swings. These extreme values were deliberately omitted in the final discussion of the test results.

In Figure 5.7 the mean immersion values of the animated film show a higher average immersion i.e., at 55s running time: This may be caused by the visual action

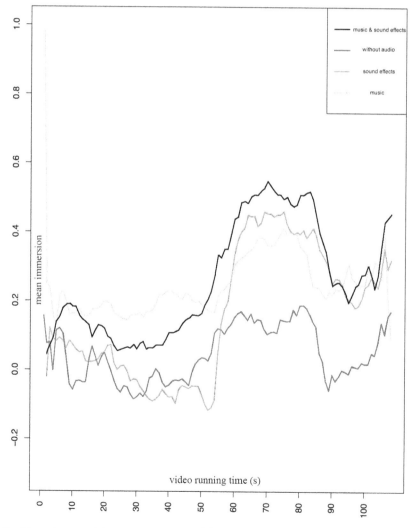

Figure 5.7   Mean (average) immersion animated film *Goldenberg*.

described in this special part of the video. The dark gray line representing the silent version depicts the same gradient with lower values. The black (music and sound effects) and the gray (sound effects) lines show much more dynamic immersion ratings than the light gray (music) and the dark gray (without audio) graphs. In general, music (light gray line) seems to balance the induced immersion, reducing its dynamic range compared to the other conditions.

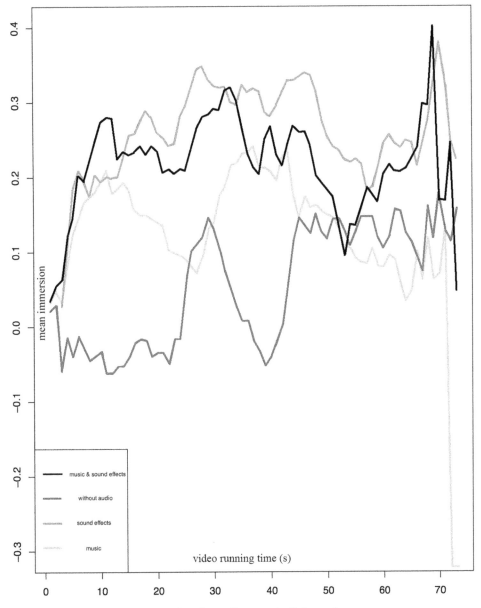

Figure 5.8 Mean (average) immersion, live action movie *Catacombs*.

Figure 5.8 displays the mean immersion values of the live action movie, indicating a higher immersion of the audio versions in contrast to the silent one. All four graphs in this figure are obviously less in sync compared to the ones of the animated film *Goldenberg* in Figure 5.7. The versions without audio (dark gray line) and with music (light gray line) produce the lowest immersion.

Figure 5.9 depicts the mean suspense values of the animated film *Goldenberg*. Like in Figure 5.7, figure 5.9 also displays a similar perceived suspense in the audio versions compared to the silent version. All four graphs in Figure 5.9 representing the animated film are more in sync compared to the ones in Figure 5.10 representing the live action movie. The music graph in Figure 5.10 (light gray line) produces, on average, the lowest suspense values during the video timeline.

Figure 5.10 illustrates the mean suspense values of the live action movie *Catacombs*: the light gray graph (music), as said before, has the lowest level of perceived suspense on average, whereas the dark gray graph (no audio) surprisingly indicates nearly always higher values of suspense. Sometimes this dark gray graph rises even higher than the gray graph (sound effects) and the black graph representing the full audio version (music and sound effects). This can be noted especially towards the end of the video with its final scene designed dramaturgically as a 'cliffhanger' (see section 2.2, Stimuli).

4 Discussion

In contrast to the silent version nearly all participants perceived enhanced levels of immersion when their headphones were filled with sound. However, this effect depended on the visual information and the type of audio. A mere music track rather tended to flatten the perceived dynamic range of immersion. Sound effects alone and in a full audio mix, i.e., music and sound effects, extended the dynamic range.

On the contrary, the 'perceived suspense' correlated more with the visual information based on the film plot. Sound effects alone did not enhance the perceived suspense in general. Only the full audio mix with music and sound effects had a stronger suspense effect on the participants. Therefore, the following conclusions concerning the practice of sound design can be derived.

Apparently, the immersive power of music alone and its induced suspense, triggered by audio, should not be overestimated (see also the sections Stimuli and Procedure). These are the principal findings concerning the viewers' perception. These findings can be reasoned from Figure 5.8 (immersion) and from Figure 5.10 (suspense) regarding the live-action movie *Catacombs*. An explanation could be that

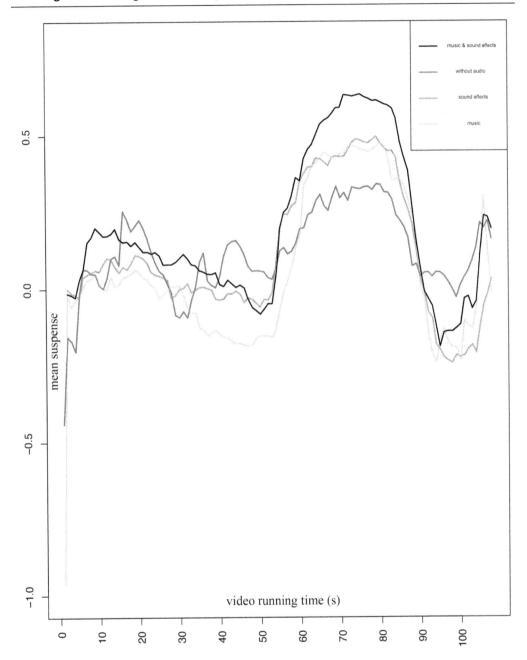

Figure 5.9   Mean (average) suspense, animated film *Goldenberg*.

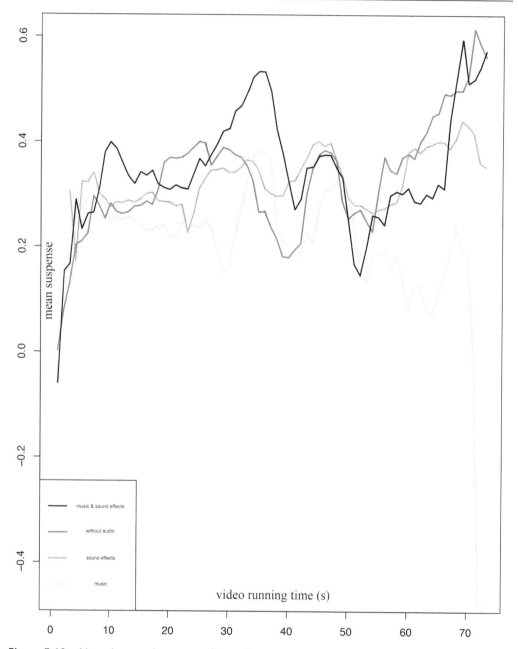

Figure 5.10 *Mean (average) suspense, live action movie* Catacombs.

the participants might have perceived incongruence between their visual and musical perception channels (see Cohen's Congruence-Association Model, CAM).[16] This incongruent sensation perhaps might have led to a lower emotional response[17] concerning immersion and suspense. Further studies have to be accomplished in order to estimate to what extent these findings might apply to other films. The researcher sees these findings of the study in accordance with his experience of more than 20 years of working in the sound design business. Therefore, the impact of sound effects should be taken more into account in the future: film sound producers and film composers perhaps ought to consider more intensely and more precisely the possibilities and the power of a well-designed audio track.

As in this study's test videos, it should be the well-balanced and congruent audio mix of music and sound effects that led to higher values of the participants' immersion and suspense in this survey. However as said before, it is difficult to easily derive these study's special conclusions to other films and videos. For that, the influence of the plot design and the dramaturgy is too important.

But what is in general the essence of a *well-balanced* audio mix? First of all, in audio postproduction, sound designers and composers take care of the different possible effects of the combinations and blending of different audio types like dialogue, sound effects and music. For example, when listening to a shattering of glass in its full impact, usually there should be no music at the same instant. It is even more impressive to mute the musical soundtrack a few seconds before the shattering event. That means that each audio track has to be worked on meticulously regarding the elements dialogue, music and sound effects: important aspects like the frequency curve, the loudness (audio compression) and the timing have to be taken into consideration very precisely. Only very experienced sound designers are able to fulfill these extraordinary tasks profoundly and completely.

Obviously, the sound design of a movie should follow a concept which is rather detached from the visual plot: the idea of a visually independent radio play.[18] After implementing this standard of quality in professional audio production, the auditory channel can really reveal its power in audiovisual media: the power of sound design. This kind of an accurately elaborated sound track, for example, has been attempted to realize in the study's test videos with non-diegetic sound effects: i.e., the sound layers of reverb and drones in the video *Catacombs*. Or sound effects like fighting noises caused by actions, which were not seen on screen in the video *Goldenberg*. In doing so, there is a higher chance of enveloping the participants and inducing them to feel the emotion (originated by the film plot) more intensely, because they feel even more immersed through the sound track. Due to the immersive power of these elaborated audio tracks, the participants perhaps could more easily match their perceived emotions to their own "neural representations"[19] or "blocks of cognition"[20] which they have built or developed in their imagination prior to the test procedure. That means the sound design of the test videos might have induced the participants to reactivate more vividly the neural pictures stored in their brain. Therefore, by watching an audiovisual movie, the participants felt the emotional impact in two ways: through the continuous actual perception, and additionally through the reactivation of their stored aural memories. Levitin asserted that there are two cerebral processes taking place in our brain simultaneously when we are watching audiovisual media: the analysis of audiovisual events and the concurrent and immediate comparison with our neural representations.[21] Hence, the findings of this study confirm Levitin's assertion.

However, the results of this study might raise a new aspect to discuss: To what extent does a participant's aural pre-experience change or manipulate his or her own current audiovisual perception? Although the post-test questionnaire did not show any significant effects, film experts presumably might react in a worldlier-wise and more experienced manner to the different audio versions than film novices. [22] The vast majority of the participants in this study already had considerable media expertise (82.5 percent were students of media production and media technology). Meaning, their experience already consisted of producing and consuming media content: this generation of web 2.0 users are often called 'prosumers.' Hence, the majority of the 240 participants formed a rather homogeneous and professionally experienced test group. This could be one possible explanation for the missing significant interaction between their answers in the post-test questionnaire and the measured perception in this study. Furthermore, the questionnaire consisted of merely eight dichotomous points (mostly yes/no). This might not have been detailed enough and should be further investigated in following studies.

5 Post-test Questionnaire

Immediately after the experiment the participants were asked eight dichotomous (mostly yes/no) questions which they could answer spontaneously without thinking about them too long:

1. Do you listen to music (e.g. radio, streaming audio) more than one hour per day?
2. Do you go to the cinema more than once a month?
3. Did or do you play a musical instrument?
4. Does an audio track generally help to understand the movie plot?
5. Which audio type helps more to understand the film plot: sound effects or music?
6. Which audio type lets you feel more deeply and more emotionally?
7. Can you describe the impact of the audio track on your audio-visual perception with one of the following four terms? Immersive – a new spatial awareness – 3D – emotional.
8. Does an audio track change the perception of time? Does a video with audio seem to be faster in perceived time than the same video without sound?

Notes

1 Chion, 2012; Görne, 2017; Holman, 2010; Lynch, 1998; Raffaseder, 2010; Steinitz, 2015; Sonnen-schein, 2001
2 Creswell, 2014, 200ff.
3 Brauch, 2012; Cohen, 2006, 2010, 2015; Flückiger, 2007; Hoekstra, 2012
4 Kock, 2018b
5 Auer et al., 2012
6 Bullerjahn, 2008
7 Scholle & Louven, 2013, 2015
8 Creswell, 2010
9 Russell, 1980; Schmidt-Atzert, 1996; Juslin et al., 2010; Todorov, 2017
10 Brauch 2012; Chion, 2012; Flückiger, 2007; Görne, 2017; Sonnenschein, 2001; Steinitz, 2015
11 Cohen, 2010
12 Russell, 1980
13 Scholle & Louven, 2013, 2015, 2017
14 Russell, 1980
15 Goldmark, 2013
16 Cohen, 2006
17 Boltz, 2004
18 Flückiger, 2007
19 Levitin, 2007
20 Rumelhart, 1980
21 Levitin, 2007
22 Filimowicz, 2012; Donnelly, 2015

References

Auer, K., Vitouch, O., Koreimann, S., Pesjak, G., Leitner, G., Hitz, M. (2012). When Music drives Vision: Influences of Film on Viewers' Eye Movements. *Proceedings of the 12th International Conference on Music Perception and Cognition*, 73–76. Retrieved from: http://icmpc-escom2012.web.auth.gr/proceedings. html.

Boltz, M.G. (2004). The cognitive processing of film and musical soundtracks. *Memory & Cognition*, 32 (7), 1194–1205. doi:10.3758/BF03196892.

Brauch, M. (2012). *Das Sounddesign im deutschen Spielfilm*. Marburg, Germany: Tectum.

Bullerjahn, C. (2008). Musik und Bild. In H. Bruhn, R. Kopiez, A.C. Lehmann (eds.). *Musikpsychologie – Das neue Handbuch*, (pp. 205–222), Hamburg, Germany: Rowohlt.

Chion, M. (2012). *Audio-Vision*. Berlin, Germany: Schiele & Schön.

Cohen, A.J. (2010). Music as a source of emotion in film. In P. Juslin & J. Sloboda (eds.). *Music and Emotion* (pp. 879–908), Oxford: Oxford University Press.

Cohen, A.J. (2015). Congruence-Association Model and Experiments in Film Music: Toward Interdisciplinary Collaboration. *Music and the Moving Image*, 8 (2), 5–24. doi:10.5406/musimoviimag.8. 2.0005.

Cohen, A.J., MacMillan, K., Drew, R. (2006). The role of music, sound effects & speech on absorption in a film: The congruence-associationist model of media cognition. *Canadian Acoustics*, 34 (3), 40–41. Retrieved from: https://jcaa.caa-aca.ca/index.php/jcaa/article/view/1812.

Creswell, J.W. (2010). *Mis 611: Research Design,* Retrieved from: http://mis611.wikidot.com/creswell-chapter-1.

Creswell, J.W. (2014). *Research Design: Qualitative and Quantitative, and Mixed Methods Approaches* – 4th ed. Los Angeles, USA: Sage

Donnelly, K.J. (2015). Accessing the Mind's Eye and Ear: What might Lab Experiments tell us about Film Music? *Music and the Moving Image,* 8 (2), 25–34. doi:10.5406/musimoviimag.8.2.0025.

Eklund, A., Nichols, T., Knutsson, H. (2016). Cluster failure: Why fMRI interferences for spatial extent have inflated false positive rates. *Proceedings of National Academy of Sciences,* 113 (28), 7900–7905. doi:10.1073/pnas.1602413113.

Filimowicz, M. (2012). The audio affect image: Five hermeneutic modalities of sound design. *The Soundtrack,* 5 (1), 29–36. doi:10.1386/st.5.1.29_1.

Flückiger, B. (2007). *Sound Design* (3rd ed.). Marburg, Germany: Schüren.

Goldmark, D. (2013). Pixar and the animated Soundtrack. In J. Richardson, C. Gorbman, C. Vernallis (eds.). *The Oxford Handbook of New Audiovisual Aesthetics* (pp. 213–226), Oxford: Oxford University Press.

Görne, T. (2017). *Sounddesign: Klang, Wahrnehmung, Emotion.* Munich. Germany: Carl Hanser Verlag.

Hoekstra, N. (2012). How to engineer a mood: A study of sound in audiovisual contexts. Bachelor thesis. Karlstad University, Sweden. Retrieved from: http://urn.kb.se/resolve?urn=urn:nbn:se:kau:diva-14120.

Holman, T. (2010). *Sound for Film and Television.* Burlington, MA: Focal Press (Elsevier).

Juslin, T., Liljeström, S., Västfjäll, D., Lundqvist, L. (2010). How does music evoke emotions? *Handbook of Music and Emotions* Oxford (pp. 605–642), Oxford: Oxford University Press. doi:10.1093/acprof:oso/9780199230143.003.0022.

Kock, M. (2018a). *Der Einfluss unterschiedlicher Audiogestaltung bei gleichem Bewegtbild – Der Weg zu einem effektiven Sounddesign.* Berlin, Germany: Schiele & Schön.

Kock, M. (2018b). The Power of Sound Design in a Moving Picture: an Empirical Study with emoTouch for iPad. *Empirical Musicology Review,* 13 (3/4) doi:10.18061/emr.v13i3–4.

Kock, M. (2019). *Wie der Ton zum Bild passt – Wege zu effektivem Sounddesign.* Berlin, Germany: Schiele & Schön (New Title Release of 2018a in 2019)

Levitin, D. (2007). *This is your brain on music: Understanding a Human Obsession.* London: Atlantic Books.

Lynch, D. (1998). The monster meets … David Lynch. Interview in *Home Theater Buyer's Guide.* Retrieved from: http://www. lynchnet.com/monster.html.

Nagel, F., Kopiez, R., Grewe O., Altenmüller, E. (2007). EMuJoy: Software for continuous measurement of perceived emotions in music. *Behavior Research Methods,* 39 (2), 283–290. doi:10.3758/BF03193159.

Raffaseder, H. (2010). *Audiodesign.* Munich, Germany: Carl Hanser Verlag.

Rumelhart, D.E. (1980). Schemata: The Building Blocks of Cognition. In R.J. Spiro, B.C. Bruce, W.F. Brewer *Theoretical Issues on Reading Comprehension* (pp. 33–58), New York, NY: Routledge.

Russell, J.A. (1980). A Circumplex Model of Affect. *Journal of Personality and Social Psychology,* 39 (6), 1161–1178. doi:10.1037/h0077714.

Schmidt-Atzert, L. (1996). *Lehrbuch der Emotionspsychologie.* Stuttgart, Germany: Kohlhammer Verlag

Scholle, C. & Louven, C. (2013). emoTouch for iPad: A New Multitouch Tool for Real Time Emotion Space Research. *Proceedings of the 3rd International Conference on Music and Emotion,* 70, Retrieved from: http://urn.fi/URN:ISBN:978-951-39-5250-1.

Scholle, C. & Louven, C. (2015). The consistency of continuous ratings and retrospective overall judgements for live performances. International Conference of Students of Systematic Musicology (SysMus15). doi:10.1037/pmu0000153.

Scholle, C. & Louven, C. (2017). *emoTouch application.* Retrieved from: https://www.musik.uni-osnabrueck.de/forschung/musikpsychologie_und_soziologie/forschungsprojekte/emotouch_web_en.html.

Sonnenschein, D. (2001). *Sound Design.* Studio City, CA: Michael Wiese Productions.

Steinitz, D. (2015, 10 23) Robert Zemeckis über Hollywood *Süddeutsche Zeitung,* weekend edition, Retrieved from: http://www.sueddeutsche.de/leben/robert-zemeckis-ueber-hollywood-1.2702043.

Todorov, A. (2017). *Face Value – The irresistible influence of first impressions.* Princeton, NJ: Princeton University Press
All figures and tables are taken from:
Kock, M. (2018a). *Der Einfluss unterschiedlicher Audiogestaltung bei gleichem Bewegtbild – Der Weg zu einem effektiven Sounddesign.* Berlin, Germany: Schiele & Schön. ©
Figure 2 is a video still. The original video was produced by Sebastian Bockisch © 2013.
Figure 3 is a video still. The original video was produced by Robert Neuber © 2013.

6

Using a Semiotic Approach to the Practice of Sound Design

Leo Murray

1 Introduction

We might take as a starting point in the practice of sound design the question: what is the ultimate aim of the sound design? We could equally reframe it as a question of the effect it has on the listener: what is the effect produced in the mind of the person who will experience the sound design? From the perspective of sound design we might begin by asking ourselves questions such as what information an audience knows, or should know, or might come to know, through experiencing the sound design. It may be that what we wish to instill in the audience is a particular feeling or sensation rather than particular sets of facts. Whatever the case, the work involved is determining the end result, and working to that end. All decisions are geared around working to create the combination of sounds or sounds/images/objects that will produce that desired effect/affect.

We could make a list of related questions for the implementation of the overall sound design in a particular project at the macro scale, or for a particular sonic element at the micro scale:

- How realistic or fantastical or abstract should it be?
- What connections should be suggested by a particular sound? What balance of elements would achieve this?, How obvious should this be?
- How complex or simple should it be? How loud or quiet? What to highlight, and what to reduce or remove?
- Should a particular sound be repeated or reintroduced to make a connection to another person, object, time or event? How obvious should this be?

Signs were of interest to many of the philosophers of the ancient world, particularly Plato and Aristotle. The founders of modern semiotics, Ferdinand de Saussure in Switzerland and Charles S. Peirce in the US, independently described the fundamental

DOI: 10.4324/9780429356360-7

nature of signs, although their work appeared almost simultaneously. Saussure in the *Course in General Linguistics* (1960) proposed a simple dyadic (two-part) structure of the sign, namely the relationship between that which is signified and its signifier. Importantly, he analyzed the sign as not merely the 'name' of a 'thing' but instead as a concept along with its sound-image (Saussure et al. 1960, 66). Saussure also described a fundamentally important characteristic of linguistic signs, that they are arbitrary; for example, the English and French words 'dog' and 'chien' refer to the same thing. Saussure (1960, 68) viewed language as the ideal example of the arbitrary sign system:

> That is why language, the most complex and universal of all systems of expression, is also the most characteristic; in this sense linguistics can become the master-pattern for all branches of semiology although language is only one particular semiological system.

Charles S. Peirce created a quite different, triadic (three-part) sign model. For Peirce, all of our understanding about the world comes to us via signs. Our senses give us signs from the world around us through which we learn to make some sense. For researchers applying this model there are implications in terms of an epistemological view – the study of knowledge, and how we obtain facts – as well as the ontological – study of being, and how we define reality, and to what degree perception and values are important to an inquiry of human existence.

As a methodology the Peircean semiotic model is pragmatic in essence in that it adopts an approach that evaluates the usefulness of a particular theory or belief in terms of the success of its practical application.[1] As a tool for analyzing sound design this particular semiotic approach also has some other benefits. It takes into account any type of sign rather than being linguistically based. It takes account of individual differences in coming to an understanding, and the modification of both the object and meaning of signs in light of context or supplementary information. Finally, it can be put to use analyzing both the practice and the product of sound design. In order to explore some of the ways we can make use of this semiotic model it is worth examining some of the principles of the model and its most fundamental elements.

2 The Universal Categories

In order to understand what Peirce meant by his system of 'Universal Categories' it is worthwhile mentioning the categories of Aristotle and Kant which Peirce's system of categories were designed to replace.[2] Aristotle classified the world (the classes of things that exist or, rather, that we can conceive) into ten categories. Kant also suggested a list of categories in four classes of three, but thought of them as properties or characteristics of any object. While acknowledging that there may be others,

Peirce sought to define three basic phenomenological categories, which he called
the Universal Categories: Firstness, Secondness, and Thirdness. These Universal
Categories are from three phenomenological conceptions: First, only through qualities of its own; Second, with reference to a correlate (a *something else*); and Third,
mediation, bringing a first and second into relation to each other.

At first glance, Peirce's categories seem so imprecise as to have little use, with the
names of the three categories – Firstness, Secondness, Thirdness – only adding to the
sense of vagueness. Yet they are fundamentally important to the system he developed.
Peirce first formally outlined them in his article "On a New List of Categories" [1867]
and continued to work on them throughout his life. In "One, Two, Three: Fundamental
Categories of Thought and of Nature" [1885] Peirce further defined these categories:

> It seems, then, that the true categories of consciousness are: first, feeling, the
> consciousness which can be included with an instant of time, passive con-
> sciousness of quality, without recognition or analysis; second, consciousness
> of an interruption into the field of consciousness, sense of resistance, of an
> external fact, of another something; third, synthetic consciousness, binding
> time together, sense of learning, thought.
>
>   (Peirce, Hartshorne and Weiss 1960, 1.377)[3]

As such, firstness is an unreflected feeling, immediacy or potentiality. Secondness is
a relation of one to another, a comparison, an experience or action. Thirdness is a
mediation, synthesis, habit or memory (Noth 1990, 41).

Thirdness is the sense made from the interaction of the three elements in a sign:
the object, its signifier and the interpreting thought of the sign: "In its genuine form,
Thirdness is the triadic relation existing between a sign, its object, and the interpreting thought, itself a sign, considered as constituting the mode of being of a sign"
(Peirce, Hartshorne and Weiss 1960, 8.328). Therefore, in the Peircean model, signs
are a phenomenon of thirdness. If we were to imagine the simplest forms of life millions of years ago we might imagine them to have the beginnings of consciousness.
For the simplest organisms, life is composed of firstnesses only; raw feelings with
nothing being related to anything. Perhaps eyes are still at the simple light or dark
stage in their evolution, hearing has not yet developed, and a sense of touch, taste
or smell does not yet yield any information that can be acted upon.

3 The Structure of a Sign

For Peirce, everything we can imagine, every object, every thought is the result of
signs. He defined the sign as consisting of three inter-related parts: a signifier, an
object (that is being represented) and an interpretant (the resultant effect in the

mind): "A sign is an object which stands for another to some mind" (Peirce et al. 1982, 3.66).

For our purposes a signifier is usually a sound. It is the relationship between the signifier, the object and the interpretant that determines how the sign will be interpreted. A sound example might be the sonic output of the Geiger counter in identifying the presence of radioactivity. The presence of radioactivity (the object) is inferred through the audible click sound or visual readout of the counter (the signifier), in order to create the interpretant (the presence of radioactive material nearby). The object is what is represented in the sign. The object of the sign need not be a physical object, but rather is simply whatever is being represented or signified, which could be an idea, a person, an inanimate object, a film or anything else: "An object is anything that can be thought" (Peirce, Hartshorne and Weiss 1960, 8.184).

The object determines the signifier. The interpretant itself acts as a signifier for a further triad, and thus semiosis is the never-ending process in which each interpretant can act as a signifier for the next sign. The link between object and signifier (and here a signifier is a sound) can be one of three levels from first to third: icon, index and symbol. Iconic sounds have no links to anything else – they exist only in themselves; indexical sounds act as evidence of an object, such as the event that caused the sound; symbolic sounds go beyond a mere causal link and instead are a learned link to the object borne of habit or rule.

4 Reasoning – The Role of Abduction

The ability to make sense of a sign requires some reasoning by some mind. Peirce suggested a new class of reasoning to add to the classifications of deductive and inductive reasoning, which he named "abductive reasoning." Abduction is the first step in logical reasoning, in that it sets a hypothesis on which to base further thought:

> All the ideas of science come to it by the way of Abduction. Abduction consists in studying facts and devising a theory to explain them. Its only justification is that if we are ever to understand things at all, it must be in that way.
>
> (Peirce, Hartshorne and Weiss 1960, 5.145)

Abduction is not something useful only in science. It is how we make sense of the world from the moment we are born. Abduction is also something that is so naturalized that we take it for granted. In 1868 Peirce set about examining some core beliefs about human faculties, in particular the idea of intuition. While it may seem common sense to us to feel as we instinctively know a particular fact without any prior knowledge, there is little evidence to support this view. There is often substantial difference between what we have seen or heard for ourselves and what we have inferred.

A child has, as far as we know, all the perceptive powers of a man. Yet question him a little as to how he knows what he does. In many cases, he will tell you that he never learned his mother-tongue; he always knew it, or he knew it as soon as he came to have sense. It appears, then, that he does not possess the faculty of distinguishing, by simple contemplation, between an intuition and a cognition determined by others.

(Peirce 1868, 105)

Rarely do we consciously acknowledge that we have a working or provisional understanding of events or what we think will happen, yet we are continually using provisional understandings and hypotheses; otherwise we would never be able to make a decision or judgement about anything. For example, any social interaction or everyday action such as driving a car, riding a bicycle or even walking down a street in any city requires us to make a multitude of these abductions in order to go about our business. We might make a guess at other people's motives, movements or predict their future actions, seemingly based on little more than a hunch. In terms of scientific inquiry there are three stages of belief: from possibility (abduction), to probability (induction), to certainty (deduction). Abduction is the hypothetical stage of reasoning that comes before the others.

In order to determine the object of a sign we must determine its use, or rather the way it is being used, since it may have multiple uses. Even the simplest indexical sound – a knock at the door – is evidence of someone knocking at a door, or rather evidence that some action has happened to create the distinctive sound of tapping on wood. For it to be understood as knocking on a door, some other process needs to have taken place. For the sound of knocking on wood to have the intended effect, it needs to be understood by the listener: *that sound means that someone is on the other side of the door and would like me to open the door*. The sound of knocking is a deliberate communication, and its meaning has to be learned. There is no natural law that determines that the particular sound of knuckle on wood means anything. It is understood as the result of habit, or a rule or law, which governs that shared understanding. The two basic forms of sign relations – icon and index – alone would be relatively useless without some way of linking similar concepts or events. In the Geiger-counter example we may know that the crackling sound of the counter indicates the presence of radioactivity, but the sound that is used to convey that information has absolutely no link whatsoever to the concept it represents. It is a symbol – a learned association or convention.

This process of icon to index to symbol happens so frequently and so seamlessly that we do not often realize it is a process at all. Imagine when you are introduced to someone for the first time. You are already using a fairly complicated sign system (language itself) to communicate and are now being given a new sound which will represent a new object – in this case a person you have just met. Unravelling the various layers of signs being used shows the massive complexity of sound signs we use every day. Interpretation frequently requires collateral information such as knowledge of the language being used, with which to interpret the sign. If we slow down the process we may see its components more clearly. If, for example, one is in discussion with a stranger speaking in an unfamiliar language, the simple task of using the words "me" and "you" are aided by pointing at each person in turn. Such collateral information helps to clarify the meaning of the word and its related concept.

5 Meaning-Making

Peircean semiotics takes account of the process of meaning-making, the role of the interpreter and of context, and so can provide a useful theoretical basis and a language for describing some of the many elements of sound practice and the ways sound can be used to add meaning. Just as words and visual signs have meanings, sounds and music also have meanings. Obviously, dialogue, being spoken language, has an overt meaning that is understandable in its own right, but it can also have multiple or additional, coded meanings, depending on the speaker of the dialogue, the situation or the intonation. Sounds and music have the potential for meanings or associations. For example, a simple sound like the knocking at a door indicates that there is someone at a door. Orchestral music such as Tchaikovsky's *Romeo and Juliet* is recognizable to an audience as a romantic piece of music, possibly because of its inherent romanticism, or from the orchestral strings providing the love themes which are most recognizable, or because we may know that it comes from a classic romantic ballet, which itself is based on a classic romantic play (Shakespeare's *Romeo and Juliet*), or because it has been used previously in a similar way in other media to represent a romantic moment or situation. In any case, as an audience we may recognize in the music signifying a sense of romance, regardless of how we came to think that.

These cultural meanings of sounds have fundamental implications for practitioners who wish to understand how they can be used and manipulated and how they might in turn be interpreted by an audience. Many of the 'standard practices' and most fruitful methods of working in the creation of sound can be analyzed from a semiotic perspective. Existing music brings with it associations of a specific time, genre, performer or feeling. Even music that is unfamiliar will have familiar aspects, whether they are the instrumentation or style that is reminiscent of another piece of music. Sounds or voices that are unfamiliar will nevertheless have similarities to people or things we have heard before, which we use to create new knowledge or adapt to our existing knowledge.

6 Ambiguity

For Peirce, ambiguity is the result of a process that begins with a hypothesis to determine the origin or meaning of a sound that is not recognized, or is partially recognized, or where it does not appear to fit its context. In a soundtrack for a film there are sounds that we attribute to the diegetic world of the story, and others that we can recognize as being non-diegetic, as is often the case with the musical score. Where there is an obvious visible (or at least diegetic) sound source, we can attribute the sound to its origin without fuss. What happens when listeners are confronted with a situation where insufficient context or information is provided in order to ascertain the origin of the sound source? We may hear a sound that does not match what we can see, or other sounds we can hear or what we might expect to hear.

Here the dominant reading is less clear. The rules that govern how a set of signs are to be interpreted and the conditions for the production and interpretation of meaning, are incomplete. The sound sign we are presented with does not always quickly tell us what it refers to. Here a lack of context can provide an opportunity to create tension, since incomplete understanding invites active participation on the part of the audience to create meaning for themselves. This 'creation of meaning' can be the application of experience of similar signs through various semiotic processes or concepts: as an iconic, symbolic or indexical sound sign; as having connotations that are implied by the sound; or as a trope such as a metaphorical, metonymic[4], ironic[5] or synecdochical[6] sound sign. In the absence of sufficiently clear evidence a process of *assimilation* or *accommodation* of the new sound sign into the code may take place through which the story can be understood (Blauert and Jekosch 2005, 200–201).

Ambiguity is one of the most useful techniques in the production of a soundtrack with which to engage an audience. Walter Murch advocates a stretching of the sound-image relationship in order to create a tension between the literal evidence of what can be seen and heard and the relationship that is created in the mind of the audience (Murch 2000). Indeed, for Murch it is this ambiguity, this incompleteness that is the source of the power of films, or any other representative art form:

> The danger of present-day cinema is that it can suffocate its subjects by its very ability to represent them: it doesn't possess the built-in escape valves of ambiguity that painting, music, literature, radio drama and black-and-white silent film automatically have simply by virtue of their sensory incompleteness -- an incompleteness that engages the imagination of the viewer as compensation for what is only evoked by the artist.
>
> By comparison, film seems to be "all there" (it isn't, but it seems to be), and thus the responsibility of filmmakers is to find ways within that completeness to refrain from achieving it. To that end, the metaphoric use of sound is one of the most fruitful, flexible and inexpensive means: by choosing carefully what to eliminate, and then adding back sounds that seem at first hearing to be somewhat at odds with the accompanying image, the filmmaker can open up a perceptual vacuum into which the mind of the audience must inevitably rush.
>
> (Murch 2000)

In Antonioni's *Blow Up* (1966), in which a photographer suspects he has witnessed a murder in a park, the sound of a click can be interpreted as a snapping twig, a camera shutter or the cocking of a gun (Weis 2007, 3). Here the sound-image relationship becomes deliberately atypical rather than the usual one of sound supporting or reinforcing the images. The photographic blow-ups – normally indications of exactitude and reliable evidence – gradually become more blurred and less trustworthy. While the soundtrack provides increasing clarity with multiple meanings, and therefore multiple interpretations. Used in such a way, deliberate ambiguity forces all but the least engaged audience member to find or create their own interpretation of the sounds and images presented to them.

7 Conventional Sound Codes

For films, codes of narrative like characterization, themes, setting, iconography and film techniques create consistent representational norms that we use to make sense of the signs that are presented to us. Whether the codes are received in their entirety is dependent on the *widespreadness* or otherwise of the code itself. In musical examples the specific operatic reference such as the use of Wagner's "Ride of the Valkyries" (1851) in *Apocalypse Now* may well be missed by the majority of the audience where a popular music reference such as "Satisfaction" (Jagger and Richards 1965) in the same film – being one which is more widely experienced, rather than formally acquired or learned, will have a greater chance of being attributed specific extra-textual meaning. This broadcast or narrowcast differentiation of codes can be applied to sounds that are more or less literal, specific or accentuated in terms of competing sound-signs (Chandler 2007, 170), such as foregrounding in the soundtrack.

The interpretation of a particular sign might develop over time. The *immediate* (initial) might give way to a *dynamical* (subsequent) interpretation. As with other types of sound-signs, the division of the object and interpretant in Peirce's model can be applied. The immediate object of the sound-sign is the music, and the immediate interpretant may well be recognition of the music or its performer. In the case of the "Valkyries" the subsequent dynamical interpretant may end up as a recognition that the music is 'classical', or 'cinematic' or it may simply be that it 'fits' the film because of the characteristics and properties of the musical instruments in the arrangement, which meld well with the images portrayed on screen. For others, the "Valkyries" music may be recognized to a lesser or greater extent: the name of the music, or Wagner as its composer, or its origin in his Ring Cycle. It may be that then the viewer/listener hears that music and brings them it way of interpreting the music as a symbolic representation, which informs the film, offering parallels from another text to the film in question.[7]

It should also be noted that TV shows, films and games are routinely produced with multiple soundtracks, or foreign language versions (FLVs), which have different dialogue, and often (as a by-product of the removal of the original language dialogue) Foley recordings replace what has been removed. In addition, FLVs may contain different music from the original language version. *2001: A Space Odyssey* (Kubrick 1968) contains the song "Daisy Bell," which is sung by HAL as he is shut down. In the FLVs this song is replaced variously by "Au Clair de la Lune" for the French version, "Hänschen Klein" for the German version and "Giro Giro Tondo" for the Italian version. In each case the song is chosen to convey a sense of a return to childhood for HAL as he becomes aware of his own mortality. The culturally specific nursery rhymes or folk songs achieve a similar effect in each of the versions of the film since the content of the song is largely irrelevant, with the song acting as a symbolic link to childhood and perhaps one of HAL's earliest memories.[8] As the music is a deliberate choice across the multiple language versions of the film, it is clear that it is not the music itself that is important, but rather the desired outcome from the use of the music. In Peircean terms, the music acts as a sound-sign that leads to the eventual dynamical interpretant, which is the emotional link to the sense of a return to childhood.

8 Learning What Sounds Mean

In the video game *Galaxian* (1979) – a variation on *Space Invaders* – shooting the enemy aliens as they periodically swoop down is rewarded with bonus points, and this is signified through a sound motif and a visible number indicating points won where the alien was destroyed. In amongst lots of shooting and swooping of aliens it might be difficult to determine initially what each sound referred to but eventually it comes to have meaning, which would render the numerical display slightly redundant as visual attention was needed elsewhere. In *Bomb Jack* (1984) different signature sounds accompanied game events – the birth of a new enemy, and their fall to earth to enable them to move, the player's jump, player's death, and so on. When a 'power pill' became available it was accompanied by a sound, and when the pill was used, it was both signified and timed by a musical motif that played for its duration. This enabled very clear feedback about the power pill's duration without requiring any countdown timer or visual representation to indicate the moment the pill would abruptly wear off.

In each case there was little or no need of instructions or rules to introduce these concepts as they became self-evident with play. What the sound designer considers is how that sound is going to be interpreted by the player, and how it might change or react to player actions, or respond to changing game dynamics, which can also give clues to the player. For example, when an object is picked up, should there be some musical acknowledgement or sense of reward? At other times the feedback to the player might be more subtle. If the ambience or music changes in some way immediately before the appearance of an enemy boss, that association may initially go unnoticed but might eventually be recognized as being somehow significant as the player becomes more accustomed to the game.

In order to play a game a player brings with them a set of existing paradigms – familiarity with other games or genres of games which function in a broadly similar way: side-scrollers, platforms games, first-person shooters, etc. Each player builds up a conceptual map that helps them understand the grammar of the game, and what they are required to do in order to progress through the game. The sound that accompanies an action is partly an aesthetic choice. The fact that the sound actually accompanies an action or event in the game is part of the game mechanics itself. For a sound designer, the sounds themselves, and the way the sounds interact or change as a result of the player's actions, and the game context, are all interrelated and part of the sound design process.

The semiotic concept of abduction as the preliminary stage of logical reasoning is a very useful one to apply to gameplay. We literally have to *learn how to play*, and learn what things are significant in the world that is presented to us as we interact with it. Sounds become linked with objects, actions and events. The exact choice for both synchronous event sounds and non-synchronous sounds might be entirely new, or might bring with them metaphorical echoes of previously heard sounds from the natural world, or from other games or other media. As each sound usage becomes more familiar it symbolizes its object and can be recognized and interpreted.

Sounds that accompany events or actions have a built-in indexical link to the event/action. Once that action is repeated and a similar sound is heard there is a very strong link made between the object (action or event) and its signifier (sound). By the same token there may be some deliberate ambiguity about the sound that is accompanying an action, or whichever object is being denoted by a particular sound. Some classic arcade games used this link between sounds and game actions to allow visual attention to remain elsewhere. The process of learning how to play a game is so naturalized that it might evade analysis, but it is a process nonetheless. At the start of a game we are presented with minimal information and work out the most reasonable explanation. We carry on with this provisional hypothesis until a better one emerges or is forced on us. Each new piece of information might be assimilated into our understanding or might modify it.

9 Conclusions

The seemingly different elements of the soundtrack may appear at first glance to be difficult to analyze in the same way that a written text or a photographic image can be analyzed. Yet, just as a moving image can be analyzed using tools adapted from areas of Saussurean semiotics, Peircean semiotic tools can be turned toward the soundtrack in order to uncover some of its uses, its strengths and its abilities to represent in ways that are too difficult, too simplistic or too cumbersome to be done through visual means. In addition, the analysis can inform the sound design rather than simply the reception and analysis by the audience. Using Peircean semiotics as a theoretical tool, the traditional practices, artistic hunches or techniques that are passed down through generations of practitioners can be examined to determine the fundamental theoretical underpinnings for their practical use. Filmmakers, game designers and other audiovisual producers can then get closer to harnessing the abilities and peculiarities of the visual and aural senses. As a means of deciphering the meanings in the audiovisual realm, semiotics is singularly useful since it "allows us to separate ideas from their representation in order to see how our view of the world, or a film, is constructed" (Turner 1993, 48).

The Peircean semiotic model is particularly useful in the analysis of sound since it takes us beyond the linguistic model and the content of a sound to the process of sound creation and sound reception. It highlights the role of the interpreting mind, and the wealth of other signs, particular contexts and experiences into which new signs are to be accommodated. What comprises the Peircean model? In addition to providing an overarching framework and a language with which to describe sound, the Peircean model has a number of specific benefits:

- it can be used as a means of analyzing individual sounds and sound-image relationships, as well as the soundtrack's role in the narrative
- it helps uncover and explain the creative processes involved in the creation of sound designs by illustrating a coherent theoretical basis for the practice
- it facilitates the critical examination of the practice by providing a language to explain processes and the theory embedded in the practice, which explain how sound is used to create meaning
- it takes into account the role of the audience and of individual interpretation in creating meaning
- it takes into account the possibility of meaning becoming modified over time or through collateral experience

The development of a conceptual framework that can be applied to all kinds of sounds, regardless of their function or positioning on the hierarchy of the soundtrack, allows collaborative and informed discussion between individuals and departments, practitioners and theorists alike. Adopting a semiotic model allows for sound to be conceptualized as a system of signs rather than simply sound types, such as dialogue, sound effects or music. It provides a means of describing how sound can be used to fulfil its many functions. It also provides the language tools to help explain not only the sound itself but what happens when it is heard by the listener, the process of understanding what happens when sounds are listened to in a particular context.

The task of creating the soundtrack or sound design can be described in terms of a series of questions about what the audience should know, feel or think as they experience it. With the overall goal of sound design as 'serving the needs of the story,' viewed in terms of Peirce's model of the sign, the choices in its creation are geared around creating the desired effect in the listener through a set of interpretants (the story, the game, the product, etc.).[9] Therefore, the sound-signs are manipulated in such a way that the interpretants can be understood or felt by the audience. This type of approach focuses attention on the decisions that influence how well the story is told, or rather how the story will be understood by the audience. It shifts the focus from the classification of the sound to its function in the soundtrack or sound design, where each element is selected and manipulated to serve the needs of the production.

Notes

1 Peirce was one of the founders of the philosophical school of Pragmatism, but after some disagreements with his colleagues later announced his coinage "pragmaticism", saying that it was "ugly enough to be safe from kidnappers" (Peirce, 1960, 5.414).

2 Aristotle's ten categories are substance, quantity, qualification, relation, place, date, posture, state, action, passion. Kant's table of categories contains four classes of three: Quantity (unity, plurality, totality); Quality (reality, negation, limitation); Relation (Inherence and subsistence, causality, community); Modality (possibility, existence, necessity) (see Thomasson 2013).

3 Many of Peirce's works are brought together in edited collections, primarily Collected Papers of Charles Sanders Peirce (volumes 1–8) and Writings of Charles S. Peirce (volumes 1–6 and 8). Here the citation is followed by the volume and paragraph number. For other sources standard referencing is used.

4 A sound metonym denotes one thing but refers to a related thing. Just as Canberra acts as a metonym for the Australian federal government, a sound can be used to represent an event or place, with which it is linked such as the use of a horn to represent an unseen car, or a national anthem standing in for a country.

5 Ironic sounds carry a contradictory meaning to what is actually heard. Here we could point to much of Michael Hordern's voiceover in *Barry Lyndon* or Michel Chion's idea of anempathetic music such as the use of "Stuck in the Middle with You" in *Reservoir Dogs*.

6 In a sonic synecdoche the part is substituted for the whole. In visual terms this might be a close-up, where a sound equivalent is a focus on one element. Individual sounds can be used to substitute for the greater mass of sounds to describe a place, such as a single explosion or whine of an incoming bomb for the sound of warfare.

7 Valkyrie ("Chooser of the Slain") in Norse mythology is "any of a group of maidens who served the god Odin and were sent by him to the battlefields to choose the slain who were worthy of a place in Valhalla. These foreboders of war rode to the battlefield on horses, wearing helmets and shields; in some accounts, they flew through the air and sea. Some Valkyries had the power to cause the death of the warriors they did not favour; others, especially heroine Valkyries, guarded the lives and ships of those dear to them." http://www.britannica.com/EBchecked/topic/622196/Valkyrie.

8 In addition, some of the songs chosen have a historical or cultural significance. Daisy Bell was the first song sung by the IBM 7094 computer at the Bell Laboratories. Au Clair de la Lune was the subject of

an 1860 phonautograph paper recording of a human voice. This was recently recovered and predates Edison's first sound recording. See: http://www.nytimes.com/2008/03/27/arts/27soun.html.

9 In the case of sound designs for objects, or simulators, or non-narrative films. Whilst not having a story as such, these could still be supposed to have a purpose. That purpose would be to create some sensation or response in the audience, player or user. The goal of the soundtrack in this case would be to serve that end.

References

Antonioni, Michelangelo. 1966. *Blow-Up*. UK: MGM. Motion Picture.

Blauert, Jens, and Ute Jekosch. 2005. "Assigning Meaning to Sounds – Semiotics in the Context of Product-Sound Design." In *Communication Acoustics*, 193–221. Berlin; Heidelberg: Springer.

Chandler, Daniel. 2007. *Semiotic : the basics*. 2nd ed., Oxford: Routledge.

Jagger, Mick, and Keith Richards. 1965. (*I Can't Get No*) *Satisfaction*: Decca. Song.

Kubrick, Stanley. 1968. *2001: A Space Odyssey*. USA: Warner Brothers Pictures. Motion Picture.

Murch, Walter. 2000. "Stretching Sound to Help the Mind See." *New York Times*, Oct. 1: 21.

Noth, Winfried. 1990. *Handbook of semiotics, Advances in semiotics*. Bloomington: Indiana University Press.

Peirce, C.S. 1868. "Questions concerning certain Faculties claimed for Man." *The Journal of Speculative Philosophy* 2 (2):103–114.

Peirce, Charles S., Max Harold Fisch, Edward C. Moore and Christian J.W. Klousel. 1982. *Writings of Charles S. Peirce : a chronological edition*. Bloomington: Indiana University Press.

Peirce, Charles S., Charles Hartshorne and Paul Weiss. 1960. *Collected papers of Charles Sanders Peirce*. Cambridge, Mass.: Belknap.

Saussure, Ferdinand de, Charles Bally, Albert Riedlinger and Albert Sechehaye. 1960. *Course in general linguistics*. 1st British Commonwealth ed. London: Owen.

Thomasson, Amie. 2013. Categories. In *The Stanford Encyclopedia of Philosophy*, edited by Edward N. Zalta.

Turner, G. (1993). *Film as social practice*. London, New York: Routledge.

Wagner, Richard. 1851. *Ride of the Valkyries* (*Walkürenritt* or *Ritt der Walküren*). In *The Valkyrie* (*Die Walküre*). Opera.

Weis, Elisabeth. 2007. "Tati, Hitchcock, Antonioni: Three Approaches to Sonic Creativity." *Offscreen* 11 (8–9): Forum 1.

Audi(o) Branding and Object Sounds in an Audio-Visual Setting

The Case of the Car

Nicolai Jørgensgaard Graakjær

1 Introduction

Please begin by watching the YouTube video "Audi – Vorspung durch Technik (Werbeslogan) – Audiologo" (see: http://bit.ly/AudiSonicLogo; Langhammer 2016). Then consider: what are we listening to? Obviously, the visual accompaniment strongly suggests that you are listening to the German car manufacturer Audi's sonic logo – that is, a specific subcategory of sound branding which embodies a short-lasting, rounded sound event deployed to identify and represent a given brand alongside a visual presentation of other continuous brand elements, such as a logo and a slogan. Generally, sound branding appears to play a pivotal role for the promotional strategy of Audi – although Audi is of course not the only car brand focused on sound; for example, "Every sound made by a BMW is analyzed by a team of over 200 acoustic engineers to ensure they are both mechanically and aesthetically correct" (Jackson and Fulberg 2003, 106; see also Bijsterveld 2014 et al., 145). Not only is Audi's focus on sound insinuated already by the brand name: the surname of the founder, August Horch, translates from German to "listen" in English and "Audi" in Latin, and the brand name "Audi" was introduced in 1910, when Horch withdrew from his own, original company (founded in 1899) and was unable to continue to use the brand name "Horch" (Erdmann et al. 2013, 12). Audi has also promoted their focus on sounds directly through various external communication channels. For example, "The Audi Corporate Sound unites recognition and creative freedom as regards content for the first time" (Bochmann 2011, 51). Also, within the practice field of sound branding, Audi has been highly praised among practitioners for its initiatives related to sound branding. For example, the launch of the "AUDI Sound Studio" – including the sound production referred to at the beginning of the chapter – has been viewed as a "historic moment [...] in the field of Sound Branding" representing a "benchmark case" (Illner 2010).

DOI: 10.4324/9780429356360-8

However, whereas the sounds can be straightforwardly identified as a(n allegedly successful) case of sound branding from Audi, what we are listening to in terms of the characteristics and potential meanings of the sounds is less obvious. On the face it, the sonic logo includes sounds conventionally associated with music defined as sound structures, which embody a discernible melodic, harmonic and/or rhythmic structure. Most significantly, the sonic logo begins by introducing the tone of A-flat (pitched a major third below the middle C) articulating a synth-keyboard tone quality. Right after its introduction, the tone comes to function as a continuous background for a foregrounded rhythmic configuration. The rhythmic impulses seem to group in two and establish a 3/8 meter with an accentuated upbeat and a fade away of the last impulse; that is: "THREE / ONE, two, THREE / ONE, two...". The tempo is relatively fast paced and approximate 270 beats (eighth notes) per minute. Generally, the rhythmic impulses do not embody sound qualities conventionally associated with musical instrumentation. Also, the impulses are not verbal and they do not seem to originate from the human voice. Rather, the impulses can be heard as object sounds, a term here suggested to account for those sounds that are apparently non-verbal and non-musical, and derive from the motion of physical objects – alternative terms for objects sounds include "sound effects" (Miller and Marks 1992), "auxiliary sounds" (Yorkston 2010) or "environmental sounds" (Özcan and van Egmond 2009). The accentuated impulses embody two low registered thumping sounds that differs slightly as the second "thump" is marginally more forceful than the first one. The thumping sounds are accompanied by clicking sounds, three of which are heard rather indistinctively at different pitches in a relatively high register. However, the presence of the clicking sounds becomes clearer as one of them occupies the first of the unaccented beats. The second, unaccentuated impulse embody a longer-lasting, bright swoosh sound that fades away and thereby ends the sonic logo.

The sonic logo's inclusion of what appears to be object sounds exemplifies a challenge for current research on sound branding. There is a scarcity of literature on how to perform a systematic examination of the characteristics and potential meanings of object sounds in the context of sound branding. There are a number of reasons for this. First, the field of sound branding is heavily practitioner oriented in the sense that insights are often offered by practitioner for practitioners (see, e.g., Bronner and Hirt 2009; Groves 2011; Jackson and Fulberg 2003; Lucensky 2011; Treasure 2011). This orientation has more to offer in terms of presenting anecdotes and advices than in providing procedures for systematic examination. Second, the research that actually does exist has focused more on other types of sounds than on object sounds. This could seem to misrepresent the fact that object sounds are ubiquitous and appear as an integral, more or less pronounced dimension of all brand objects as amply illustrated by the car. By comparison, other types of sound – music and verbal sounds – are possible but not necessary ingredients of sound branding. Nevertheless, research on sounds in the contexts of branding has focused predominantly on music (see, e.g., Graakjær, 2015; Gustafsson, 2015). Third, when focusing on the research on object sounds that actually does exist, experimental methods seem to predominate. For example, within the product category of the car, there exists a body of publicly accessible experimental studies which generally illustrates that different variants of the same type of object sound produces different effects. Among other aspects, the studies have examined the sounds of turn indicators (Wagner and Kallus 2015) and the closing of car doors (see, e.g., Parizet, Guyader and Nosulenko 2008). However, while the experimental research suggests that object sounds *can* affect the impression of the object – although the effects appear ambiguous – this line of research has less to offer with respect to illustrating *why* and *how* the sounds hold the potential to do so. With this background, this chapter aims to present an application of close reading methodology to systematically examine why and how object sounds in audio-visual settings holds the potential to influence the experience of brand objects. By relying on a case of a sonic logo, the chapter also wishes to make contributions to the scant literature on how to analyze the characteristics and potential meanings of this particular example of sound branding which appear particularly disposed for the use of object sounds because of its short duration.

The product category of the car seems particularly well-suited for present purposes. Generally, the car can be viewed as an aesthetic object in the sense that aesthetics "deal with the experience of objects which provide the consumer with an element of beauty, or which are emotionally and/or spiritually moving" (Charters 2006, 239). If this is indeed the case, that "[c]ar consumption is never simply about rational economic choices, but is as much about aesthetic, emotional and sensory responses to driving" (Sheller 2004, 222), then is seems reasonable to suggest that the sounds of the car can help produce such responses and somehow perform the role of identifying and representing the brand in audio-visual settings as well. Obviously, when a car is seen on display in a showroom, it does not normally emanate sound. However, when managed – handled, started and driven – the car produces a wide variety of sounds. The exemplary case of the car thereby clearly illustrates how brand objects of all kinds more or less pronouncedly embody sounds alongside other sensory characteristics (visual, olfactory, tactile and gustatory). An indication of the brand value of the sounds of cars can be found in the practice of making the sounds available for focused listening with the purpose to offer a *autonomous sensory meridian response* (see, e.g., RSG 2016). Moreover, in addition to the alleged central position of the brand in the context of sound branding, the sonic logo of Audi is (expectedly) well known and accessible; the latter feature is particularly relevant for a close reading of sounds, as it cannot rely on stills or 'frozen' sound events but must be based on "repeated hearings of a single sound" (Chion 1994, 32). The chapter uses the case of the sonic logo as a springboard to present relevant theories for a close reading of sounds which can hopefully be generalized to other types of object sounds in audio-visual settings of sound branding.

The chapter's approach is informed by semiotics, as it aims to show how sound embodies part of the potential meanings that can be considered to be built into brand objects (inspired by Danesi 2007, 3). The approach is hence based on the premise that brand object sounds influence the brand, given that a brand is defined as "a semiotic enterprise of the firm" (Tybout and Calkins 2005, 41) including "a set of associations linked to a name, mark, or symbol associated with a product or a service" (Tybout and Calkins 2005, 1). Specifically, the chapter is stimulated by Michel Chion's (1994) identification of three "modes of listening" (introduced in the context of film sound) as well as Charles Sanders Peirce's (1955) tripartite classification of modes of relationships between a representamen and its object – or signifier and signified (following Chandler 2007, 36). Also, this perspective bears resemblance to the approach offered in Gaver (1986) in the context of analyzing sounds in computer interfaces – a context relevant for the present chapter's focus on object sounds exemplified by the sounds of cars – where three kinds of mapping between data and the auditory means used to represent it are presented. Whereas the three modes of listening can help organize and sharpen the auditory analytical attention, the modes of relationships can help conceptualize and describe ways in which sounds produce potentials of meanings (for more on the relationships between the perspectives of Peirce and Chion in the context of sound analysis, see Capeller 2018). Overall, this approach to the sounds of brand objects offers the advantage of being based on the way people actually listen to objects and events in their everyday life (for a similar argument, see Gaver 1986, 176).

After a short introduction to the sounds of branding, the chapter proceeds to present and illustrate the three modes of listening which will be addressed one at a time. The three modes overlap and combine, but they are presented separately for educational reasons. Subsequently, the object sounds from the sonic logo are profiled as specific examples of interior car sounds. The chapter ends with a summary of the presented analytical perspectives including suggestions for their wider applicability.

2 The Sounds of Branding

A simple typology of connections between sounds and brand objects can provide an initial conceptualization of sound branding. Two perspectives inform the suggested typology. On the one hand, the typology is structured according to different kinds of sounds. Specifically, a widely used and accepted tripartite distinction (inspired by van Leeuwen 1999) between music, verbal sounds and object sounds has been adopted. Whereas music and object sounds have already been introduced above, verbal sounds include the prosody of the human voice (i.e., speech intonation, speed of delivery and accent) as well as phonetic symbolism (i.e., associational meaning caused by the phonological structure and fluency). The tripartite distinction is here

Table 7.1. Possible relations between types of sound and connections between the sounds and the brand object.

Type of sound	Intrinsic	Extrinsic
Object sounds	*Examples: product and packaging sounds*	*Examples: sonic logos*
Verbal sounds	*Examples: voice of GPS system*	*Examples: prosody of names and slogans*
Music	*Examples: melodic motifs as indicators*	*Examples: music in advertising*

proposed for practical, analytical reasons. While it should be obvious that each of the three types of sound can exist in their own right, it is equally obvious that, occasionally, the different types overlap and blend as exemplified by the case sounds. On the other hand, the typology is informed by different kinds of connections between sounds and objects. The connection can be established and performed in two different ways (following Graakjær and Bonde, 2018). In one way, the branding process addresses sounds that connect intrinsically to the object in the sense that they embody and emanate from the handling of object: they are *the sounds of the object*. In another way, the branding process focuses on sounds that are extrinsically connected to the object. By existing "on behalf of" the object, that is, they are *the sounds for the object*, they basically serve representative and authentication purposes. Table 7.1 provides an overview of the possible relations between, on the one hand, types of sound and, on the other hand, connections between the sounds and the brand object.

The table indicates that the same type of sound (e.g., object sounds) can be connected to a brand in different ways. Conversely, different types of sound can be connected to the object in the same way. Additionally, the typology implies that not all sounds in an audio-visual setting are connected to a brand object. For example, the sounds of a car can appear as intrinsic when they are associated with a particular brand in a car commercial. However, the same or similar sounds might not appear as sound brands by failing to qualify for any of the six categories when they appear, for example, as inconspicuous parts of a commercial for some other brand object.

3 Reduced Listening and Iconic Signification

The perspective of reduced listening addresses what the sounds *sound like*. According to Chion, reduced listening "takes the sound [...] as itself the object to be observed instead of as a vehicle for something else" (1994, 29). However, in this context, the study of the sounds themselves will inspire an examination of iconic signification – that is, a relationship between signifier and signified based on imitation or resemblance (Chandler 2007, 36). As we are habitually inclined to listen to sounds from the perspective of their origin (see more below on causal listening), we are faced with some "difficulty of paying attention to sounds in themselves" (Chion 1994, 29). Yet, reduced listening has the advantage of "opening up our ears and sharpening our power of listening" (ibid., 31). Consequently, reduced listening facilitates a specification of the distinctive features of the given object sounds.

The analysis of the sounds' quality and structure includes an analysis of "some distinctive, recognizable, and recurring physical form" (Danesi 2007, 29). Accordingly, the temporal and dynamic form of sounds can be analyzed from the perspective of *sound envelope*, which specifies a sound events' attack, decay, sustain and release (Tagg 2013, 277ff.). As already indicated, the sonic logo includes several distinct object sounds. The clicking sounds embody an abrupt attack with no extended decay, sustain nor release. The thumping sounds are slightly more rounded as the attack and release are both rather abrupt (which allows them to play a distinct rhythmic role) yet "soft" with respect to both the attack and release. By comparison, the swoosh sound at the end of the sonic logo evolves through a more extended attack (the sound initially emerges with an increase of intensity) followed by a short-lasting decay (a decline in the intensity of the sound) which sustains only very shortly (regarding the continuation of the sound) before a fading release (the ending of the swoosh sound).

Further aspects of what the sounds sound like can be approached from the perspective of sound modality, as presented by van Leeuwen (1999). Modality refers to what extent the sounds embody "a true representation of the people, places and/or things represented" (ibid., 180). Van Leeuwen offers eight articulatory parameters (1999, 172ff.) for the analysis of various degrees of modality. Already, the parameters *durational variety* and *dynamic range* have been included when examining the case sounds from the perspective of envelope. Additionally, the *pitch range* of the case sounds is rather static, and embodies only a minor increase and decrease in the pitch (or melody height) of the swoosh sound. The thumping sounds are statically low registered whereas the clicking sounds reside in a high register. There is no significant *degree of friction*, as the sound qualities of the object sounds do not appear distorted. Also, there is no significant *degree of fluctuation* as the sounds embody no obvious vibrato. As regards *perspectival depth*, there is clear differentiation of background and foreground. The musical tone, as identified previously, sets the background for the rhythmic impulses, of which the thumping sounds are clearly foregrounded. With respect to the *degree of directionality*, the thumping sounds appear in extreme "close up," as if we were listening to the sounds with our ears in very close proximity to the given object. Moreover, *absorption range* is significant, as the sounds generally appear with some degree of reverberation – the general quality of the sounds are thereby somewhat "wet" as opposed to entirely "dry" (van Leeuwen 1999, 181).

From the specification of these articulatory parameters, the case sounds resemble – and hence have an iconic relationship to – specific sounds as they are (conceivably) known from outside the particular setting of the sonic logo. The foregrounded position of the thumping sounds makes them particularly noteworthy. These sounds might be considered to bear resemblance to sounds produced by a human heart. Thus, actually, each heartbeat includes two sound events: "a low, slightly prolonged "lub" (first sound) occurring at the beginning of ventricular contraction […] and a sharper, higher-pitched "dup" (second sound), caused by closure of aortic and pulmonary valves" (Britannica 1998). The average listeners might not have actually heard the sounds themselves, unless having experienced *objective, pulsatile tinnitus* (Snow 2004, 1). However, listeners could be familiarized with the sounds through various media settings (e.g., documentaries, TV series and films), where the sounds occasionally emerge as heard from, for example, a doctor's "point of audition" (Chion 1994, 89ff.) when checking the heart rhythm of a patient through a stethoscope. From this perspective, the sonic logo includes two heartbeats. However, an alternative association based on the articulation of the thumping sounds is the sounds of a car running over an unevenness on the road surface. Perhaps most illustratively, the sonic logo's representation resembles the sounds from a moving car's tires against the small gaps in the road surface of a bridge (e.g., expansion joints). From this perspective, the two slightly varied accentuated beats resemble the small difference in the sound quality from, first, the sounds of the front wheel tires, and, second, the (rapidly succeeding) sounds of the back wheel tires.

As regards the swoosh sound, obvious resemblances to other sounds do not seem to emerge. Arguably, the sounds could be considered to appear "metallic" and hence to resemble the sounds of friction between metal materials. For example, the "schwing" sound from a sword being unsheathed could come to mind, but compared to the "schwing" – possibly a conventionalized sound quality produced for the purpose of effect in films – the swoosh of the sonic logo appears in a lower register (less "bright") and more prolonged in terms of its attack and release. As to the less pronounced clicking sounds, the associations to other sounds are likewise less obvious. Arguably, the sounds could be heard to bear some resemblance to the sounds from switching or pressing buttons.

The concept of anaphones – offered by Tagg (2013, 486ff.) in the context of music analysis – can be used to examine even further aspects of resemblances. Tagg identifies anaphones as *"homologous* sign types" that usually appear as composites of "three main categories – *sonic, kinetic* and *tactile*" (Tagg 2013, 487; italics in original; Tagg explicitly excludes tastes and smells [see 2013, 496ff.]). For a similar approach, though focused more on perception than on sound structure and including a focus on tastes, see the examination of cross modal correspondences in Spence and Wang (2015). Anaphones can be used to address the sounds' possible homologous relations to not only other sounds but also other sensory phenomena and thereby the main categories can be seen as specifications of iconic signification. Non-auditory resemblances have already been touched upon when, for example, identifying the sounds as "wet" and not "dry" – illustrating a tactile anaphone. Also, the swoosh sound can be seen from the perspective of a tactile anaphones illustrated by the metallic, "cold" sound quality. These examples can be seen to illustrate how "[h] earing is a way of touching at a distance" (Schafer 1977, 11).

4 Causal Listening and Indexical Signification

From the perspective of the causal mode of listening, the analysis addresses *from what and where the sounds originate*. Correspondingly, by way of indexical signification – that is, a relationship between signifier and signified based on direct (causal or physical) connection (Chandler 2007, 37) – sounds offer the listener an impression of the characteristics of the "object that has the sound" (Pasnau 1999, 316) as well as aspects of the context of that object, that is, "an *interaction* of *materials* at a *location* in an *environment*" (Gaver 1993, 6; italics in original). Often, reduced and casual listening emerge in tandem as the sounds' quality and structure form the basis for the assessment of from what and where the sounds originate: "What leads us to deduce the sound's cause if not the characteristic form it takes?" (Chion 1994, 32). Illustratively, the previous examination of the sounds' resemblance to other sounds has already implied perspectives on the (possible) origin of the sounds – for example, the thumping sounds have been likened to the sounds from the human heart, and the swoosh sounds have been described as the sounds from the interaction of metallic materials. The latter example illustrates how the quality of the sounds offer information of the physical materiality of the sounds' origin.

To examine this perspective in further detail, Gaver (1993) proposes a tripartite classification of object sounds based on differences between the attributes of objects and their interactions. First, *aerodynamic sound* is characterized by a "direct introduction and modification of atmospheric pressure differences of some source" (Gaver 1993, 13). Aerodynamic sounds can appear continuously (e.g., the whistling sound from a tea kettle) or explosively, as illustrated by a champagne bottle pop sound. Second, *liquid sounds* are generally "determined by the formation and change of resonant cavities in the surface of the liquid" (Gaver 1993, 15). For example, it is normally "easy to tell whether a liquid gurgling out of a bottle is water, or a thicker syrup or oil" (ibid., 16). Third, sounds from 'vibrating solids' emerge as the result of vibrating objects, that is "when a force is exerted upon and then removed from a system that is otherwise at equilibrium" (Gaver 1993, 6).

From this perspective, the object sounds of the sonic logo predominantly exemplify the sounds of vibrating solids. For example, in addition to the metallic quality of the swoosh sounds, the clicking sounds likewise indicate a friction between small and swiftly moving metallic elements. Also, when approached as the sounds of car tires running over gaps in the road surface, the thumping sounds could be considered to include dimensions of vibrating solids and hence indicate a relatively fast, firm and stable car ride. However, the thumping sounds represent a rather ambiguous sound quality. When heard as the sounds of the human heart, the sound quality could be considered to include aspects of aerodynamic sounds in the form of pressure differences (a pumping heart). The particular organization of the heartbeat sounds further indicates a relatively excited heart: the powerfulness and foregrounding of the sounds combine with a relatively fast-paced pulse – that is, the sonic logo expresses a heart rate of approximately 90 beats per minute compared to the average adult rate of 70 beats at rest (Britannica 1998) – to indicate a relatively excited human heart.

Basically, from the perspective of causality, the question arises as to whether the sounds originate from what they seem to originate. For example, the sounds might be produced by other means than what would seem obvious – for example, synthesizers or produced from other objects. The practice of Foley artists in the context of film sound production further illustrates the point (see, e.g., Ament 2014). Indeed, the causal mode of listening has been referred to as "the most easily influenced and deceptive mode of listening" (Chion 1994, 26). However, no matter how the sounds are actually produced, what matters from the perspective of a close reading of sounds is to what extent the sounds resemble the sounds in question and whether this resemblance is perceivable to listeners.

As to the more specific question of whether or not the sounds represent a "true representation" of the resembling sounds, the thumping sounds appear intensified and decontextualized. They appear hyperbolic, that is, somewhat exaggerated in extreme close up. This feature can be seen to reflect the context of the sounds as they are made available for purposes of promotion and branding. In addition to help arrest the attention of viewers – illustrating the indexical function of the sounds, given that "[a]nything that focusses the attention is an index" (Peirce in Chandler 2007, 42) – the hyperbolic appearance of the sounds mirrors conventions of advertising, where brand objects are typically represented looking (or, indeed, sounding) their best and not necessarily the way they will precisely appear in all settings of everyday life. Van Leeuwen provides the comparable example of a lemonade commercial, where the "bubbles fizz and the ice cubes clink against the glass in crystal-clear and hyper-real fashion" (1999, 176). Furthermore, this corresponds to the *sensory coding orientation* that listeners can adopt when they review the modality of sounds. In this orientation, "what matters is emotive effect, the degree to which the sound event has an *effect* of pleasure or its opposite" (van Leeuwen 1999, 179).

A further aspect of the sounds' origin concerns the issue of whether or not the sounds can be seen (or are implied) in the audio-visual setting in which they appear. Generally, the setting of a promotional video for the car manufacturer Audi motivates a reading of the sounds as somehow related to cars and Audi. Also, the detailed visualization of the sounds encourages specific readings. The beginning of the visual sequence is marked by bright red lights intermixed with flickering white beams of light. Initially, the lights are vivid and blurred, but soon, as they move towards the right of the screen, they become increasingly focused. Ultimately, the brand name is presented in a saturated red color while the slogan "Vorsprung durch Technik" ("Progress through technology") is presented in white letters. At the beginning of the sequence, the particular (quality and movement of the) colors seem to suggest that the viewer is positioned inside a moving car. This reading is supported by the presence of reverberation (as mentioned previously), which suggests that the sounds emerge in a particular location – that is, reverberation is the result of the size of a space and the acoustically absorptive quality of the space's finishes (for an examination of the acoustical properties of car cabins, see, e.g., Kaplanis et al., 2017). This perspective supports an interpretation of the thumping sounds as originating from the cabin: as the sounds of car tire friction or the sounds of one own's (excited) heart while driving Audi. Likewise, the clicking sounds could seem to originate from the handling of car switches and buttons, and, especially motivated by the transition from blurring to focused lights, the clicking sounds could be heard as associated with the operation of the windshield wipers – as if the screen is wiped clean.

However, these readings are based on the implied presence of a car cabin, as the aforementioned car sounds do not originate from any explicitly visualized objects. Therefore, the sounds are *acousmatic*, as coined by Chion (1994, 32). Only the swoosh sound seems to originate from a specific visual object, namely the animated introduction of the brand logo in the form of the four merged rings. As the dynamic envelope of the swoosh sound corresponds synchronically to the gradual illumination of the four rings, the interaction of the visuals and the sound event exemplify how not only music (as originally referred to in the following citations) but also object sounds can be seen as "a bundle of generic attributes in search of an object" (Cook 1998, 23), and that the sounds "readily finds an object" (Cohen 2001, 263). Incidentally, the quality of the swoosh sounds seems to correspond to the metallic, aluminum colored presentation of the logo, and the sounds help stimulate an association between Audi and a certain (aluminic, advanced) built quality. To further the understanding of this process, it might prove fruitful to include the perspective of information linking, that is, an examination of "how items of information [...] can be and are *meaningfully linked* to other items of information" (van Leeuwen 2005, 219ff.; italics in original). Specially, although Roland Barthes' (1977) focus on relations between elements was primarily concerned with linguistic messages and pictures, this approach can arguably be broadened to include the examination of relations between sounds and visuals. From this perspective, the swoosh sound performs a denotative function as it "replies [...] to the question: what is it?" and "helps to identify purely and simply the elements of the scene"; consequently, the function of the swoosh is principally that of anchorage (Barthes 1977, p. 39) and thus "very close" to the concept of elaboration as a form of specification (van Leeuwen 2005, 229).

5 Semantic Listening and Symbolic Signification

The perspective of semantic listening examines *to what the sounds refer* beyond their resemblance to other sounds and the object and location of their origin. The relationship between signifier and signified in the symbolic mode is "fundamentally *arbitrary* or purely *conventional*" (Chandler 2007, 36; italics in original), and based on learning and convention. As was the case with the relationship between causal and reduced listening, semantic listening can overlap: "Obviously one can listen to a single sound sequence employing both the causal and semantic modes at once" (Chion 1994, 28).

The total selection and combination of sounds of the sonic logo hold the potential to act as a symbol for Audi. Based on continuous and consistent distribution of the sonic logo in the setting of Audi promotion, listeners might have 'learned' to associate the sonic logo with Audi. However, each of the included object sounds do not seem to possess a high degree of distinctiveness when compared to sounds from outside the promotional setting. From the perspective of branding, the association between the thumping sounds – viewed as car tires friction sounds – and Audi can thus be considered to exemplify an attempt at "connotative hijacking" (Tagg 2013, 184) of a generic object sound. This would exemplify a possible variant of intrinsic sound branding. Rather than exemplifying the design of a distinctive and possibly rights-protected variant of the generic sound in question – a strategy exemplified by the crunching sound of Kellogg's cornflakes (Lindström 2005, 16f) and the engine roar of the Harley Davidson motorcycle (Sapherstein 1998) – the association between the sounds and Audi is possibly established through the promotional audio-visual setting of the sounds. Because the sounds represent an actual, possible feature of (also) Audi, the relationship between the sounds and Audi would be partly motivated or "restrained" (ibid., 38). Therefore, such association with Audi could be seen to include indexical aspects as well: the sounds would appear "as a member of a class" that has come to "stand for the class," thus representing a subcase of a synecdoche, namely "*species for genus*" (Chandler 2007, 133; italics in original). This illustrates how relationships between a signifier and a signified can be more or less motivated and conventional – for example, "[w]ithin each form signs also vary in their degree of conventionality" (ibid., 38). However, as the sounds are only ambiguously (thumping sounds) and vaguely (clicking sounds) associated with Audi in the audio-visual setting of the sonic logo, it is doubtful whether the process of connotative hijacking can develop as a result of the sonic logo alone. A proper hijacking would imply that listeners would be disposed to associate the generic car sounds with Audi even when heard in the setting of another (or unidentified) car mark.

When listening to the thumping sounds as heartbeats, they arguably play a pivotal role in symbolizing the drivers' possible relation to the brand. In comparison with the thumping sounds as heard as car tires friction sounds – and with the examples of Kellogg's and Harley Davidson's focus on intrinsic branding – the sounds heard as heartbeats would represent a process of extrinsic sound branding. The object sounds of the heartbeats are not features of the product, although the car could perhaps be considered to be anthropomorphized as having a beating heart – a reading supported by the potential conventional symbolic meaning (e.g., blood and warmth, see Kress and van Leeuwen, 2002) associated with the red color in which the brand name is presented in clear contrast to the otherwise white, aluminum and black colors. However, rather than suggesting that the/a heart is in the Audi, the sonic logo seems to suggest that Audi is in the heart: As indicated above, the sounds could be seen to symbolize an emotional engagement on behalf of the listeners and possible drivers of an Audi; as if the sounds of the beating heart suggests that driving an Audi makes the blood run faster through your veins. From the perspective of information linking, the function of the thumping sounds in relation to the visuals is mainly that of relay, and thus "very close" to the concept of extension (van Leeuwen 2005, 229): the sounds stand in a "complementary relationship" to the visuals, and they invoke meaning potentials "not to be found in the image itself" (Barthes 1977, 39 and 41).

As regards the swoosh sound, it might invoke interpretations in addition to its previously described denotative function of identifying the presentation of the visual logo. The metallic sound quality could thus be viewed to symbolize advanced technology as hinted at through the slogan. The metallic swoosh is offered as the sound of "Technik," and because the sonic logo does not include combustion engine sounds, it is possible to further associate the swoosh with advanced technology other than combustion engines.

6 The Sounds of the Car

Obviously, the sonic logo's inclusion of sounds does not reflect the whole range of possible sounds of the car. In this section, the examination aims to further profile the object sounds from the sonic logo by viewing them in light of what can, more broadly, be said to characterize the sounds of the product category of the car.

First, the sonic logo's inclusion of car sounds represents (a selection of) sounds as heard from what appears to be a *moving* car. Of course, the stationary, untouched car does not produce sounds (as indicated at the beginning of the chapter). However, for example, the sound of a door closing would normally emerge from a non-moving car. Incidentally, the sound of the door closing embodies a "very important" sound because "closing the doors is one of the operations a customer can do while he is examining a car in a seller's hall" (Parizet, Guyader and Nosulenko 2008, 12). Additionally, unlike other types of car sounds (e.g., alert sounds, see more below), the sounds of the closing door are practically inevitable *first* (and *last*) *impression sounds* when driving a car. Indeed, the sound of the car door closing has long been the subject of branding (see, e.g., Bijsterveld et al. 2014, 21).

Second, the sounds are seemingly heard from *inside* a car. The moderate reverberation and lack of directionality of the sounds of the sonic logo help produce the impression of a cocoon-like environment. By comparison, a moving car as heard from the outside would normally produce a wider variety of reverberations and more radical fluctuations in terms of the sounds' directionality and frequencies. An examination of the moving car as heard from the outside has for example led to the following observations (amongst others): "much of the sound that reaches us from a source has reflected off various other objects in the environment, which color the spectrum of reflected sound" and "[a]s sources move with respect to a potential observation point, their frequencies shift" (Gaver 1993, 4ff.). These observations imply that the car's sounds are not only relevant for the driver situated in the cabin, but also to other road users. The engine sound is particularly relevant in this respect – at least when the car travels slowly enough so that the engine sound can still be heard over the sounds from the friction of tires touching the road surface (which will predominate at higher speeds). Illustratively, electric cars have recently been required to include an acoustic vehicle alert system, which produces sounds when the car is reversing or travelling below a certain speed. For example, in the US, the limit has been set to 18.6 mph (to be implemented no later than September 2020; see Stenquist, 2019). The artificial sounds are meant to compensate for the almost silent movement of the electric car at slow speeds, which could otherwise cause mishaps as the car could be overlooked by pedestrians, cyclists and visually impaired people.

Third, the car sounds from the sonic logo can be profiled further by positioning them as particular examples of *sounds from inside* the moving car. While it is beyond the scope of this chapter to specify all possible examples of interior car sounds, the sounds of the sonic logo can be viewed in light of general categories. In this context, the categories exclude the sounds produced by, for example, deliberately strumming or scratching the cabin material (as exemplified through the ASMR video referred to previously) and the sounds produced by mechanical malfunction (see, e.g., Cole 2015). Also, the mediated sounds that do not originate from the car, for example in the form of music through the car's sound reproduction technology, are not addressed. Based on these premises, two broad categories of sounds can be distinguished (inspired by Özcan et al., 2020), namely operating sounds and indicator sounds.

Operating sounds is a term here suggested to account for the sounds from, for example, the engine, the rustle of the wind and the friction of tires touching the road surface. Similar to the concept of keynote sounds offered by Murray Schafer – a perspective applied here with the reservation that Schafer did not focus on sound branding – operating sounds are continuous and "ubiquitous" (1977, 9) and vary only according to exterior conditions (e.g., wind, rain and road surface).

Operating sounds thereby represent a kind of auditory background on which indicator sounds emerge. When the thumping sounds are viewed as originating from the car tire friction against gaps in the road surface, the sounds thus exemplify operating sounds.

The term indicator sounds is here proposed to identify the sounds that basically perform the function of informing and signaling to the driver that something specific relating to the handling and status of the car has happened or is about to happen. Because the sounds will emerge as auditory figures "on top of" the background sounds constituted by the operating sounds, the sounds can be likened to signal sounds in the sense offered by Schafer. Indicator sounds can be subdivided based on what the sounds indicate. One subtype of sounds indicates that an action on behalf of the driver has been successfully performed – this type here includes both "sound as *brand carriers*" and "sound as *indicators*" as coined by Özcan et al. (2020, 168; italics in original). These sounds function as *receipts* to confirm, for example, that the car has been started, a door has been successfully closed, a button has been pressed, or the turn signal has been activated. Another subtype of sounds indicates that an action on behalf of the driver has not been successfully performed and/or is now needed – this type corresponds to "sounds as *prompters*" as coined by Özcan et al. (2020, 168; italics in original). These sounds function as *alerts* to warn or remind the driver that, for example, the car is soon running out of resources (electricity or fuel), the seatbelt has not been fastened, a door has not been properly closed or the car is reversing toward a wall (as possibly detected by the proximity sensor). Alert sounds can be likened to signal sounds "which must be listened to because they constitute acoustic warning devices" (Schafer 1977, 10). The car horn exemplifies an extraordinary type of alert sound. Actually, the car's horn is not activated to indicate something for the driver, but rather (and normally) to call another driver or pedestrian's attention to some hazard.

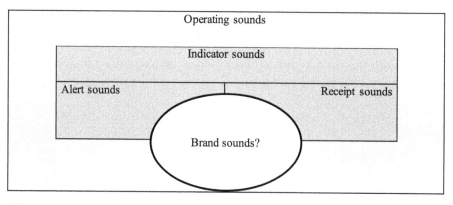

Figure 7.1   An overview of the interior sounds of a car.

Inspired by the perspective offered by Schafer, specific car sounds could be heard as brand sounds to the extent that they emerge as so-called sound marks which are "unique or possesses qualities which make it specially regarded or noticed" (Schafer 1977, 10). It might be the case that "[c]urrently, design is based on drawing attention without considering brand value," and that "auditory messages are often designed to be merely functional and technical, devoid of brand values" (Özcan et al. 2020, 155 and 166). Also, arguably, some types of car sounds are more disposed for processes of sound branding than others, as the exterior conditions (as mentioned previously), legal regulations and the function of the sound all affect the range of possible designs. For example, an alert sound has to be distinguishable from indicator sounds, and it would normally emerge relatively forceful and highly registered to ensure that the attention of the driver is arrested. By comparison, the design of the sound to indicate that the car has been started is less restricted – this is arguably why Özcan et al. (2020) highlights this particular sound as a "brand carrier." However, it should be stressed that no sound is neutral or wholly insignificant, as, potentially, all sounds assist in producing sensory responses and associations relevant to the branding of a car. Therefore, the circle in figure 1 covers both operating and (subtypes of) indicator sounds to illustrate that, principally, both types of sounds can emerge as sounds brands.

7 Conclusions

The sonic logo of Audi can be seen to include examples of interior sounds of the moving car: an operating sound and an indicator receipt sound. As previously indicated, it is doubtful whether these car sounds can play a role in processes of intrinsic sound branding. The sounds could perhaps be seen to hold the potential to influence a connotative hijacking process of *intrinsic generic sound branding* rather than a connotative building process of *intrinsic exclusive sound branding* (in which original sounds appear distinctive based on the sound quality). However, the sounds are arguably not distinctive enough for such processes to actually happen. Moreover, the fact that the object sounds play a role as impulses in a musical structure seem to tone down the potential for them to act as brand object sounds.

Consequently, the sonic logo does not seem to appeal to consumers to listen for the *sounds of Audi* in the sonic logo. Rather, consumers seem encouraged to listen for and acknowledge the *sounds for Audi*, and from this perspective the thumping sounds' resemblance to two heartbeats exemplify a case of *extrinsic generic sound branding*. While the characteristics of the sounds of the heartbeat do not represent an intrinsic part of the product of Audi – and might not be unique for Audi – the specific way in which the sounds are audio-visually integrated could result in an effective sound brand. For example, the sonic logo appears to embody the three most common functions of a sonic logo: 1) a reveille or heraldic function to attract attention; 2) a mnemonic function to facilitate memorization, recognition and identification; and 3) a preparatory or identity function to express the characteristics and values of the brand (see, e.g., Tagg and Clarida 2003; van Leeuwen 2017; Graakjær 2019) – and the close reading has indicated how these functions are specifically structured. It is beyond the scope of this chapter to explore how the sonic logo of Audi relates to other sound branding initiatives by Audi (for example, with respect to the "AUDI Sound profile" mentioned previously). Also, the chapter must leave it for future research to, for example, compare the sonic logo (and total sound profile) of Audi with the ones offered by other car brands.

The chapter has referred to the case of the sonic logo for Audi to illustrate a method for the analysis of object sounds in the setting of audio-visual branding. The method embraces the following three questions: What do the sounds sound like? From what and where do the sounds originate? To what do the sounds refer? Supplementary questions, which are particularly relevant in the setting of audio-visual branding, include: To what extent do the sounds embody a "true" representation? How does the sounds relate to the visuals? The examination of the sonic logo has furthermore inspired an exemplification of specific theories that can be relied upon when examining these questions from the perspective of a close reading methodology. The particular case has illustrated how object sounds can blend with other types of sound (illustrated by the sonic logo's inclusion of music sounds; see table 1), that the potentials of meaning associated with object sounds are not always unambiguous (illustrated by the thumping sounds), and that the object can be associated with the brand in different ways (illustrated by both intrinsic and extrinsic connections). More generally, the case indicates – by the example of the car (see figure 1) – that particular objects embody a particular range of possible sounds from which the audio-visual brand setting displays a selection. Hopefully, the presented analytical framework can inspire further research on how to listen to and explore why and how the sounds of brand objects embody potentials of meanings, and how these meanings can be systematically examined from the perspective of a close reading methodology.

References

Ament, Vanessa. 2014. The Foley Grail: The Art of Performing Sound for Film, Games, and Animation. Abingdon, UK: Focal Press.

Barthes, Roland. 1977. Image, Music, Text. New York: Hill & Wang.

Bijsterveld, Karin, Eefje Cleophas, Stefan Krebs and Gijs Mom. 2014. Sounds and Safe. A History of Listening behind the Wheel. Oxford: Oxford University Press.

Bochmann, Margarita. 2011. "Audi Corporate Sound." In Audio Branding Academy Yearbook, 2010/2011, edited by Kai Bronner, Rainer Hirt and Cornelius Ringe, 51–53. Baden-Baden: Nomos.

Bodden, Markus and Torsten Belschner. 2011. "Sound design for silent vehicles. Security – identity – emotion." In Audio Branding Academy Yearbook, 2010/201, edited by Kai Bronner, Rainer Hirt and Cornelius Ringe, 67–83. Baden-Baden: Nomos.

Britannica. 1998. "Heart." Accessed January 15, 2020. https://www.britannica.com/science/heart#ref174626.

Bronner, Kai. 2009. "Jingle all the way? Basics of audio branding." In Audio branding. Brands, Sound and Communication, edited by Kai Bronner and Rainer Hirt, 77–87. Baden-Baden: Nomos.

Bronner, Kai and Rainer Hirt, eds. 2009. Audio branding. Brands, sound and communication. Baden-Baden: Nomos.

Capeller, Ivan. 2018. "Sounds, Signs and Hearing: Towards a Semiotics of the Audible Field." Athens Journal of Philology 5, no. 1: 45–60.

Chandler, Daniel. 2007. Semiotics. The basics. Abingdon: Routledge.

Charters, Steve. 2006. "Aesthetic Products and Aesthetic Consumption: A Review." Consumption, Market and Culture 9, no. 3: 235–255.

Chion, Michel. 1994. Audio-vision – Sound on Screen. New York: Columbia University Press.

Cook, Nicholas. 1998. Analysing Musical Multimedia. Oxford: Oxford University Press.

Cohen, Annabel. 2001. "Music as a source of emotion in film." In Music and Emotion: Theory and Research, edited by Patrik Juslin and John Sloboda, 249–272. Oxford: Oxford University Press.

Cole, Craig. 2015. "10 Car Noises to be Concerned About." AutoGuide.com. Accessed January 15, 2020. https://www.autoguide.com/auto-news/2015/04/10-car-noises-to-be-concerned-about.html.

Danesi, Marcel. 2007. The Quest for Meaning. A Guide to Semiotic Theory and Practice. Toronto: University of Toronto Press.

Erdmann, Thomas, Falf Friese, Peter Kirchberg and Ralph Plagmann. 2013. Four Rings. The Audi Story. Bielefeld: Delius Klasing.

Gaver, William. 1986. "Auditory Icons: Using Sound in Computer Interfaces." Human-Computer Interaction 2, no. 2: 167–177.

Gaver, William. 1993. "What in the world do we hear? An ecological approach to auditory source perception." Ecological Psychology 5, no. 1: 1–29.

Graakjær, Nicolai. 2015. Analyzing Music in Advertising: Television Commercials and Consumer Choice. New York: Routledge.

Graakjær, Nicolai and Anders Bonde. 2018. "Non-musical Sound Branding – A Conceptualization and Research Overview." European Journal of Marketing 52, no. 7/8: 1505–1525.

Graakjær, Nicolai. 2019. "Sounding out *i'm lovin' it* – a multimodal discourse analysis of the sonic logo in commercials for McDonald's 2003–2018," *Critical Discourse Studies* 16, no. 5: 569–582.

Groves, John. 2011. Commusication. From Pavlov's Dog to Sound Branding. Cork: Oak Tree Press.

Gustafsson, Clara. 2015. "Sonic branding: a consumer-oriented literature review." Journal of Brand Management 22, no. 1: 20–37.

Illner, Karlheinz. 2010. "Benchmark Case: A New AUDI Sound Branding." Sound Branding Blog. Accessed January 15, 2020: https://soundbrandingblog.com/2010/05/07/benchmark-case-new-audi-sound-branding/.

Jackson, Daniel and Paul Fulberg. 2003. Sonic Branding. London: Palgrave.

Kaplanis, Neofytos, Søren Bech, Sakari Tervo, Jukka Pätynen, Tapio Lokki, Toon van Waterschoot and Søren H. Jensen. 2017. "Perceptual aspects of Reproduced Sound in Car Cabin Acoustics." Journal of Acoustical Society of America 143, no. 3: 1459–1469.

Kress, Gunther and Theo van Leeuwen. 2002. "Colour as a semiotic mode: notes for a grammar of colour." Visual Communication 1, no. 3: 343–368.

Langhammer, Marc. 2016. Audi - Vorsprung durch Technik (Werbeslogan) – Audiologo. YouTube, June 8, 2016. Accessed January 15, 2020: https://www.youtube.com/watch?v=X3HR1YBNlM8.

Lindstrom, Martin. 2005. Brand Sense: How to Build Powerful Brands through Touch, Taste, Smell, Sight & Sound. New York: Free Press.

Lucensky, Jakob. 2011. Sounds like Branding: Use the Power of Music to turn Customers into Fans. London: A & C Black Publishers Ltd.

Miller, Darryl and Lawrence Marks. 1992. "Mental imagery and sound effects in radio commercials." Journal of Advertising 21, no. 4: 83–93.

Özcan, Elif and René van Egmond. 2009. "The effect of visual context on the identification of ambiguous environmental sounds." Acta Psychologica 131, no. 2: 110–119.

Özcan, Elif, René van Egmond, Alexandre Gentner and Carole Favart. 2020. "Incorporating Brand Identity in the Design of Auditory Displays. The Case of Toyota Motor Group." In Foundations in Sound Design for Embedded Media. A Multidisciplinary Approach, edited by Michael Filimowicz, 155–193. New York: Routledge.

Parizet, Etienne, Erald Guyader and Valery Nosulenko. 2008. "Analysis of car door closing sound quality." Applied Acoustics 69, 12–22.

Pasnau, Robert. 1999. "What is sound?" The Philosophical Quarterly 49, no. 196: 309–324.

Peirce, Charles Sanders (edited by James Hooper). 1955. Peirce on Signs. Writings on Semiotic by Charles Sanders Peirce. Chapel Hill: University of North Carolina Press.

RSG. 2016. "ASMR Random Interior Car Sounds." YouTube. Accessed January 15, 2020: https://www.youtube.com/watch?v=LIw_SUHqTt4.

Sapherstein, Michael. 1998. "The Trademark Registrability of the Harley-Davidson Roar: A Multimedia Analysis." Intellectual Property & Technology Forum. Boston College Law School.

Schafer, Murray. 1977. The Soundscape. Our Sonic Environment and the Tuning of the World. Rochester, VT: Destiny Books.

Sheller, Mimi. 2004. "Automotive emotions. Feeling the car." Theory, Culture & Society 21, no. 4/5: 221–242.

Snow, James. 2004. Tinnitus. Theory and Management. Hamilton, Ont.: BC Decker Inc.

Spence, Charles and Qian Wang. 2015. "Wine and Music (I): On the Crossmodal Matching of Wine and Music." Flavour 4, no. 34: 1–14.

Stenquist, Paul. 2019. Why Quiet Cars Are Getting Louder. New York Times, October 24, 2019. Accessed January 15, 2020: https://www.nytimes.com/2019/10/24/business/electric-vehicle-noises-nhtsa.html.

Tagg, Philip. 2013. Music's Meaning. A Modern Musicology for Non-musos. New York: MMMSP.

Tagg, Philip and Clarida, Bob. 2003. Ten little Title Tunes. New York, NY: MMMSP.

Treasure, Julian. 2011. Sound Business. How to use Sound to Grow Profits and Brand Value. Gloucestershire, UK: Management Books 2000 Ltd.

Tybout, Alice and Tim Calkins. 2005. Kellogg on Branding. Hoboken, NJ: Wiley.

van Leeuwen, Theo. 1999. Speech, Music, Sound. London: Macmillan Education.

van Leeuwen, Theo. 2005. Introducing Social Semiotics. London: Routledge.

van Leeuwen, Theo. 2017. "Sonic logos." In Music as Multimodal Discourse. Semiotics, Power and Protest, edited by Lyndon Way and Simon McKerrell, 119–134. London: Bloomsbury.

Wagner, Verena and Konrad Kallus. 2015. "Sound Quality of Turn Indicator Sounds – Use of a Multidimensional approach in the Automobile Product Development." Journal of Traffic and Transportation Engineering 3, 158–165.

Yorkston, Eric. 2010. "Auxiliary auditory ambitions: assessing ancillary and ambient sounds." In Sensory Marketing: Research on the Sensuality of Products, edited by Aradhna Krishna, 157–167. New York: Routledge.

Designing and Reporting Research on Sound Design and Music for Health

Methods and Frameworks for Impact

Kjetil Falkenberg and Emma Frid

1 Background

Research overviews demonstrate that there is a long history of scholars from different areas reporting on the benefits of music-related interventions within therapy, medicine and health. In particular, music therapy and music medicine interventions have been studied widely, as can be seen in numerous publications in the Cochrane Database of Systematic Reviews (https://www.cochranelibrary.com/cdsr/reviews/topics). However, this research in sound and music is seldom put under the same scrutiny by evaluators as more typical medical studies such as interventions including pharmaceutics. The current chapter focuses on evaluating and making strategic choices for conducting sound design– and music-related research for health, accessibility and disability. The aims are to provide an introduction to study design and guidelines for reporting of research studies in order to achieve impact, and to describe the common frameworks that systematic reviews are conducted in accordance with.

In this chapter, we use the term 'sound and music research' to refer to work focusing on sound design and music interaction within *Sound and Music Computing* (SMC) and *Sonic Interaction Design* (SID), areas which in the context of this chapter are applied to health. For the sake of clarity, we will not include *music therapy* or *music medicine interventions* in our reading of sound and music computing applied to health; the main reason is that those two areas are already frequently featured in the systematic review databases. Furthermore, we will in this text suggest a broadened definition of "health" to include studies on accessibility and disability.

Sound and Music Computing research "approaches the whole sound and music communication chain from a multidisciplinary point of view. By combining scientific, technological and artistic methodologies it aims at understanding, modelling and generating sound and music through computational approaches" (Bernardini et al. 2007, 9). Sonic Interaction Design is positioned at the intersection of auditory display,

DOI: 10.4324/9780429356360-9

ubiquitous computing, interaction design and interactive arts (Rocchesso et al. 2008, 1). SID focuses on emergent research topics related to multisensory, performative and tactile aspects of sonic experiences, exploring how sounds can be used to convey information, meaning and aesthetic and emotional qualities in interactive contexts (Franinović and Serafin 2013, vii). One example of clinical application of SMC and SID is *sonification*, or data representation through sound (Hermann et al. 2011, 1).

Music therapy (MT) is defined as "a systematic process of intervention wherein the therapist helps the client to promote health, using music experiences and the relationships that develop through them as dynamic forces of change" (Bruscia, 1998, 20). *Music medicine interventions*, or intervention using music as medicine, is defined as listening to pre-recorded music, offered by medical staff (Bradt et al. 2016, 1). The methods mentioned in this chapter are expected to be well-known to researchers in both MT and other clinical use of music. It is thus worth noting that while both the SMC and SID communities would typically consider that MT and music as medicine belong to their communities, music therapists and researchers in clinical use of music might not consider themselves to be part of either SMC or SID. Therefore, in this chapter the areas are separated.

Previous work focusing on evaluating the quality of research on sound and music in health contexts include the article "Reporting quality of music intervention research in healthcare: A systematic review" by Robb and colleagues (2018, 24), in which reporting quality of music intervention research was examined using the *Checklist for Reporting Music-based Interventions*. The checklist or guidelines were introduced in previous work carried out by Robb et al. (2010, 271–279). Out of 860 articles considered, 187 met the review criteria. Results revealed a number of areas in which reporting could be improved; the authors reported concerns about inadequate intervention reporting (e.g., information about the music used, decibel levels or materials) and inconsistent terminology, which limits validity, replicability and clinical application of findings. Most interventions were delivered by credentialed music therapists or registered nurses. The authors concluded that problems with reporting quality impedes meaningful interpretation and cross-study comparisons; improved reporting quality and creation of a shared language would advance scientific rigor and clinical relevance of music intervention research.

The following chapter will present practical suggestions for effectively increasing the impact of research done within the field of sound and music research. Such suggestions include:

1. to define distinct research questions that can easily be evaluated for inclusion in meta-studies according to existing guidelines
2. to avoid unnecessary research bias
3. to design studies with appropriate control groups
4. to report results that are possible to assess in a comparison with other studies.

These suggestions are acquired from existing methods commonly used in health research, such as in Cochrane's published meta-studies (see Higgins and Wells 2011). The methods are not novel, but, we argue, are not used extensively within the field of sound and music research. The ambition is to guide researchers to be prepared for meeting the inclusion criteria in scientific evaluations and have their works assessed in systematic reviews and meta-analyses, thus gaining a larger impact. We focus on recommendations that may strengthen the research conducted within the field of sound and music research by emphasizing relevant aspects when conducting and reporting a systematic review. For the purpose of understanding how quantitative research is evaluated and assessed, we present a number of tools and frameworks used to evaluate systematic reviews. An understanding of these frameworks may provide a better commencement of improving the impact of quantitative research. The methods, frameworks and guidelines that will be described are: PICO, ROBIS, GRADE, QATQS, AMSTAR and PRISMA, in addition to some alternatives to or variations of these.

1.1 Terminology

The material presented in this chapter discusses research methods, employing a range of different terminology commonly used in clinical and health research practice. Since these terms are perhaps not as common in the sound and music research field, and some have even been used haphazardly, a short introduction of the terminology is presented below. For clarity, the terms are presented in alphabetical order.

Allocation concealment: there are two corollaries to double-blinding (see definition of *blinding* below), allocation concealment and blind statistical analysis. If an allocation algorithm (i.e., the process of allocating participants to experimental groups) is completely random, then the allocation of participants to groups is concealed. If someone were to be unblinded, then whoever knew about the allocation system could trace back and forth from this participant and find out about the group allocation of other participants. For blind statistical analysis, data are input by automatic means or by assistants who are blind to group assignment of participants. The analysis is normally run with a database that is still blinded in the sense that the groups are named "A" and "B" and only after this analysis has been conducted and documented is the blind broken (Walach 2010, 387–389).

Bias: a systematic error in data collection to address a specific research question. Biases are errors that are systematically related to people, groups, treatments or experimental conditions. As such, they are different from random errors, which are randomly distributed, as the name suggests. Bias results in overestimation or underestimation of the measurement of a behavior or trait. Therefore, bias can endanger the ability of researchers to draw valid conclusions regarding whether one variable causes a second variable or whether the results generalize to other people. There are several different types of bias: sampling bias, selection bias, experimenter expectancy effects and response bias (Kovera 2010, 84–85).

Blinding: studies can be single-blind, double-blind or triple-blind. In a single-blind study, participants are deliberately kept ignorant of the group to which they have been assigned or key information about the materials they are assessing (Stuart-Hamilton 2010, 1384–1386). In a double-blind study, both experimenters and participants are without knowledge of crucial aspects of the study (Walach 2010, 387–389). In a triple-blind study, also the individual(s) who assess the outcomes of the study are unaware of crucial aspects of the study (Dawes 2010, 1542–1543). The purpose of these approaches is to prevent the risk for biasing results.

Confounders: confounding occurs when two variables systematically covary. Understanding the relationship between or among variables and whether those relationships are causal can be complicated when an independent or predictor variable covaries with a variable other than the dependent variable. The confounding variable is a variable that systematically varies with the independent variable. It provides an explanation other than the independent variable for changes in the dependent variable (Kovera 2010, 221–222).

Control group versus experimental group: in the simplest case of experimental research, a study contrasts two groups. The independent variable is present in one group but not the other. For example, one group may receive a treatment while the other receives no treatment. The group in which the treatment occurs is called the experimental group and the group in which treatment is withheld is called the control group (Gill & Walsh 2010, 252–252).

Effect size: a description of the magnitude of observed effects, independently of the possibly misleading influences of sample size. Studies with different sample sizes and the same basic descriptive characteristics (e.g., means and confidence intervals) will differ in statistical significance but not in effect size estimates. Effects that are large but nonsignificant can suggest further research with greater power. Effects that are small but significant, caused by large sample sizes, can warn researchers against possibly overvaluing the observed effect (Fritz, C. et al. 2012, 2).

Evidence-based practice: evidence-based practice has its origins in medicine. It is essentially a methodology which claims a neutral stance. As such, the process appears capable of expansion to a range of different disciplines involving human services (Trinder 2008, 4). Evidence-based decision making is defined as "using the findings from the statistical measures employed and correctly interpreting the results, thereby making a rational conclusion." An evidence-based approach ensures that the methodology and logic to arrive at conclusions are reasonable. Evidence compiled by researchers is viewed scientifically through the use of a defined methodology that values systematic and replicable methods for production (Mirtz & Greene 2010, 434–437).

Impact factor: a journal's impact factor (JIF) is based on two elements: a numerator consisting of the number of citations in the current year to items published in the previous two years, and the denominator, the number of articles and reviews published in the same two years (Garfield 2006, 90). The best journals within a field are usually those in which it is most difficult to have an article accepted, and these are the journals that have a high impact factor (Garfield 2006, 92).

Inclusion/eligibility criteria: a set of predefined characteristics used to identify subjects who will be included in a research study. The criteria should respond to the scientific objective of the study and proper selection of inclusion criteria will optimize external and internal validity. Good selection criteria will ensure homogeneity (i.e., the state of being uniform or similar, see Cramer & Howitt 2004, 75), of the sample population, reduce confounding and increase the likelihood of finding the true association between intervention and outcomes (Velasco 2010, 629–591).

Independent versus dependent variable: independent variables can be manipulated by the researcher, whereas dependent variables are the responses to the effects of independent variables (Fan 2010, 592). The dependent variable, or the outcome variable, is the result of an action of one or several independent variables (the treatment or intervention) (Salkind 2010, 348).

Meta-analysis: meta-analysis uses statistical techniques to integrate and summarize results of a set of studies. It is common that systematic reviews also contain meta-analyses. Meta-analyses can provide more precise estimates of effects than those derived from individual studies included in a review (Liberati et al. 2009, 2). Common summary measures for meta-analyses with continuous variables are mean difference (when outcome measurements are made on the same scale) and standardized mean difference (when the studies assess the same outcome but do not yield directly comparable data, e.g., using different scales to measure depression), also referred to as *Cohen's d* (Liberati et al. 2009, 12). The most common summary measures for binary outcomes are the risk ratio, odds ratio and risk difference (Liberati et al. 2009, 12).

MeSH: Medical Subject Headings, the US National Library of Medicine's controlled vocabulary thesaurus (US National Library of Medicine 2019a). MeSH is a controlled and hierarchically organized vocabulary used for indexing articles in/ from the MEDLINE®/PubMED® database. It is used for cataloguing and searching of biomedical and health-related information; each article citation is associated with a set of MeSH terms that describe the content of the citation, which enables search on such entry terms instead of keyword searching (US National Library of Medicine 2019a).

Power: the ability to detect significant results if they exist, or, probability that a test appropriately rejects that a hypothesis is true. The power value refers to the likelihood that the test will not make a false negative (Type II error). A false negative refers to not finding a significant difference when one actually exists. If the likelihood of a false negative is $\beta = 80$, and power is equal to $1 - \beta$, then power is 20. In general, the power to detect small effects is low, but it can be increased by manipulating other parameters, e.g., the sample size. Power is influenced by statistical significance, effect size and sample size (Vo & James 2010, 1066–1067).

PubMed and *MEDLINE*: PubMed comprises more than 30 million citations for biomedical literature from the MEDLINE database, life science journals and online books (US National Library of Medicine 2020). MEDLINE is the US National Library of Medicine premier bibliographic database. It contains more than 25 million references to journal articles in life sciences with a concentration on biomedicine. An important feature of MEDLINE is that the records are indexed with NLM Medical Subject Headings (MeSH) (US National Library of Medicine 2019b).

Quasi-experimental design: scientific experiments consist of a controlled set of observations aimed at testing whether two or more variables are causally related. Two types of scientific experiments can be distinguished: randomized experiments and quasi-experiments. For randomized experiments, study units are randomly assigned to observational conditions. In quasi-experiments, study units are not randomly assigned to observational conditions, due to ethical or practical constraints. Randomized experiments have an advantage over quasi-experiments since alternative explanations (e.g., confounding variables) are equally likely across conditions and thus can be ruled out. Due to the lack of random assignment between conditions, the same cannot be said about quasi-experiments (Baldwin & Berkeljon 2010, 1172–1176).

Randomized controlled trial (RCT): a study design in which the effects of the intervention (treatment) are compared with those of a control treatment. The participants are randomly assigned to one of the two groups (control or treatment). Randomization is done to ensure that all potential confounding factors are divided equally among the groups (Kabisch et al. 2011, 664).

Reliability: reliability quantifies the precision of measurement over numerous consistent administration conditions or replications and corresponds to the trustworthiness of the recorded data. Observed responses are viewed as imperfect representations of an unobserved variable. Often, it is this unobserved characteristic that is of interest to the researchers (Gushta & Rupp 2010, 1238–1242).

Sampling: when researchers examine a portion (sample) of a larger group of potential participants and the results are used to make statements that apply to a broader group or population (Fritz, A. & Morgan 2010, 1303).

Statistical significance: the difference between two measurements that results from more than randomness (Michaelson & Hardin 2010, 1362–1366). A statistically significant difference is caused by an actual difference; it is not the result of random variation (Michaelson & Hardin 2010, 1362–1366). A commonly used significance level is $\alpha = .05$. This signifies that there is only a 5 percent probability that a researcher will incorrectly detect a significant effect, when such an effect does not actually exist (Vo & James 2010, 1066–1067).

Systematic reviews: systematic reviews attempt to collect all empirical evidence that fits a pre-specified eligibility criterion to answer a specific research question. They have a clearly stated set of objectives with an explicit and reproducible methodology, a systematic search that attempts to identify all studies that meet the eligibility criteria, an assessment of the validity of the findings of included studies, and systematic presentation and synthesis of findings and characteristics of included studies (Liberati et al. 2009, 2; see also Grant & Booth 2009, 91–108).

Validity: there are two types of validity, external and internal. Results from a research study that possesses external validity can be generalized to a wider group of people, or a population, not only those originally included in the study (Leighton 2010a, 467–470). Internal validity, on the other hand, refers to the accuracy of statements made about the causal relationship between two variables, i.e., the manipulated variable (the treatment or independent variable) and the measured variable (the dependent variable) (Leighton 2010b, 620–622).

2 Tools and Frameworks

In the following sections, we present a number of tools and frameworks. The first section focuses on the Cochrane organization and their risk of bias tool, which is used to evaluate the risk of bias in Randomized Controlled Trials (RCTs). The subsequent section introduces ROBIS, Risk of Bias in Systematic Reviews, a tool used for assessing risk of bias. The next section presents PICO, Patient/Problem, Intervention, Comparison/Control and Outcome, a framework that can be used to form questions and facilitate literature search. This is followed by GRADE, Grading of Recommendations Assessment, Development and Evaluation, a system for rating quality of evidence and strength of recommendations. We also describe QATQS, the Quality Assessment Tool for Quantitative Studies, and AMSTAR, A Measurement Tool to Assess Systematic Reviews, which can be used to assess methodological quality. Finally, we present the PRISMA (Preferred Reporting Items for Systematic reviews and Meta-Analyses) system for reporting systematic reviews and meta analyses, and derivatives.

2.1 Cochrane

Cochrane (previously known as the Cochrane Collaboration) is a British organization with the mission to "promote evidence-informed health decision-making by producing high-quality, relevant, accessible systematic reviews and other synthesized research evidence" (https://www.cochrane.org/about-us). They publish 12 issues of systematic reviews per year, and in 2018 Cochrane published 629 reviews, with an impact factor of 7.755. Chapter 3 of their handbook for authors, "Defining the criteria for including studies and how they will be grouped for the synthesis," includes a number of guidelines which are, naturally, helpful for improving chances of being assessed in meta reviews.

Cochrane uses PICO (see below) to determine the elements of the review question and eligibility criteria. Moreover, they emphasize that a predefined, unambiguous eligibility criterion is a fundamental prerequisite for a systematic review. This criterion should be "sufficiently broad to encompass the likely diversity of studies, but sufficiently narrow to ensure that a meaningful answer can be obtained when studies are considered in aggregate." Moreover, systematic review authors should make sensible post-hoc decisions about exclusion of studies, which should be reported in the review. The authors also emphasize that randomized trials is the best study design for evaluating the efficacy of treatments (McKenzie et al. 2019, 33--65).

2.1.1 Cochrane Risk of Bias Tool

Cochrane has developed a tool for assessing risk of bias in randomized controlled trials called the *Cochrane Risk of Bias Tool*. This tool separates a judgment about risk of bias from a description of the support for that judgment, for a series of items covering different bias domains (Higgins et al. 2011, 9). The tool covers the following domains: selection bias, performance bias, detection bias, attrition bias, reporting bias and other bias. Bias is assessed as low, unclear, or high risk for each domain (Higgins et al. 2011, 2–3). The following sources of bias are listed: random sequence generation, allocation concealment, blinding of participants and researchers, blinding of outcome assessment, incomplete outcome data, selective reporting or other reasons. Selection bias may occur due to inadequate randomized sequence generation or inadequate concealment of allocations before assignment; performance bias due to knowledge of the allocated interventions by participants and personnel during the study; detection bias due to knowledge of the allocated interventions by outcome assessment, attrition bias due to amount, nature, or handling of incomplete outcome data; and reporting bias due to selective outcome reporting (Higgins et al. 2011, 6). An example of a risk of bias table from a Cochrane review is presented in Table 8.1.

2.2 ROBIS

ROBIS (Risk of Bias in Systematic Reviews) is the first rigorously developed tool designed specifically to assess the risk of bias in systematic reviews (Whiting et al. 2016, 225). Bias occurs if systematic limitations or flaws in the design, conduct or analysis of the work distort the results or conclusions (Whiting et al. 2016, 226). ROBIS evaluates 1) the extent to which the research question addressed in the review

Table 8.1 Example of risk of bias table from a Cochrane review

Bias	Judgment	Support for judgment
Random sequence generation (selection bias)	Low risk	
Allocation concealment (selection bias)	High risk	
Blinding of participants and researchers (performance bias)	Unclear risk	
Blinding of outcome assessment (detection bias)	...	
Incomplete outcome data (attrition bias)		
Selective reporting (reporting bias)		
Other bias		

Adapted from table 2 in Higgins et al. 2011

matches the research question defined by its user, and 2) the degree to which the review minimizes the risk of bias in summary estimates and presented conclusions (Whiting et al. 2016, 226). The target audience of ROBIS consists of guideline developers, authors of overviews of systematic reviews and review authors (Whiting et al. 2016, 226). The ROBIS tool is made up of three different phases: *assess relevance*, *identify concerns with the review process* and *judge risk of bias* (Whiting et al. 2016, 227), which are described in the following sections.

2.2.1 Phase 1 – Assess Relevance

Phase 1 is optional and involves assessing if the target question, i.e., the question that the review is trying to answer, and the systematic review question, i.e., the question defined by the user, match (Whiting et al. 2016, 229). Results may be "yes", "no" or "partial."

2.2.2 Phase 2 – Identify Concerns with the Review Process

Phase 2 covers four domains that may introduce bias into a systematic review: 1) *study eligibility criteria*; 2) *identification and selection of studies*, 3) *data collection and study appraisal* and 4) *synthesis and findings* (Whiting et al. 2016, 227). The domains aim to assess the following: 1) whether primary study eligibility criteria were prespecified, clear and appropriate; 2) whether any studies that would have met the inclusion criteria were not included in the review; 3) whether bias may have been introduced through the data collection or risk of bias assessment processes; and 4) whether the reviewers have used appropriate methods to combine data from the included studies (Whiting et al. 2016, 230–231). Each domain is composed of three sections that should be considered sequentially: i) *information used to support judgement*, ii) *signaling questions* and iii) *judgement of concern about risk of bias* (Whiting et al. 2016, 230). Signaling questions for respective domains in Phase 2 (and Phase 3) are presented in Table 8.2. Possible answers for the signaling questions are "yes" "no" "probably yes/no" and "no information" (only used when there is not enough data to permit a judgement), where "yes" indicates low concerns (Whiting et al. 2016, 230). If the signaling questions for a domain all result in "yes" or "probably yes" the level of concern is judged as "low", but if there are some answers that are "no" or "probably no" there is a potential concern about bias (Whiting et al. 2016, 230). In this manner, the level of bias for each domain is judged as "low" "high" or "unclear" (Whiting et al. 2016, 230).

2.2.3 Phase 3 – Judge Risk of Bias

Phase 3 focuses on assessing the overall risk of bias in the interpretation of review findings and if this interpretation considers limitations identified in Phase 2 (Whiting et al. 2016, 227). The same structure as for the domains described for Phase 2 is used for Phase 3, with signaling questions and information used to support judgement (Whiting et al. 2016, 232), see Table 8.2. However, the judgement regarding concerns about bias is replaced with an overall judgement of risk of bias (Whiting et al. 2016, 232). Table 8.3 shows an example of ROBIS results presented in tabular format.

Table 8.2 Phases 2 and 3 of the ROBIS tool for assessing risk of bias in systematic reviews, presented with signaling questions

Signaling questions	Phase 2				Phase 3
	1. Study eligibility criteria	**2. Identification and selection of studies**	**3. Data collection and study appraisal**	**4. Synthesis and findings**	**Risk of bias in the review**
	Did the review adhere to predefined objectives and eligibility criteria?	Did the search include an appropriate range of databases/ electronic sources for published and unpublished reports?	Were efforts made to minimize errors in data collection?	Did the synthesis include all studies that it should?	Did the interpretation of findings address all of the concerns identified in domains 1–4 of Phase 2?

Were the eligibility criteria appropriate for the review question?	Were methods additional to database searching used to identify relevant reports?	Were sufficient study characteristics available for both review authors and readers to be able to interpret the results?	Were all predefined analyses reported or departures explained?	Was the relevance of identified studies to the review's research question appropriately considered?
Were the eligibility criteria unambiguous?	Were the terms and structure of the search strategy likely to retrieve as many eligible studies as possible?	Were all relevant study results collected for use in the synthesis?	Was the synthesis appropriate given the nature and similarity in the research questions, study designs, and outcomes across included studies?	Did the reviewers avoid emphasizing results on the basis of their statistical significance?
Were all restrictions in eligibility criteria based on study characteristics appropriate?	Were restrictions based on date, publication format, or language appropriate?	Was the risk of bias (or methodologic quality) formally assessed using appropriate criteria?	Was between-study variation minimal or addressed in the synthesis?	
Were any restrictions in eligibility criteria based on sources of information appropriate?	Were efforts made to minimize errors in the selection of studies?	Were efforts made to minimize error in the risk of bias assessment?	Were the findings robust, for example, as demonstrated through funnel plot or sensitivity analyses? Were biases in primary studies minimal or addressed in the synthesis?	
Judgment Concerns regarding…	Concerns regarding…	Concerns regarding…	Concerns regarding…	Risk of bias in the review

Adapted from table 2 in Whiting et al. 2016

Table 8.3 Example of ROBIS results presented in tabular format

Review	Phase 2				Phase 3
	1. Study eligibility criteria	2. Identification and selection of studies	3. Data collection and study appraisal	4. Synthesis and findings	Risk of bias in the review
1	low risk	high risk	unclear risk	high risk	high risk
2					
....					

Adapted from table 1 in Whiting et al. 2016.

2.3 PICO

A key aspect of high-quality research is to ask adequate questions. Without well-focused questions, identifying appropriate resources and searching for relevant evidence can be difficult and time consuming (Schardt et al. 2007, 2). PICO is a specialized framework that is used to form questions and facilitate literature search. The acronym stands for *Patient/Problem* (sometimes *Population* is also used, see e.g., Huang & Demner-Fushman 2006, 359), *Intervention, Comparison/Control* and *Outcome(s)* (Richardson et al. 1995, A12). The PICO framework originates in medicine and is the most frequently cited framework for asking well-built questions (Schlosser et al. 2007, 229). Using PICO to frame questions serves three purposes: 1) it forces the questioner to focus on the single most important issue and outcome; 2) it facilitates the next step in the process, i.e., the search process, by selecting language or key terms to be used; and 3) it directs you to clearly identify the problem, results and outcomes (Miller and Forrest 2001, 136). This will in turn allow you to determine the type of evidence required to solve the defined problem, and to measure the effectiveness of the intervention (Miller and Forrest 2001, 136). PICO may also be used to formulate titles of systematic reviews, as suggested by Liberati (Liberati et al. 2009, 5). An example of such a review from the Cochrane Database of Systematic Reviews is "Music education for improving reading skills in children and adolescents with dyslexia" (see Cogo-Moreira et al. 2012). A schematic representation of the PICO template is shown in Table 8.4. A more detailed description of the PICO components is presented in the succeeding section.

The *P* component of PICO relates to the patient and her/his membership in a population, age, gender, ethnic group, risk profile as well as other traits judged to be important (Armstrong 1999, 25–28). In the review example presented above by Cogo-Moreira and colleagues (2012), this component would be children and adolescents with dyslexia. The *I* component is the intervention, or the exposure under consideration. It applies not only to therapy but also to prevention, diagnostic

Table 8.4 The PICO template for asking questions

	Component	Definition
P	Patient or Problem	Describe the person who is affected by the intervention and the problem to be solved
I	Intervention	Describe the standard intervention, i.e. the treatment
C	Comparison/Control	Describe the condition comparisons
O	Outcome	Describe the expected results

testing and exposure; and thus resembles the course of the clinical action under consideration (Schlosser et al. 2007, 228). In the example by Cogo-Moreira et al., the intervention corresponds to music education. The *C* component deals with considering an intervention relative to another (Schlosser et al. 2007, 228). A comparison condition can also be a baseline case or the equivalent of "doing nothing" (Armstrong 1999, 25–28). For the Cogo-Moreira et al. example, the *C* component is not explicitly defined in the title; the control group would possibly be children and adolescents who have not participated in music education, but such studies were not found. Finally, the *O* component targets what the practitioners seek to accomplish in terms of treatment goals (Schlosser et al. 2007, 228). This may also include unwanted outcomes, e.g., the probability of side effects that should be avoided, costs or effort associated with achieving an outcome (Armstrong 1999, 25–28). In the example above, the outcome is the improved reading skills. There are several derivatives of PICO available: PECO, or *Population, Exposure, Comparator* and *Outcomes* (Morgan et al. 2018, 1027); PICOS, where *S* refers to study design (Methley et al. 2014, 2); SPICE, or *Setting, Population, Intervention, Comparison* and *Evaluation* (Booth 2006, 363); CIMO, or *Context, Intervention, Mechanism, Outcome* (Denyer et al. 2008, 393); ECLIPSE, or *Expectation, Client group, Location, Impact, Professionals, Service* (Wildridge & Bell 2002, 113–114); and SPIDER,or *Sample, Phenomenon of Interest, Design, Evaluation, Research type* (Cooke et al. 2012, 1435).

2.4 GRADE

GRADE stands for *Grading of Recommendations Assessment, Development* and *Evaluation*. It is a system that is primarily used for grading evidence when submitting a clinical guidelines article, used by many different organizations (Guyatt et al. 2008c, 924). GRADE provides a system for rating quality of evidence and strength of recommendations. The system classifies the quality of evidence into four levels: high, moderate, low and very low. It also provides two grades of recommendations: weak versus strong (Guyatt et al. 2008c, 926).

The quality of evidence is classified as "high quality" if further research is very unlikely to change the confidence in the estimate of effect. If further research is likely to have an important impact on the confidence in the estimate of effect and may change the estimate, it is classified as "moderate quality." If further research is very likely to have an important impact on the confidence in the estimate of effect and is likely to change the estimate, it is classified as "low quality." The "very low quality" label is used if any estimate of effect is very uncertain. Evidence based on randomized controlled trials (RCTs) begins as high quality evidence. Observational studies (e.g., case-control studies) start with a low quality rating, but grading upwards may be warranted for example if the magnitude of the treatment effect is very large or if all plausible biases would decrease the magnitude of an apparent treatment effect. The confidence in the evidence can be decreased for several reasons, such as *study limitations, inconsistency of results, indirectness of evidence, imprecision* and *reporting/publication bias* (Guyatt et al. 2008c, 926).

Study limitations are limitations that may bias estimates of the treatment effect. This includes lack of allocation concealment, lack of blinding, large losses to follow-up (subjects who at one point in time were actively participating in a research trial, but have become lost at the point of follow-up in the trial), failure to adhere to an intention to treat analysis, stopping early for benefit or failure to report outcomes (Guyatt et al. 2008a, 996). When it comes to *inconsistent results*, heterogeneity or variability of results across studies suggest true differences in underlying treatment effects. Existence of heterogeneity without a plausible explanation decreases the quality of the evidence (Guyatt et al. 2008a, 996). There are two different types of *indirectness of evidence*: the first one occurs when considering the use of one of two active drugs and if both drugs are compared to a placebo case (Guyatt et al. 2008a, 996). This is an indirect comparison of the magnitude of effect, and thus the evidence is of lower quality than if the two drugs were compared head-to-head (Guyatt et al. 2008a, 997). The second type of indirectness of evidence relates to differences between the population, intervention, comparator to the intervention, and outcome of interest (Guyatt et al. 2008a, 997). *Imprecision* occurs when there are few subjects and few events, which results in wide confidence intervals (Guyatt et al. 2008a, 997). Finally, *publication bias* relates to bias in terms of what is reported. Failure to report studies, for example, studies that show no effect, reduces the quality of evidence. Another situation that should elicit suspicion about publication bias occurs when published evidence is limited to a small number of trials, and all of them are funded by industry (Guyatt et al. 2008, 997).

In clinical research, the strength of a recommendation reflects "the extent to which we can be confident that the desirable effects of an intervention outweigh the undesirable effects" (Guyatt et al. 2008b, 1049). Examples of desirable effects are improvement in quality of life or reduction in the burden of treatment, whereas undesirable effects include adverse effects that have negative impact on quality of life or increase the use of resources (Guyatt et al. 2008b, 1049). Factors that affect the strength of a recommendation are *quality of evidence, uncertainty about the balance between desirable and undesirable effects, uncertainty or variability in values and preferences* and *uncertainty about whether the intervention represents a wise use of resources*. When desirable effects of an intervention outweigh the undesirable effects (or do not), strong recommendations are given. However, when trade-offs are not as certain, for example because of low quality evidence or if the desirable and undesirable effects are balanced, weak recommendations are mandatory (Guyatt et al. 2008c, 926).

2.5 QATQS

QATQS stands for *Quality Assessment Tool for Quantitative Studies*. It is a tool developed in Canada by the Effective Public Health Practice Project (EPHPP), which can be used to evaluate quantitative studies. It consists of the following eight components: 1) *selection bias*, 2) *study design*, 3) *confounders*, 4) *blinding*, 5) *data collection methods*, 6) *withdrawals* and *drop-outs*, 7) *intervention integrity* and 8) *analysis*. The components are considered to be universally relevant to any health topic. In the assessment, each examined component receives a mark ranging between "strong," "moderate" and "weak." A global rating is then computed for the paper.

Component 1, *selection bias*, focuses on whether the individuals selected to participate in the study are likely to be representative of the target population, and what percentage of selected individuals that agreed to participate. Component 2, *study design*, focuses on the following aspects: which type of study design was used, if the study was described as randomized and if the method for randomization was described, as well as if the method was appropriate. The following categories of study designs are given: randomized controlled trial, controlled clinical trial, cohort analytic (two group pre+post), case-control, cohort (one group pre+post/before and after), interrupted time series, others or if the design cannot be specified (EPHPP 2009, 1). Component 3, *confounders*, assesses if there were important differences between groups prior to the intervention and the percentage of relevant confounders that were controlled, either in the design or analysis. Examples of how confounders can be controlled are stratification and matching. The following examples of confounders are given: race, sex, marital status/family, age, socioeconomic status (SES, i.e. income or class), education, health status and pre-intervention score on outcome measure (e.g., a measured health metric before a treatment is carried out). For Component 4, *blinding*, assessment focuses on whether the outcome assessor(s) were aware of the intervention or exposure status of participants, and if the study participants were aware of the research question. Component 5, *data collection methods*, focuses on whether data collection tools were shown to be valid and reliable (EPHPP 2009, 2). Component 6, *withdrawals* and *drop-outs*, focuses on whether withdrawals and drop-outs were reported in terms of numbers and/or reasons per group, as well as the percentage of participants that completed the study. If participation varied in different groups, the lowest percentage should be considered. Component 7, *intervention integrity*, poses questions about the percentage of participants that received the allocated intervention or exposure of interest, if the consistency of the intervention was measured, and if it is likely that the participants received an unintended intervention such as contamination or co-intervention that may have influenced the results. Finally, Component 8 is *analysis*. This considers whether the unit of allocation is a community, organization/institution, practice/office or individual. The appropriateness of the employed statistical methods are also evaluated (EPHPP 2009, 3). A summary of the QATQS components is presented in Table 8.5.

Table 8.5 Summary of QATQS components for evaluation of quantitative studies

Component	Description
Selection bias	If the individuals selected to participate in the study are likely to be representative of the target population and what percentage of selected individuals that agreed to participate.
Study design	Which type of study design that was used, if the study was described as randomized and if the method for randomization was described, as well as if the method was appropriate.
Confounders	If there were important differences between groups prior to the intervention and the percentage of relevant confounders that were controlled, either in the design or analysis.
Blinding	If the outcome assessor(s) were aware of the intervention or exposure status of participants and if the study participants were aware of the research question.
Data collection methods	If data collection tools were shown to be valid and reliable.
Withdrawal and drop-outs	If withdrawals and drop-outs were reported in terms of numbers and/or reasons per group, as well as the percentage of participants that completed the study.
Intervention integrity	The percentage of participants that received the allocated intervention or exposure of interest, if the consistency of the intervention was measured, and if it is likely that the participants received an unintended intervention such as contamination or co-intervention that may have influenced the results.
Analysis	If the unit of allocation and analysis was a community, organization/institution, practice/office, or individual.

2.6 AMSTAR

AMSTAR stands for *A Measurement Tool to Assess Systematic Reviews* and is a tool used to assess the methodological quality of systematic reviews (some other sources use the phrase *Assessment of Multiple Systematic Reviews*, see Pieper et al. 2018). The tool includes 11 items for evaluation (Shea et al. 2009, 1018): 1) Was an "a priori" design provided? 2) Was there duplicate study selection and data extraction? 3) Was a comprehensive literature search performed? 4) Was the status of the publication used as inclusion criteria? 5) Was a list of studies (included/excluded) provided? 6) Were the characteristics of the included studies provided? 7) Was the scientific quality of the included studies assessed and documented? 8) Was the scientific quality of the included studies used appropriately in formulating conclusions? 9) Were the methods used to combine the findings of studies appropriate? 10) Was the likelihood of publication bias assessed? and 11) Was the conflict of interest included? Important aspects to consider for each item are described below.

For (1), the research question and inclusion criteria should be established before the review is conducted. For (2), there should be at least two independent data extractors as well as a consensus procedure for disagreements in place. Regarding search (3), at least two electronic sources should be searched. Moreover, years and databases used should be reported, as well as keywords and/or MeSH terms. Where feasible, the search strategy should also be provided. Concerning status of publication (4), authors should state that they searched for reports regardless of publication type, and whether or not they excluded any reports based on publication status or language. A list of the included versus excluded studies should also be provided (5). Data from the original studies should be provided in aggregated form, e.g., in a table explaining participants, interventions and outcomes (6). Characteristics in all of the studies (age, race, sex, socioeconomic data, etc.) should also be reported. Considering scientific quality (7), "a priori" methods should be provided. For example, if only randomized, double-blind, placebo-controlled studies, or allocation concealment were included, this should be specified in an inclusion criterion. Moreover, results of the methodological rigor and scientific quality should be considered (8). When it comes to combining findings from different studies (9), tests should be done on the pooled results to ensure that the studies were combinable and to assess homogeneity (e.g., Chi-squared test for homogeneity). However, if heterogeneity exists, the appropriateness of combining the data should be taken into account, or a random space effects model could be used. Regarding bias (10), assessment of publication bias should include a combination of statistical tests (e.g., Egger's regression test) and graphical aids (e.g., funnel plots). Finally, regarding conflict of interest (11), potential sources of financial support should be clearly acknowledged both in the systematic review and the reviewed studies (Shea et al. 2009, 1018).

2.7 PRISMA

PRISMA stands for *Preferred Reporting Items for Systematic reviews and Meta-Analyses* (Liberati et al. 2009, 1; Moher et al. 2009, 1). It was developed as an evolution of the QUOROM statement (*Quality of Reporting of Meta-analysis*), a reporting guideline published in 1999 (Liberati et al. 2009, 1). PRISMA focuses on how authors can ensure transparent and complete reporting of systematic reviews and meta-analyses (Liberati et al. 2009, 4). The PRISMA statement includes a 27-item checklist and a 4-phase flow diagram (Liberati et al. 2009, 2). It is important that information for each of the 27 items is provided somewhere in the report (Liberati et al. 2009, 4). The following phases of a systematic review are defined: *identification* (records identified through database search and other sources), *screening* (records to be screened, after duplicates have been removed), *eligibility* (full-text articles assessed for eligibility after exclusion of articles), and *inclusion* (studies included in the qualitative synthesis, leading to inclusion in the quantitative synthesis, i.e., the meta-analysis) (Liberati et al. 2009, 4).

PRISMA's first item focuses on the *Title* and involves identifying the report as a systematic review, meta-analysis, or both. The second item is the *Abstract* and involves producing a structured summary (Liberati et al. 2009, 5. A structured abstract could include the following headings: context/background), objective/purpose, data sources, study selection/eligibility criteria, study appraisal and synthesis methods/data extraction and data synthesis, 6) results, limitations, and conclusions/implications (Liberati et al. 2009, 6). The third and fourth items focus on the *introduction* of the reported results. The third item is the *rationale* for the review in the context of what is already known (Liberati et al. 2009, 6). The fourth addresses the *objectives* of the work, which involves providing an explicit statement of PICOS (Liberati et al. 2009, 6). Items 5 to 16 focus on *Methods*, and items 17 to 23 focus on *Results*. Finally, items 24–26 address *Discussion*; the included items are *summary of evidence, limitations*, and *conclusions*. The last item, 27, focuses on *Funding* used to carry out the systematic review. A summary of the items defined by the PRISMA statement is presented in Table 8.6.

Apart from PRISMA, there are a number of other reporting guidelines for different types of research designs (Tate & Douglas 2011, 1). For example, *The Consolidated Standards of Reporting Trials* (CONSORT) statement is the reporting guideline for RCTs. This was published in an updated version in 2010, both in the form of a checklist (Schulz et al. 2010, 1–11) and an explanation and elaboration document (Moher et al. 2010, e1–e37). CONSORT is a checklist of 25 items, some of which also have subcomponents, resulting in a total of 37 points to consider. The items cover the following aspects of reporting: *Title* and *Abstract* (2 items), *Introduction* (2 items), *Methods* (17 items), *Results* (10 items), *Discussion* (3 items), and *Other information* (3 items) (Tate and Douglas 2011, 4). An alternate statement to improve

Table 8.6 PRISMA checklist of items to include when reporting a systematic review

Section/Topic	No	Item
TITLE		
Title	1	Identify the report as a systematic review, meta-analysis, or both.
ABSTRACT		
Structured summary	2	Provide a structured summary including, as applicable: background, objectives, data sources, study eligibility criteria, participants, interventions, study appraisal and synthesis methods, results, limitations, conclusions and implications of key findings, systematic review registration number.
INTRODUCTION		
Rationale	3	Describe the rationale for the review in the context of what is already known.
Objectives	4	Provide an explicit statement of questions being addressed with reference to participants, interventions, comparisons, outcomes, and study design (PICOS).
METHODS		
Protocol and registration	5	Indicate if a review protocol exists, if and where it can be accessed (e.g., web address), and, if available, provide registration information including registration number.
Eligibility criteria	6	Specify study characteristics (e.g., PICOS, length of follow-up), and report characteristics (e.g., years considered, language, publication status) used as criteria for eligibility, giving rationale.
Information sources	7	Describe all information sources (e.g., databases with dates of coverage, contact with study authors to identify additional studies) in the search and date last searched.
Search	8	Present full electronic search strategy for at least one database, including any limits used, such that it could be repeated.
Study selection	9	State the process for selecting studies (i.e., screening, eligibility, included in a systematic review, and, if applicable, included in the meta-analysis).
Data collection process	10	Describe the method of data extraction from reports (e.g., piloted forms, independently, in duplicate) and any processes for obtaining and confirming data from investigators.
Data items	11	List and define all variables for which data were sought (e.g., PICOS, funding sources) and any assumptions and simplifications made.
Risk of bias in individual studies	12	Describe methods used for assessing risk of bias of individual studies (including specification of whether this was done at the study or outcome level), and how this information is to be used in any data synthesis.
Summary measures	13	State the principal summary measures (e.g., risk ratio, difference in means).

(Continued)

Table 8.6 (Continued)

Section/Topic	No	Item
Synthesis of results	14	Describe the methods of handling data and combining results of studies, if done, including measures of consistency for each meta-analysis.
Risk of bias across studies	15	Specify any assessment of risk of bias that may affect the cumulative evidence (e.g., publication bias, selective reporting within studies).
Additional analyses	16	Describe methods of additional analyses (e.g., sensitivity or subgroup analyses, meta-regression), if done, indicating which were pre-specified.
RESULTS		
Study selection	17	Give numbers of studies screened, assessed for eligibility, and included in the review, with reasons for exclusions at each stage, ideally with a flow diagram.
Study characteristics	18	For each study, present characteristics for which data were extracted (e.g., study size, PICOS, follow-up period) and provide the citations.
Risk of bias within studies	19	Present data on the risk of bias of each study and, if available, any outcome-level assessment (see Item 12).
Results of individual studies	20	For all outcomes considered (benefits or harms), present, for each study: (a) simple summary data for each intervention group and (b) effect estimates and confidence intervals, ideally with a forest plot.
Synthesis of results	21	Present results of each meta-analysis done, including confidence intervals and measures of consistency.
Risk of bias across studies	22	Present results of any assessment of risk of bias across studies (see Item 15).
Additional analysis	23	Give results of additional analyses, if done (e.g., sensitivity or subgroup analyses, meta-regression, see Item 16).
DISCUSSION		
Summary of evidence	24	Summarize the main findings including the strength of evidence for each main outcome; consider their relevance to key groups (e.g., health care providers, users, and policymakers).
Limitations	25	Discuss limitations at study and outcome level (e.g., risk of bias), and at review level (e.g., incomplete retrieval of identified research, reporting bias).
Conclusion	26	Provide a general interpretation of the results in the context of other evidence and implications for future research.
FUNDING		
Funding	27	Describe sources of funding for the systematic review and other support (e.g., supply of data) and role of funders for the systematic review.

Adapted from table 1 in Liberati et al. 2009

the reporting of non-intervention evaluation studies using non-randomized designs is presented in the TREND guidelines (*Transparent Reporting of Evaluations with Non-randomized Designs*) (Des Jarlais et al. 2003, 362).

3 Discussion

It seems that sound and music research, as it is reported in the Sound and Music Computing and Sonic Interaction Design communities, is underrepresented in systematic reviews on music and health. To test this assumption, we searched the Cochrane Database of Systematic reviews for reports with the term "music" in the title (limiting the search to "Record Title").[1] The search gave 14 hits, out of which 2 were marked as "withdrawn" in the database. From the remaining 12 reports, there are strikingly few studies included from authors that are active in the sound and music research area; similarly, only a few studies that are included (or even considered but excluded) have been published in related academic journals and conference proceedings (e.g., *Musicae Scientiae* and *Psychology of Music*). About half of the included studies are found in clinical research journals, and the other half in music therapy journals. -An extended search to find the term "music" in title, abstract and keywords results in 61 hits (3 were marked as "withdrawn" in the database) and revealed a similar tendency, in terms of included studies.-

Commonly, all of the 12 above-mentioned systematic reviews have search strategies that include searching (at least) MedLine and Cochrane's databases. Most state that manual searches of specific music therapy journals and reference lists were performed. No reports mention a more extensive search that would include typical outlet channels for sound and music research, such as proceedings from the Sound and Music Computing conferences, *Journal of the Acoustic Society of America*, *Organised Sound*, *Journal of New Music Research*, *Journal on Multimodal User Interfaces*, or *Music Perception*. Even general interest journals such as *Applied Sciences* and *PLoS ONE* are absent from the search strategies. With this consideration, it is interesting to note that in the conclusions sections and in the implications for research, it is often commented that further research involving studies that go beyond music therapy and the music listening of music medicine interventions are needed.

From the above, an immediate interpretation is that little emphasis is put on designing randomized control studies in sound and music research. In other words, there appears to be a tendency towards use of quasi-experimental design. Another interpretation is that the Sound and Music Computing and Sonic Interaction Design communities are unaware that their studies will have little or no impact in comparison to studies reporting results in line with the music therapy or music medicine intervention traditions. Yet another possibility is that the search strategies and criteria as defined in the Cochrane reviews are too narrow and should benefit from also looking at the sound and music research outlets.

While we argue in this chapter that researchers benefit from adopting a strategy like following the PICO framework to increase impact, we admit that conducting studies under proper controlled conditions might be impractical. In the book *Case Study Research*, For many case studies, the objective is not to discover something about a broader population of cases, but rather to investigate one unit in an attempt to explain a single outcome within that unit. In the book Case Study Research, Gerring discusses approaches for single-outcome analysis with singular case observations and contrast these to generalizations across populations in other case study methods. Single-outcome single-case methods are generally problematic because of the situations in which they are deployed (one example being the Watergate robbery), but nonetheless indispensable (Gerring 2007, 187–210). This consideration is also relevant for various types of studies within sound and music research and can challenge the assumption of having a controlled experiment.

4 Implications for Research

The foundation of evidence-based practice is the use of the best scientific evidence to support decision making. Identification of best evidence requires appropriate research questions, i.e., questions that can lead to something that can be measured. As stated earlier, a well-written research question will have many benefits and be helpful not only to the researcher in question, for example by feeding into the development of factors of interest for analysis, but also to the research community as a whole. In order to be considered for inclusion in systematic reviews, it is essential that the titles and abstracts are following accepted formats. The most common in clinical reviews is PICO. Employing the PICO framework makes the research more accessible to search engines and databases, and thus more likely to be included in systematic reviews. A made-up example of a title that successfully employs the PICO framework could be: "Actively improvising with musical instruments reduces depression symptoms more than passive music listening among long-term depressed patients." This title clearly describes the patient (P: long-term depressed patients), intervention (I: actively improvising on a musical instrument), comparison (C: control group in the form of patients who were only exposed to passive music listening) and outcome (O: a reduction in the number of depression symptoms).

With a clearly defined research question, the risk of having unnecessary bias can be reduced. The problem of bias in the studies was common among all the 12 above-mentioned reviews with "music" in the title from Cochrane's Database of Systematic

Reviews. Although the ROBIS framework was developed for systematic reviews, the guidelines are also applicable for experimental design of other types of studies. Another recurring problem among the 12 found studies was power: none of the reviews could conclude that the quality of evidence was high. Of course, the use of different study designs has impact in this regard. Even if it is more difficult to draw causal inferences from quasi-experiments than from randomized experiments, adequate planning can allow for strong causal inferences. This can be achieved for example by identifying and studying possible threats to internal validity, designing controls that limit such threats and by defining hypotheses that limit the number of alternative explanations. Moreover, when it comes to reporting results of systematic reviews and meta-analyses, PRISMA, CONSORT or TRADE can be applied as tools for ensuring that the presented results are comparable to other work.

To conclude, it was apparent from the search strategies adapted for Cochrane reviews that the impact of research is dependent on finding the study in MEDLINE or other dedicated databases. A journal may be indexed in MEDLINE if it is related to life science. Thus, authors active in Sound and Music Computing or Sonic Interaction Design in areas of health, accessibility and disability who aim for impact and inclusion in the relevant systematic reviews would benefit from targeting their output towards such indexed journals.

Note

1   https://www.cochranelibrary.com/advanced-search?q=*music&t=6.

References

Armstrong, E.C. (1999). The well-built clinical question: The key to finding the best evidence efficiently. *Wisconsin Medical Journal*, 98, 25–28.

Baldwin, S. & Berkeljon, A. (2010). Quasi-experimental design. In N. J. Salkind (ed.), *Encyclopedia of research design,* 1172–1176. Thousand Oaks, CA: SAGE. DOI: 10.4135/9781412961288.n353.

Bernardini, N., Serra, X., Leman, M. & Widmer, G. (2007). *A roadmap for sound and music computing.* The S2S2 Consortium.

Booth, A. (2006). Clear and present questions: Formulating questions for evidence based practice. *Library Hi Tech*, 24 (3), 355–368. DOI:10.1108/07378830610692127.

Bradt, J., Dileo, C., Magill, L. & Teague, A. (2016). Music interventions for improving psychological and physical outcomes in cancer patients. Cochrane Database of Systematic Reviews, (8).

Bruscia, K. E. (1998). *Defining music therapy*. Barcelona Publishers.

Cramer, D., & Howitt, D., eds. (2004). The SAGE dictionary of statistics. London: SAGE. DOI: 10.4135/9780857020123.

Cogo-Moreira, H., R.B. Andriolo, L. Yazigi, G.B. Ploubidis, C.R. Brandão de Ávila and J. J. Mari. (2012). "Music education for improving reading skills in children and adolescents with dyslexia." Cochrane Database of Systematic Reviews,(8). DOI: 10.1002/14651858.CD009133.pub2.

Cooke, A., Smith, D. & Booth, A. (2012). Beyond PICO: the SPIDER tool for qualitative evidence synthesis. *Qualitative health research*, 22 (10), 1435–1443.

Dawes, M. (2010). Triple-blind study. In N.J. Salkind (ed.), *Encyclopedia of research design,* 1542–1543. Thousand Oaks, CA: SAGE. DOI: 10.4135/9781412961288.n471.

Denyer, D., Tranfield, D. & Van Aken, J.E. (2008). Developing design propositions through research synthesis. Organization Studies, 29 (3), 393–413. DOI: 10.1177/0170840607088020.

Des Jarlais, D.C., Lyles, C. & Crepaz, N. (2004). Improving the Reporting Quality of Nonrandomised Evaluations of Behavioral and Public Health Interventions: The TREND Statement. *American Journal of Public Health*, 94 (3), 361–366. DOI: 10.2105/ajph.94.3.361.

Effective Public Health Practice Project (EPHPP). (2009). Quality Assessment Tool for Quantitative Studies. Available online: https://www.ephpp.ca/PDF/Quality%20Assessment%20Tool_2010_2.pdf. See also: https://www.ephpp.ca/PDF/QADictionary_dec2009.pdf for full descriptions of the terms.

Fan, S. (2010). Independent variable. In N.J. Salkind (ed.), *Encyclopedia of research design,* 592–593. Thousand Oaks, CA: SAGE. DOI: 10.4135/9781412961288.n184

Franinović, K. & Serafin, S., eds. (2013). *Sonic Interaction Design. Introduction.* MIT Press.

Fritz, C.O., Morris, P.E. & Richler, J. J. (2012). Effect size estimates: current use, calculations, and interpretation. *Journal of experimental psychology: General*, 141 (1), 2.

Fritz, A. & Morgan, G. (2010). Sampling. In N.J. Salkind (ed.), *Encyclopedia of research design,* 1303–1305. Thousand Oaks, CA: SAGE. DOI: 10.4135/9781412961288.n398.

Garfield E. (2006) The History and Meaning of the Journal Impact Factor. *JAMA*, 295 (1), 90–93. DOI:10.1001/jama.295.1.90.

Gerring, J. (2007). *Case Study Research*. Cambridge: Cambridge University Press.

Gill, J. & Walsh, J. (2010). Control group. In N.J. Salkind (ed.), *Encyclopedia of research design,* 252–252. Thousand Oaks, CA: SAGE. DOI: 10.4135/9781412961288.n76.

Grant, M.J. & Booth, A. (2009). A typology of reviews: An analysis of 14 review types and associated methodologies. *Health Information and Libraries Journal*, 26, 91–108.

Gushta, M. & Rupp, A. (2010). Reliability. In N.J. Salkind (ed.), *Encyclopedia of research design,* 1238–1242. Thousand Oaks, CA: SAGE. DOI: 10.4135/9781412961288.n377.

Guyatt, G.H., Oxman, A.D., Kunz, R., Vist, G.E., Falck-Ytter, Y. & Schünemann, H.J. (2008a). Rating Quality of Evidence and Strength of Recommendations: GRADE: What is "quality of evidence" and why is it important to clinicians? *BMJ*, 336 (7651), 995–998.

Guyatt, G.H., Oxman, A.D., Kunz, R., Falck-Ytter, Y., Vist, G.E., Liberati, A. & Schünemann, H.J. (2008b). Going from evidence to recommendations. *BMJ*, 336 (7652), 1049–1051.

Guyatt, G.H., Oxman, A.D., Vist, G.E., Kunz, R., Falck-Ytter, Y., Alonso-Coello, P. & Schünemann, H.J. (2008c). GRADE: an emerging consensus on rating quality of evidence and strength of recommendations. *BMJ*, 336 (7650), 924–926. DOI: 10.1136/bmj.39489.470347.AD.

Hermann, T., Hunt, A. & Neuhoff., J.G. (2011). *The sonification handbook*. Berlin: Logos Verlag.

Higgins, J.P.T., Altman, D.G., Gøtzsche, P.C., Jüni, P., Moher, D., Oxman, A. D. & Sterne, J. A. (2011). The Cochrane Collaboration's tool for assessing risk of bias in randomized trials. *BMJ*, 343, d5928.

Higgins, J.P.T., Thomas, J., Chandler, J., Cumpston, M., Li, T., Page, M.J., Welch, V.A. (eds.). *Cochrane Handbook for Systematic Reviews of Interventions* version 6.1 (updated Sept. 2020). Cochrane, 2020. Available from www.training.cochrane.org/handbook.

Huang, X., Lin, J. & Demner-Fushman, D. (2006). Evaluation of PICO as a knowledge representation for clinical questions. AMIA. Annual Symposium proceedings. AMIA Symposium, 2006, 359–363.

Kabisch, M., Ruckes, C., Seibert-Grafe, M. & Blettner, M. (2011). Randomized controlled trials: part 17 of a series on evaluation of scientific publications. *Deutsches Arzteblatt international*, 108 (39), 663–668. DOI: 10.3238/arztebl.2011.0663.

Kovera, M. (2010). Bias. In N.J. Salkind (ed.), *Encyclopedia of research design*. Thousand Oaks, CA: SAGE, 84–84. DOI: 10.4135/9781412961288.n29.

Kovera, M. (2010). Confounding. In N.J. Salkind (ed.), *Encyclopedia of research design*. Thousand Oaks, CA: SAGE, 221–222. DOI: 10.4135/9781412961288.n70.

Leighton, J. (2010a). External validity. In N.J. Salkind (ed.), *Encyclopedia of research design*. Thousand Oaks, CA: SAGE, 467–470, DOI: 10.4135/9781412961288.n146.

Leighton, J. (2010b). Internal validity. In N.J. Salkind (ed.), *Encyclopedia of research design*. Thousand Oaks, CA: SAGE, 620–622. DOI: 10.4135/9781412961288.n192.

Liberati, A., Altman, D.G., Tetzlaff, J., Mulrow, C., Gøtzsche, P.C., Ioannidis, J. P. & Moher, D. (2009). The PRISMA statement for reporting systematic reviews and meta-analyses of studies that evaluate health care interventions: explanation and elaboration. *Annals of internal medicine*, 151 (4).

McKenzie J.E, Brennan S.E, Ryan R.E, Thomson H.J, Johnston R.V & Thomas J. (2019). Chapter 3: Defining the criteria for including studies and how they will be grouped for the synthesis. In: Higgins J.P.T, Thomas J., Chandler J., Cumpston M., Li T., Page M.J., Welch V.A. (eds). *Cochrane Handbook for Systematic Reviews of Interventions* version 6.0 (updated July 2019). Available from www.training.cochrane.org/handbook.

Methley, A.M., Campbell, S., Chew-Graham, C., McNally, R. & Cheraghi-Sohi, S. (2014). PICO, PICOS and SPIDER: a comparison study of specificity and sensitivity in three search tools for qualitative systematic reviews. *BMC health services research*, 14 (1), 579.

Michaelson, G. & Hardin, J. (2010). Significance, statistical. In N.J. Salkind (ed.), *Encyclopedia of research design*. Thousand Oaks, CA: SAGE, 1362–1366. DOI: 10.4135/9781412961288.n417.

Miller, S.A. & Forrest, J.L. (2001). Enhancing your practice through evidence-based decision making: PICO, learning how to ask good questions. *Journal of Evidence Based Dental Practice*, 1 (2), 136–141.

Mirtz, T. & Greene, L. (2010). Evidence-based decision making. In N.J. Salkind (ed.), *Encyclopedia of research design*. Thousand Oaks, CA: SAGE, 434–437. DOI: 10.4135/9781412961288.n136.

Moher, D., Hopewell, S., Schulz, K.F., Montori, V., Gøzsche, P.C., Devereaux, P.J. & Altman, D.G. (2010). CONSORT 2010 Explanation and elaboration: Updated guidelines for reporting parallel group randomized trials. *Journal of Clinical Epidemiology*, 63 (8), e1–e37.

Moher, D., Liberati, A., Tetzlaff, J. & Altman D.G., The PRISMA Group. (2009). Preferred Reporting Items for Systematic Reviews and Meta-Analyses: The PRISMA Statement. *PLoS Med* 6 (7): e1000097. DOI: 10.1371/journal.pmed.1000097.

Morgan, R.L., Whaley, P., Thayer, K.A. & Schünemann, H.J. (2018). Identifying the PECO: a framework for formulating good questions to explore the association of environmental and other exposures with health outcomes. *Environment international*, 121 (Pt 1), 1027.

Richardson, W.S., Wilson, M.C., Nishikawa, J.A. & Hayward, R.S. (1995). The well-built clinical question: a key to evidence-based decisions. *ACP journal club*, 123 (3), A12–A13.

Pieper, D., Koensgen, N., Breuing, J., Ge, L. & Wegewitz, U. (2018). How is AMSTAR applied by authors–a call for better reporting. *BMC medical research methodology*, 18(1), 56.

Robb S.L., Burns D.S. & Carpenter J.S. (2011) Reporting Guidelines for Music-based Interventions. *Music and Medicine*, 3 (4), 271–279. DOI: 10.1177/1943862111420539.

Robb, S.L., Hanson-Abromeit, D., May, L., Hernandez-Ruiz, E., Allison, M., Beloat, A. & Polasik, S. (2018). Reporting quality of music intervention research in healthcare: A systematic review. *Complementary therapies in medicine*, 38, 24–41.

Rocchesso, D., Serafin, S., Behrendt, F., Bernardini, N., Bresin, R., Eckel, G. & Visell, Y. (2008). Sonic Interaction Design: Sound, Information and Experience. In *Proceedings of the CHI '08 Extended Abstracts on Human Factors in Computing Systems*, 3969–3972.

Salkind, N. (2010). Dependent variable. In N.J. Salkind (Ed.), *Encyclopedia of research design*. Thousand Oaks, CA: SAGE, 348–348. DOI: 10.4135/9781412961288.n109.

Schardt, C., Adams, M.B., Owens, T., Keitz, S. & Fontelo, P. (2007). Utilization of the PICO framework to improve searching PubMed for clinical questions. *BMC medical informatics and decision making*, 7 (1), 16.

Schlosser, R.W., Koul, R. & Costello, J. (2007). Asking well-built questions for evidence-based practice in augmentative and alternative communication. *Journal of communication disorders*, 40 (3), 225–238.

Schulz, K.F., Altman, D.G., Moher, D., for the CONSORT Group. (2010). CONSORT 2010 Statement: Updated guidelines for reporting parallel group randomized trials. *Annals of Internal Medicine*, 152 (11), 1–11.

Shea, B.J., Hamel, C., Wells, G.A., Bouter, L.M., Kristjansson, E., Grimshaw, J. & Boers, M. (2009). AMSTAR is a reliable and valid measurement tool to assess the methodological quality of systematic reviews. *Journal of clinical epidemiology*, 62 (10), 1013–1020.

Stuart-Hamilton, I. (2010). Single-blind study. In N.J. Salkind (ed.), *Encyclopedia of research design.* Thousand Oaks, CA: SAGE, 1384–1386. DOI: 10.4135/9781412961288.n423.

Tate, R.L. & Douglas, J. (2011). Use of reporting guidelines in scientific writing: PRISMA, CONSORT, STROBE, STARD and other resources. *Brain Impairment*, 12 (1), 1–21.

Trinder, L. (2008). Introduction: the Context of Evidence-Based Practice. In *Evidence-based practice: A critical appraisal.* John Wiley & Sons.

US National Library of Medicine (NLM), National Institutes of Health (NIH) (2019a). Medical Subject Headings. Modified: December 5, 2019. https://www.nlm.nih.gov/mesh/meshhome.html.

US National Library of Medicine (NLM), National Institutes of Health (NIH) (2019b). MED-LINE®: Description of the Database. Modified: April 10, 2019. https://www.nlm.nih.gov/bsd/medline.html.

US National Library of Medicine (NLM), National Institutes of Health (NIH) (2020). PubMed. Accessed May 4, 2020. https://www.ncbi.nlm.nih.gov/pubmed/.

Velasco, E. (2010). Inclusion criteria. In N.J. Salkind (ed.), *Encyclopedia of research design.* Thousand Oaks, CA: SAGE, 629–591. DOI: 10.4135/9781412961288.n183.

Vo, H. & James, L. (2010). Power. In N.J. Salkind (ed.), *Encyclopedia of research design.* Thousand Oaks, CA: SAGE, 1066–1067. DOI: 10.4135/9781412961288.n324.

Walach, H. (2010). Double-blind procedure. In N.J. Salkind (ed.), *Encyclopedia of research design.* Thousand Oaks, CA: SAGE, 387–389. DOI: 10.4135/9781412961288.n122.

Whiting, P., Savović, J., Higgins, J.P., Caldwell, D.M., Reeves, B.C., Shea, B. & Churchill, R. (2016). ROBIS: a new tool to assess risk of bias in systematic reviews was developed. *Journal of clinical epidemiology*, 69, 225–234.

Wildridge, V., & Bell, L. (2002). How CLIP became ECLIPSE: A mnemonic to assist in searching for health policy/management information. *Health Information and Libraries Journal*, 19, 113–115.

Practice-led and Interdisciplinary Research
Investigating Affective Sound Design

Jonathan Weinel and Stuart Cunningham

1 Introduction

In this chapter we will discuss approaches to researching sound design that utilize 'practice-led' research methodologies. Practice-led methodologies are based on the premise that the creative work itself constitutes a contribution to knowledge, recognizing that practice feeds into the generation of theory; while conversely, engagement with theory also informs practice (Smith and Dean 2009). This is an appropriate strategy for investigating many areas of research that are orientated towards the arts, such as sound design, since it allows for the development of innovative approaches through 'doing' or through the creation of original artifacts. Whether these are musical compositions, sound installations or software-based projects, practice-led research can allow novel compositional or design strategies to be uncovered, yielding original insights that could not otherwise be obtained through alternative methodologies. Practice-led methodologies are often used in music and sound design research at universities, and may also take place in industrial contexts. While our focus in this chapter is on academia, historical examples of practice-led research in the industrial context include Bell Labs' artist residencies; and conferences such as the ACM *Audio Mostly* series on interaction with sound, where we often present our research, regularly feature presentations from individuals working in industrial contexts such as the automotive industry (e.g., Fagerlönn et al. 2013; Larsson 2016).

Even though practice-led approaches are widely used in the arts, these methodologies are sometimes misunderstood within the scientific community, since they do not typically involve qualitative or quantitative measurements, which can cause confusion with regards to the means by which the efficacy of any outcomes can be judged. Nonetheless, as we shall discuss in this chapter, practice-led research can provide a substantial basis for conducting research in sound design. However, rather

DOI: 10.4324/9780429356360-10

than simply reiterating the validity and associated claims in favor of this research methodology, here we shall argue instead that practice-led approaches can be combined with qualitative and quantitative research methodologies, which are more commonly encountered in the scientific community. Whilst this may not always be appropriate, in some cases this can lead to productive, interdisciplinary outcomes that provide the best of both worlds. Education has been quick to adopt interdisciplinary approaches, as evidenced by the increasing emphasis on STEAM (Science, Technology, Engineering, Arts, and Mathematics) learning (Yakman 2008; Guyotte et al. 2014; Henriksen 2014) and thus it seems likely that there will be an increasing demand for research frameworks and approaches to deal with outcomes in STEAM higher learning activities. Furthermore, as many research funding bodies increasingly highlight the importance of interdisciplinary research, there is clearly a great deal of support for such approaches.

In order to address these methodologies as they apply to sound design, through the course of this chapter we shall provide an outline of how practice-led research can be meaningfully used to investigate sound design and how this may occur in combination with qualitative or quantitative methodologies. Finally, we shall also provide an overview of some 'worked examples' from our own research that explores issues of emotion, affectivity and subjective perception, as they apply to sound design. Here we have been able to utilize practice-led approaches to develop original sound design artifacts and software prototypes, however, we have also carried out various user studies to investigate the experience of these. Through the exploration of these examples from our own research, we aim to demonstrate how combined methodologies can yield successful insights into affective sound design, while also showing more broadly how such approaches might generalize to other areas of sound design.

2 Practice-led Research Methodologies in Sound Design

In their discussion of practice-led research methodologies, Smith and Dean (2009) highlight the importance of creative works themselves in the construction of knowledge. Creative artifacts or performances can demonstrate innovation within a given field. Their existence can expand the diversity of practice within that field, and display new aesthetic or technical innovations. In the sciences, research is a collective endeavor, in which investigations demonstrate positive or negative results, with different groups building upon each other's outcomes, leading to common insights and advances in knowledge. The arts are fundamentally similar, in that most artists borrow and develop ideas from other artists and existing works, building upon these to develop new working practices that demonstrate artistic innovations that develop the form within a particular area.

However, as Nelson (2013) discusses, artists sometimes confuse the matter because they often have a vested interest in protecting the idea that they harbor some kind of elusive 'creative genius' that cannot be quantified. Yet if we delve a little deeper, we usually find that key figures of importance in the arts emerge from scenes of related activity. Just as the iceberg remains largely hidden beneath the water, or the mushroom is supported by the mycelium below the surface, the most recognized artistic figures in a given aesthetic movement are the most visible and successful, but beneath their work lies vast networks of less visible artistic practice. These wider networks of activity, perhaps including exhibitions, concerts or installations at less prestigious venues, are of critical importance in feeding, shaping and nurturing those artists who achieve wider critical acclaim.

Understood in this way, the arts are not so different from the sciences: there is a collective process taking place, and works build on top of each other gradually to advance the field. However, the means through which knowledge advances and is transmitted is fundamentally different. While the sciences usually depend on empirical data and results, in the arts, contributions to the field often occur through artifacts and performances, and may reflect advances towards new artistic forms or modes of practice. In such cases, it may be the artistic works themselves that encapsulate these developments in the purest form. This epistemological distinction becomes an important factor to understand when considering artistic practices as research, because it is at the root of confusions that arise between disciplines. Some may doubt the validity of artistic research, because they do not see measurements or effects being verified. However, empirical approaches are not always necessary, because artifacts and performances can demonstrate innovative advances in the field, which are evident from the forms themselves.

Following this line of thought, we may understand that advances in the arts occur through collective endeavors, and may be transmitted through the materiality of artifacts or other temporal formats such as live performances. However, it would be incorrect to claim on this basis that all artistic practice is research, since most artists who have no involvement in academia would not describe their work in these terms. What then, distinguishes artistic practice as a form of research? In Nelson's (2013) discussion, he emphasizes that "practice as research" should involve a clear line of inquiry and praxis, in which the individual is engaged in both practice and critical reflection in an iterative process. We would however argue that this is still not enough to distinguish artistic activities as research, because most professional artists fulfill these criteria, and it does them a disservice to assume they do not engage in deep critical reflection on what they are doing. Instead we would argue that what really distinguishes artistic practices as research is that they take place within an academic context and with the explicit intention to make a significant contribution to knowledge within a specific domain. In doing so, the work is engaged in the forums of knowledge exchange and dissemination associated with universities and the academy. This requires serious engagement with theory, and thorough documentation of praxis and lineages of knowledge in the form of text. Practice-led research then, involves contributions to knowledge that stem from artistic practices as a key or primary factor, but also requires that these be transmitted into the academic research discourse, which means that ideally they must also be written about, discussed and communicated within academic forums, so that there is a clear and rigorous knowledge base accompanying them.

Returning to apply these ideas to sound design, we may consider that practice-led research in sound design can occur through processes of inquiry in sound design such as musical compositions, sound installations or software-based projects. Work in areas such as these can demonstrate important new approaches that advance the field. Practical work can lead to the development of new ways of making or manipulating sound, while compositional or sound design work can demonstrate ways in which these might practically be utilized. Such work should be undertaken with deep critical engagement in associated theoretical discourses. While artifacts and performances are absolutely central to the transmission of knowledge, these must also be clearly and effectively communicated in a specific academic research context. On the one hand, this can occur through concerts, sound installations software artifacts or recordings. However, supporting discussion of the work should also be provided through forms such as conference papers, book chapters and journal articles, which provide the accompanying discourse that is necessary for a more comprehensive understanding of the work in the research context. By doing this, new innovations in sound design can be realized, but they can also be fully understood and properly communicated in the academic research context, advancing knowledge within the field.

3 The Case for Interdisciplinary Approaches

Whilst the above section highlights practice-led research as an appropriate methodology for sound design research, we may also utilize empirical approaches to investigate this domain, and in some cases it can be beneficial to combine the two. Interdisciplinary research may typically involve individuals from a variety of disciplinary backgrounds and with a range of expectations and epistemological views that need to find some form of harmony in order for successful collaborations to manifest. This is noted by Serra (2019), who discusses the diverse origins and backgrounds of the Music Technology Group at the Universitat Pompeu Fabra Barcelona. While Serra notes that the structure of higher education establishments can present challenges for interdisciplinary work, he also highlights its importance for the field of music technology, emphasizing that the very nature of music requires researchers to cross disciplinary boundaries. Although Serra deals specifically with music technology, there are many similarities and overlaps with the field of sound design.

Sound design research is a particularly interesting field to consider since it has foundations in both the arts and sciences. On the one hand, sound design is almost always carried out as a creative process for the benefit of human listeners, such as visitors to exhibitions or installations, cinema audiences, music aficionados or video game players. In these types of contexts, sound design is subject to human sensory perception and the subjective interpretation of the listener, in much the same way as a piece of visual art, such as a painting. On the other hand, the tools and techniques required to manipulate and disseminate sound are grounded firmly within the domains of physics and engineering. To understand the fundamental principles behind the capturing and processing of sound requires knowledge of waveforms and their manipulation through analogue or digital electronics; and their presentation and distribution, perhaps via radio, compact disc, vinyl or Internet broadcast, draws upon similar specialisms. The combination of these two broad disciplines – that is, sound design as an art, and audio engineering as a science – requires that in many cases, research in sound design should not only utilize practice-led approaches to advance creative practices and our understanding of them, but also scientific methods, to underpin these insights with relevant technical knowledge. This approach may be advantageous, not only as a way of satisfying the discipline-specific expectations of a field inhabited by researchers from these respective backgrounds, but also because, from the perspective of the listener, there may be no immediate sign of this division – only the experience of the product crafted by the symbiotic relationship of the two.

Thus, we come to consider the case for the use of scientific approaches as complementary to practice-led investigations in sound design. To begin, it is useful to establish some of the principles and features that define these ways of working. It is commonplace to begin a scientific or empirical study with the specification of an intention, prototypically expressed as a hypothesis, or perhaps as a question, although this is usually preceded by some kind of articulation of the problem in general. Such intentions are often the product of previous research activities, be they primary (outcomes arrived at by research performed oneself) or secondary (outcomes reported by other researchers) in nature. The remainder of the approach then revolves around constructing an organized and systematic way to gather evidence that will allow the researcher to prove or disprove the stated hypothesis. It is important to stress that the formulation of hypotheses or research questions should not be rushed and that they are carefully crafted.

Hypotheses and questions will lead the researcher to make decisions about what data or information should be gathered in order to try to reach an outcome, or answer, in response. The data that is obtained is constituted of one or more *variables*, things that can be collected or measured, and which may be direct or indirect measures of the phenomenon of interest. Thus, any statement of research intent will lead the researcher to the ways in which their investigations can be conducted.

Often these variables are factors or indicators of the circumstance that we are interested in. For example, in our own work we often seek to measure the emotional response a person has whilst listening to a sound. Measuring emotion directly and reliably is not easy. There is no objective 'emotion meter' that works in the same way as a thermometer, for instance. Even though we can ask someone for their emotional response by requesting them to tick a box on a form, this can be colored by their cultural values, system of beliefs and other social and psychological factors. To deal with this, we could capture a mixture of variables that are known *indicators* of emotion, such as skin conductivity (sweat secretion) and skin temperature, alongside asking the person to tick some boxes on a form (a type of *self-report* method). Ultimately, it is the intention that analysis of these variables will allow the researcher to reach a conclusion about the validity of a hypothesis or provide an answer to a research question. Further guidance on producing research problem statements, hypotheses, questions, and determining variables of interest can be found in the work of Leedy and Ormrod (2016).

Variables are identified that serve as parameters that can be manipulated by the researcher as well as being used as indicators of an outcome (in order to detect some form of effect). These are respectively known as the *independent* and *dependent* variables. For example, an independent variable might be the volume control on a sound system and the dependent variable could be a decibel meter in the room. As we manipulate the volume control, we would discover that the readings on the decibel meter change in response – an *effect* that we could go on to explore and describe in more detail with structured tests (i.e., setting specific values on the volume control and measuring the corresponding reading on the decibel meter). The gathering of data is considered to be objective, at least from the perspective of eliminating potential bias on the part of the researcher, although it is fully acknowledged that there exist quantitative metrics that are designed to capture subjective human response phenomenon, such as the Self-Assessment Manikin (Bradley and Lang 1994), used in studies of emotional response.

Scientific research is characterized by adherence to a structure that supports rigor and control, allowing the variables and factors being scrutinized to be analyzed independently of the majority of other interfering factors. This permits data to be gathered from which a response to the original hypothesis can be inferred, often supported by statistical methods and analytical models that support objective conclusions to be reached. Furthermore, such activities are typified by their clarity and documentation in dissemination, so as to attempt to maximize transparency and support replicability of the research activities (Leedy and Ormrod 2016). The methods most often encountered when adopting this approach are quantitative in nature, involving the gathering of various measures that can be counted or quantified. For instance, one might record the heart rate of a research participant who is listening to a sound or ask them to rate the 'roughness' they perceive in the *texture* or *timbre* of the sound. The process of quantitative analysis typically then involves describing and evaluating the characteristics of the data by using statistics and machine-learning. Statistics and machine-learning provide objective and repeatable ways to interrogate the data and discover or model a range of effects.

However, there is scope for the collection and analysis of qualitative information, where descriptive information about the object(s) of investigation is sought and recorded. The information captured can take a wide variety of forms, including text, images, drawings, video, sound and so forth (Lazar, Feng, and Hochheiser 2017). For instance, one might play a research participant a sound and ask them to describe, in words, its textural qualities, or one could ask the participant to draw a picture of the texture that represents the sound they are hearing.

Both quantitative and qualitative information can be used together, often in the capacity of a mixed methods study. Mixed methods studies are designed to capture a wider variety of information, usually with the aim of producing a richer picture of the phenomenon and one that, consequently, is more robust. In such a situation the information gathered could be used to interrogate one another, known as *triangulation*, to determine where findings converge and diverge (Plano Clark and Ivankova 2016). Triangulation is the practice of utilizing multiple sources of information in order to provide a more accurate indication of the true nature of the occurrence that is being investigated. To draw upon the examples from the previous two paragraphs (of quantitative and qualitative data relating to a presented sound stimulus), a researcher would now have indicators of the participant's heart-rate response, a roughness rating, a description of the texture of the sound and a picture of the texture. These pieces of information could be collectively analyzed to determine how that listener perceives the sound in question.

As established earlier, in sound design research, we may be interested in working with sound from the perspective of: waveforms and signals; its perception, experience and affect upon its audience; or a combination of both. Each of these areas can be satisfied by a structured process of inquiry. We first of all deal with activities that focus upon working with sound at a *signal* level before then considering how its *perception* and *affect* upon the human listener can be considered.

In the case of dealing with sound as a signal, a great deal can be achieved by analyzing waveform audio using a variety of existing metrics, as well as the creation of new ones for specific purposes. These processes are almost always carried out using computer software that performs digital signal processing, although task-specific hardware devices are not uncommon in the world of sound. Dawson (2019) highlights a broad range of audio analysis techniques and applications that operate principally at the signal level, such as: *acoustic analysis*; *audio event and sound recognition*; *music information retrieval*; *sound analysis*; *semantic audio analysis*; *sound scene and event analysis*; and *speech analysis*. Such activities are typically concerned with the extraction of *features* from audio signals or with their manipulation in order to achieve a specific goal. Audio features are typically classified depending upon their degree of granularity relative to a complete audio signal and include measures indicating everything from intensity and frequency to musical mode or tempo (Giannakopoulos and Pikrakis 2014; Moffat, Ronan and Reiss 2015). Working with audio signals in this manner is typically a 'desk-based' task, meaning it is economical and fast to obtain results or try out ideas. The analysis and manipulation of audio signals can be rewarding and satisfying in their own right, and being able to explain artistic expression or meaning on this level can be beneficial in understanding and replicating the work of others or in its application in other sub-domains. But it can be very difficult, especially in an interdisciplinary context, to separate the signal from its perception by the human listener, and so attention is also required there.

When it comes to the perceptions and experience of the audience, these too can be approached by using a scientific and empirical approach. Dawson (2019) highlights the techniques of *multimodal analysis* and *psychoacoustic analysis*, which we would argue are good examples of activities that can be used to explain the relationship between the sound signal, its perception and its interpretation, notably the affective response and/or the attachment of meaning that is arrived at by the listener. As is becoming obvious, the closer we come to the human aspect of sound design, the more we move away from engineering and technology and the closer we come to exploring our understanding of neurology and the psychology of sound – also scientific domains in their own right. In terms of human perception of sound design, there is now potential to capture this information from the listener, using self-report tools or observations of the researcher, and these can be achieved using a variety of quantitative and qualitative measures, such as surveys and interviews. These data can be collected with relative ease, but the issue of subjectivity, and how it might influence the validity of findings, must be taken into account. To this end, it is worth noting that there is the opportunity to capture physiological response data from listeners, using a range of passive, as well as more intrusive, sensors. For instance, facial expressions, body language, heart rate, and perspiration might be measured in order to provide more objective indicators of listener response. These measures are not without issue either, notably the level of intrusion and the requirement to then infer specific outcomes based upon the resultant data. Both of these avenues present effective ways to capture listener perception and experience of sound designs and are relatively easy to conduct in a methodical way, subject to appropriate ethical scrutinies and approvals.

Due consideration must also be given to the environment and equipment that is being used in research where audio reproduction and perception are a core component. Where a large amount of control over the listening environment and equipment is needed, purpose-built listening rooms should be employed, which typically exhibit desirable characteristics in terms of reverberation and ambient noise, and are equipped with high-grade equipment (Corey 2016). In other circumstances, a more artificial, but more controlled acoustic space with no sound reflections and which is sealed from ambient noise, known as an *anechoic chamber*, may be used. In many cases, often due to resource or practical constraints, it may not be possible to use such spaces and quality headphones can be a useful alternative. Otherwise, it may be considered appropriate to the research in question to evaluate it *in-situ* and accept that the ambient noise and reflections of the surrounding environment are outside the control of the researcher, and thus the activities can be performed 'in the wild'.

In summary of the case for supporting practice-led research in sound design with scientific methods, there are two main arguments for the application of empirical techniques when developing and evaluating such work. First, the ability to capture data, often in a mixed methods setting, perhaps using sound designers themselves, or audiences, permits triangulation. From this comes a greater insight into how the process and outcomes of sound design work, and how these can be realized effectively. In essence, more of the 'big picture' is revealed, and the adoption of multiple methods means that greater reliability and validity of outcomes can be realized. The second argument for this approach is the ability to communicate and disseminate results relating to the underlying technologies and perceptual experiences that are inextricably connected with the inter-disciplinary work of the sound designer. By adopting approaches recognized by all interested parties, work can be easily understood and evaluated by all concerned.

4 Researching Affective Sound Design

Having outlined possible methodologies for researching sound design based on practice-led and/or interdisciplinary approaches, we will now turn to consider a series of examples of work carried out in the context of the Affective Audio research group (see Weinel et al. 2014). The Affective Audio group is a research collective that explores the application of approaches from the field of affective computing (Picard 1998) to the field of sound design. Our work broadly consists of projects investigating sound design in the context of immersive sound and audio-visual systems, in relation to aspects of human emotion and perception. A large number of the projects that we describe fall into the broad category of being a *participant study* or *user study*. It is important to stress that such studies can make use of a wide variety of research methods, information and data. However, the defining feature is that studies of this nature make use of a small number, or *sample*, of people who are representative of the larger group, the *population*, who would typically engage with the outcomes of the project in question.

In what follows, we shall discuss several areas of our research in which practice-led and interdisciplinary methods have been used in combination. Our research group engages with quite a diverse range of projects and endeavors, and what follows is certainly not a definitive account of our activities. Rather, these projects have been chosen because they exemplify the type of practice-led and interdisciplinary work described earlier. Through these, we aim to illustrate how the methodological approaches discussed in this chapter have been applied in practice. We intend this exploration to be informative for other researchers in sound design considering how to develop their work using practice-led and/or interdisciplinary approaches. These projects will also illustrate how fluidity may occur between projects, and the possible advantages of utilizing practice-led and interdisciplinary research methodologies.

4.1 Biofeedback Audio-visual Projects Based on Altered States of Consciousness

Weinel's earlier PhD work (prior to joining the Affective Audio group) investigated the composition of electroacoustic music based on altered states of consciousness (ASCs) (Weinel 2012; see also Weinel 2021). The project consisted of practice-led research composing works of electroacoustic music and audio-visual compositions that sought to represent the typical form and structure of hallucinations. In addition to the compositions themselves, the project also included software tools that were used to create sonic materials and live performances of the work. In terms of theory, the project drew significantly upon research regarding the experience of ASCs. For instance, research by Klüver (1971) discusses the typical form of hallucinations, in which visual funnel patterns ('form constants') may be observed during stages of hallucination. Research such as this, and other work such as Strassman's (2001) studies of DMT (Dimethyltryptamine, a hallucinogenic substance) hallucinations informed an original compositional process, in which the typical form, structure and features of hallucinations were used as a basis for composing electroacoustic music. The main output of this project was a portfolio of electroacoustic compositions, audio-visual compositions and software tools, accompanied by a written commentary. While drawing upon the outcomes of scientific studies that investigated the experience of ASCs, this project was firmly situated as practice-led research, with the main contribution to knowledge being the development of an approach to composing electroacoustic music based on altered states of consciousness (Weinel 2016).

Several projects developing from Weinel's work on ASCs were later developed in the context of the Affective Audio group. The first of these, *Psych Dome*, consisted of an immersive audio-visual installation utilizing biofeedback, sound and video projection (Weinel et al. 2015b). The work was originally created for presentation in a mobile fulldome (a type of immersive video projection environment in a dome; see Lantz 2006). Participants entering the dome wore a consumer-grade EEG headset (Neurosky Mindwave), which sent brainwave signals via OSC (Open Sound Control) to a Max/MSP patch that generated electronic music. This patch also sent control data to a Processing sketch, which generated a visualization based on Klüver's (1971) form constants. In this way, the project provided an artistic audio-visual installation based on representations of hallucinations, where brainwave activity influences aspects of the experience, by modulating colors and frequencies of sound.

The *Psych Dome* installation itself can be understood as a product of practice-led research, since it extended compositional methodologies derived from Weinel's earlier research in electroacoustic composition. However, this work generates questions regarding the interaction between the biofeedback (EEG) device, the user and the artwork itself, which could be investigated through empirical methods. The Affective Audio team therefore decided to use the project as a basis for a study exploring the type of interaction that brainwave signals from consumer EEG could provide for an installation such as this, and to determine the efficacy of this novel form of input. We carried out a participant study, which compared live EEG signals with a pre-recorded EEG signal, to explore the extent to which users of the installation experienced a sense of control and/or connection with the artwork via biofeedback. This was done using a pretest-posttest approach (Salkind 2010), where 10 participants experienced *Psych Dome*. In both cases, participants wore the EEG headset with the pre-recorded signal version acting as a placebo control condition. For each occasion, participants were unaware whether the EEG was live or pre-recorded, and this sequence was randomized across all of the participants, so as to avoid any kind of presentation bias. After each experience, the participants provided five ratings of the experience on a number of scales, the scores from which were then subject to paired statistical testing to determine the efficacy of the EEG input.

The outcomes of the study suggested that most people could not tell if a live EEG signal was actually being used or not, suggesting that the type of interaction experienced was what Zander et al. (2010) refer to as "passive," in that the device elicits a sense of connection to the artwork but not one that includes volitional control. In this case, the *Psych Dome* installation, developed as practice-led research, can therefore be understood as advancing the discourse regarding the creative design of biofeedback artworks; however, it also provided a basis for the empirical study, which will be informative for further practice-led work in this area, since it enriches our understanding of what artworks utilizing consumer-grade EEG can, and cannot, provide for audiences.

The outcomes of the *Psych Dome* study informed a subsequent development of one of Weinel's other creative projects, *Quake Delirium*, which similarly emerged from his practice-led research regarding ASCs. Weinel conceived of a video game system that could provide a first-person representation of a hallucination state. He created an early prototype of this system through a modification of the video game *Quake* (Weinel 2011). In this modification, the user plays the video game, whilst automation of graphical effects, game parameters and an associated real-time soundtrack are generated, in order to provide a warped version of the game that aims to simulate features of hallucination. Similar approaches would later be used in commercial video games such as *Far Cry 3* (Ubisoft Montréal 2012) to represent experiences of intoxication via psychedelic mushrooms (Weinel 2018, 143–144).

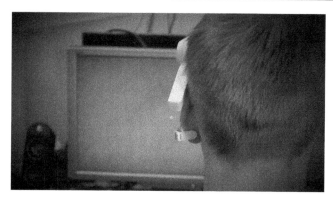

Figure 9.1 A participant wears a consumer-grade EEG headset, whilst playing the *Quake Delirium EEG* video game. Aspects of graphics and sound in the video game are linked to biofeedback signals produced by the headset.

In a later extension of this project, *Quake Delirium EEG,* the Affective Audio team added a consumer-grade EEG headset as a controller, as shown in Figure 9.1. This allowed the modifications to graphics and sound to be linked to the brainwave activity of the player, who wears the headset whilst playing (Weinel et al. 2015a). Based on the *Psych Dome* study, we hypothesized that *Quake Delirium EEG* would provide a 'passive' form of biofeedback, where changes to graphical parameters and sound were perceived non-volitionally by the user. Using similar approaches to the *Psych Dome,* we carried out an empirical study that confirmed this.

4.2 Sound Design Based on States of Intoxication

A notable body of work to come out of the Affective Audio group's activities has focused upon sound design to represent states of intoxication. Representations of such states in popular media, especially film, television and computer games, are far from scarce, but often the sound designs employed in these media rely upon stereotypes, clichés and hyper-realistic representations. As such, we set out to investigate how more accurate designs could be provided, to represent the perception of sound during intoxication with greater authenticity. To achieve this aim, we wanted these designs to be grounded in empirical and valid evidence, as far as possible. Thus, in this collection of work, we moved away from using structured research methods as a way to *evaluate* sound designs and instead use them as the *premise* upon which sound designs are based. The adoption of this approach is also one that is driven by necessity: the practicalities, ethical, social and legal issues associated with performing any kind of participant-based study related to intoxication are necessarily significant.

Our initial work (Weinel, Cunningham and Griffiths 2014) around intoxication sound designs drew upon a large database of self-reports that is freely available online (http://www.erowid.org). Using this database, it was possible to carry out a systematic search, organization and filtering of these accounts to identify reports that specifically dealt with auditory elements. The remaining reports were then subjected to a process of thematic analysis (Clarke, Braun and Hayfield 2015) in order to identify the dominant features present in the sonic experiences of the individuals lodging the reports. From this evidence base, we were then able to construct a framework of sound categories describing sonic experiences relating to intoxication, which were related to Hobson's (2002) neurologically-based model of consciousness. The framework itself was then applied in the creation of several prototypes based upon literal and metaphorical mappings of these descriptions to production and post-production sound design activities.

The framework and categories of sound intoxication were later applied in the dynamic mapping of sound signal characteristics to various stages of ASCs in a prototype game engine audio system (Weinel and Cunningham 2017). This provided several interactive mechanisms for simulating *selective auditory attention* to different sound sources, by attenuating the amplitude of unattended sources; *enhanced sounds*, by adjusting perceived brightness through filtering; and *spatial disruptions* to perception, by dislocating sound sources from their virtual acoustic origin in 3D space, causing them to move in oscillations around a central location. The prototype itself, which was created in the 3D game development environment Unity, can be considered as a piece of practice-led work in its own right, as well as being potentially used in the future as a tool for evaluation and experience testing using game players, or other participants, from whom a range of empirical metrics could be obtained and analyzed to determine the efficacy of the prototype.

More recently, the approach of utilizing self-report data to inform sound design was also applied specifically to sound design work based on communicating the perception of sound whilst under the influence of alcohol (Cunningham 2019). In the case of this study, the scientific literature was searched to find information relating to audiometry outcomes for participants who had consumed alcohol, which then informed creative practice leading to the composition of several electroacoustic pieces using soundscape and soundwalk forms. These compositions featured sound elements, in this case equalization filters, based upon the evidence from the scientific articles and then located within the sound design in response to actions taking place within the narrative of each piece. Therefore, as with the earlier examples discussed, an empirical evidence base was used to inform creative practices in sound design in order to communicate states of intoxication with improved accuracy.

4.3 Evaluation of Sound Designs Using Repertory Grid Technique

Over a number of years, we have found value in the use of the Repertory Grid Technique (RGT) as a way to interrogate sound design from a variety of angles, particularly those that are concerned with the listener perception and experience

(Cunningham 2010; Grill, Flexer and Cunningham 2011; McGregor and Cunningham 2015; Cunningham, Weinel and Picking 2016). RGT is grounded in *personal construct psychology*, which is based on the premise that a person learns about the world around them through their experiences and interactions with it (Kelly 1955). Initially devised as a theory to investigate perception of personalities, the principles have been applied to many other domains, or topics, including technological ones.

Our body of work began by highlighting the possibilities that RGT offers and demonstrating its ability to allow a study participant to be able to appraise a variety of media forms, specifically individual audio and video components (Cunningham 2010) within the context of designing sound for audio-visual media, such as computer games. Subsequently, the technique was applied to identify a range of perceptual qualities that could be used to describe textural sounds (Grill, Flexer and Cunningham 2011). This particular work began by interviewing a range of participants to elicit descriptive terms for a set of sound stimuli, before the terms and their quantitative ratings were analyzed using hierarchical clustering to determine the ten major terms used. Following this mixed methods approach, the resultant terms were validated by a large-scale online survey that involved the participants rating a range of sounds using the determined descriptive terms, to determine how consistent or alike ratings from all participants were. This is technically referred to as the level of *inter-rater agreement* or *inter-rater reliability*.

A key benefit of the RGT approach is that it uses qualitative descriptors of terms, *constructs*, to describe bipolar characteristics of objects of interest, known as *elements* (in our case sound designs). Constructs can then be placed on a semantic differential scale and a rating given for each element, providing quantitative data (Fransella, Bell and Bannister 2004). This can be achieved in several ways, but the most common approaches use individual interviews, sometimes facilitated with card sorting techniques, or online survey tools. The process of card sorting offers "techniques concerned with understanding how people classify and organize things" (Benyon 2019, 167). This provides a tactile and visual way for RGT participants to organize and order the elements being discussed using paper cards. In the case of RGT interviews, the cards typically show the names of the elements under discussion, and often the constructs too. Participants can then arrange the cards in relation to the constructs, helping to lead them toward a numeric rating on a scale (scales of 1 to 5 or 1 to 7 are common). An illustrative example of this is shown in Figure 9.2, in a RGT context, where a participant is being asked to rate several sound elements with respect to the construct of 'soothing/alarming'.

Thus, RGT is in itself a mixed methods approach, which can be quickly used to assess individual perceptions along with those of a group. Providing a complete primer on the use of RGT is beyond the scope of this chapter, but there are a range of materials in the literature that detail its use in broad contexts (Hassenzahl and Wessler 2000; Fransella, Bell and Bannister 2004). These are complementary to the

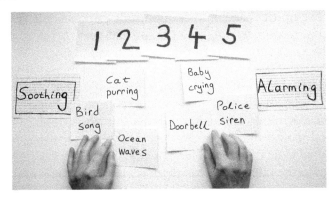

Figure 9.2 A participant using card sorting to assist them in providing ratings for a set of sounds.

comprehensive case study that we included in one of our previous projects relating to perceptions of altered states of consciousness in computer game sound designs (Cunningham, Weinel and Picking 2016).

In a particularly relevant recent endeavor, RGT was used in a study (McGregor and Cunningham 2015) that sought to validate the intentions of expert sound designers with the perceptions of listeners. Evaluation of sound designs typically involves only the judgements of expert reviewers, despite usually being produced for a mass, public audience. Using RGT, we were able to provide a way for non-experts to evaluate sound designs without being trained in formal sound design evaluation. Significantly, this method provided a way to highlight where the experts' audio designs have failed or succeeded to get their intentions across clearly to their listeners. A large, multidimensional grid was produced from the research, which was used to inform the overall outcomes of the work. The qualitative constructs utilized in extracting quantitative measures from participants were based upon a previously validated set. RGT, in this situation, provided an efficient and effective way for sound designers to test out and validate their ideas.

4.4 Reflections on Interdisciplinary Working

While the examples discussed provide only a few possible instances of projects that involve combination of practice-led and/or interdisciplinary methods, these begin to illustrate what such approaches might actually look like. On the one hand, practice-led approaches are very good at providing tangible investigations into sound design approaches, because they result in artifacts such as electroacoustic compositions or art installations that operate within the lifeblood of the discipline – the actual act of doing sound design itself. However, the experience within our group has also been that there is much to gain from complementary studies of a scientific nature, which further our understanding of underlying technologies and perceptual processes.

The adoption of empirical techniques alongside practice-led research naturally requires additional resources, notably time, on the part of the researchers. This is in addition to the knowledge, experience and skill requirements to incorporate the techniques necessary. These, of course, can be learned as part of the researcher's development or included in the context of a team of researchers. In the Affective Audio setting, the team has benefitted significantly from having researchers with different foundation disciplines and research skills, but whose interests have significant areas of intersection and significant overlap. This allows for the successful realization of projects by virtue of the different theoretical and practical skillsets that the researchers bring to the table; however, there is also a supplementary benefit in that the researchers also inevitably learn from each other, developing their expertise as these different approaches brush off on each other.

5 Conclusion

Through the course of this chapter we have explored how practice-led and interdisciplinary approaches may be applied to research in sound design. As Serra (2019) notes, practice-led research is not always well understood, and whilst it has traction in some environments such as the UK, elsewhere it is not necessarily valued as highly. Here we have reinforced the validity and utility of practice-led research methodologies. In the projects we have outlined here, as well as in other ongoing work such as Weinel's (2019a; 2019b) investigations into music visualizations in virtual reality, practice-led research is an appropriate methodology, because it allows for the construction of new prototypes that test possible design strategies and approaches, yielding insights that could not be obtained through other means. In many cases, it is not necessary, or even appropriate, to carry out empirical studies to measure or verify the outcomes of practice-led research in sound design. In our experience, carrying out user studies or listener investigations at an early stage of practice-led work is not always useful, as it can interrupt or slow down progress during phases of creative activity. However, as we have outlined in this chapter, there are often very clear benefits to adopting interdisciplinary approaches in which practice-led work is combined with complementary work investigating the underlying technologies or perceptual processes with which sound design is almost always intertwined. As our own example projects illustrate, these complementary approaches can be formulated in various ways, such as by carrying out studies that verify the outcomes of sound design, or conversely, by providing an evidence base that subsequently informs creative approaches. Lastly, from a personal perspective, we have also argued that interdisciplinary teams can benefit the outcomes of sound design research, not only for the successful completion of projects, but also due to the positive impact of interdisciplinary teams on the professional development of the researchers themselves. By its very nature, sound design takes places in the nexus of art, technology and human perception, and so, we believe there is much to gain from the intersection of research methodologies and specialists aligned with these respective areas.

References

Benyon, D. (2019) Designing User Experience: A guide to HCI, UX and interaction design. Harlow: Pearson.

Bradley, M.M., and Lang, P.J. (1994) Measuring Emotion: The self-assessment manikin and the semantic differential. Journal of Behavior Therapy and Experimental Psychiatry, 25 (1), 49–59.

Clarke, V., Braun, V. & Hayfield, N. (2015). Thematic analysis. In Smith, J.A. (ed.),

Qualitative psychology: A practical guide to research methods, (pp.222–248), SAGE. https://uk.sagepub.com/en-gb/eur/node/60648/print.

Corey, J. (2016) Audio Production and Critical Listening: Technical ear training. New York: Taylor & Francis.

Cunningham, S. (2010) Applying Personal Construct Psychology in Sound Design using a Repertory Grid, Proceedings of the 5th Audio Mostly Conference: A Conference on Interaction with Sound, Interactive Institute, Sweden. doi: 10.1145/1859799.1859807.Cunningham, S. (2019) Communicating Auditory Impairments Using Electroacoustic Composition, EVA London 2019 (Electronic Visualisation and the Arts). doi: 10.14236/ewic/EVA2019.63.

Cunningham, S., Weinel, J., and Picking, R. (2016). In-Game Intoxication: Demonstrating the Evaluation of the Audio Experience of Games with a Focus on Altered States of Consciousness. In: Garcia-Ruiz, M. (ed.) Games User Research: A Case Study Approach. Boca Raton: CRC Press, 97–118.

Dawson, C. (2019) A-Z of Digital Research Methods. Abington: Routledge.

Fagerlönn, J., Larsson, S. and Lindberg, S. (2013) An auditory display that assist commercial drivers in lane changing situations, Proceedings of the 8th Audio Mostly Conference: A conference on Interaction with Sound, Interactive Institute, Sweden. doi: 10.1145/2544114.2544120.Fransella, F., Bell, R., and Bannister, D. (2004) A Manual for Repertory Grid Technique. Hoboken: John Wiley & Sons.

Giannakopoulos, T., and Pikrakis, A. (2014) Introduction to Audio Analysis: A MATLAB® approach. Cambridge: Academic Press.

Grill, T., Flexer, A., and Cunningham, S. (2011) Identification of Perceptual Qualities in Textural Sounds Using the Repertory Grid Method, Proceedings of the 6th Audio Mostly Conference: A Conference on Interaction with Sound, University of Coimbra, Portugal. doi: 10.1145/2095667.2095677.Guyotte, K.W., Sochacka, N.W., Costantino, T.E., Walther, J., and Kellam, N.N. (2014) STEAM as Social Practice: Cultivating creativity in transdisciplinary spaces. Art Education, 67 (6): 12–19.

Hassenzahl, M., and Wessler, R. (2000) Capturing Design Space from a User Perspective: The repertory grid technique revisited, International Journal of Human-Computer Interaction, 12 (3–4): 441–459.

Henriksen, D. (2014) Full STEAM Ahead: Creativity in excellent STEM teaching practices, The STEAM journal, 1 (2): 15.

Hobson, A.J. (2002) Consciousness and Brain Science, The Dream Drugstore: Chemically Altered States of Consciousness. Cambridge: MIT Press, 44–46.

Kelly, G.A. (1955) The Psychology of Personal Constructs. New York: Norton.

Klüver, H. (1971) Mescal and Mechanisms of Hallucinations. Chicago: University of Chicago Press.

Lantz, E.J. (2006) Digital Domes and the Future of Large-format Film, LF Examiner, 9 (8): 112.

Larsson, P. (2016) Speech Feedback Reduces Driver Distraction Caused by In-vehicle Visual Interfaces, Proceedings of the 11th Audio Mostly Conference: A Conference on Interaction with Sound, Interactive Institute, Sweden. doi: 10.1145/2986416.2986435.Lazar, J., Feng, J.H., and Hochheiser, H. (2017) Research Methods in Human-Computer Interaction. Cambridge: Morgan Kaufmann.

Leedy, P.D., and Ormrod, J.E. (2016) Practical Research: Planning and design (Global; 11th ed.). Boston: Pearson.

McGregor, I.P., and Cunningham, S. (2015) Comparative Evaluation of Radio and Audio Logo Sound Designs, Journal of the Audio Engineering Society, 63 (11): 876888.

Moffat, D., Ronan, D., and Reiss, J.D. (2015) An Evaluation of Audio Feature Extraction Toolboxes, Proc. of the 18th Int. Conf. on Digital Audio Effects (DAFx-15), 30 Nov.–3 Dec., 2015.

Nelson, R. (ed.) (2013) Practice as Research in the Arts: Principles, Protocols, Pedagogies, Resistances. Basingstoke: Palgrave Macmillan.

Picard, R.W. (1998) Affective Computing. Cambridge: MIT Press.

Plano Clark, V.L., and Ivankova, N.V. (2016) Mixed Methods Research: A guide to the field. Los Angeles: SAGE.

Salkind, N.J. (2010) Encyclopedia of Research Design (Vol. 1). Thousand Oaks, CA: SAGE Publications. doi: 10.4135/9781412961288.

Serra, X. (2019) The Interdisciplinarity of Music Research: The Perspective of the Music Technology Group of the UPF, Higher Education in the World 7: Humanities and Higher Education: Synergies between Science, Technology and Humanities, 115–117. https://www.upf.edu/web/mtg/blog/-/asset_publisher/pYSKGP3RLdWn/content/id/230972346/maximized#.YAdRL-j7S70.

Smith, H. and Dean, R.T. (2009) Practice-led Research, Research-led Practice in the Creative Arts. Edinburgh, UK: Edinburgh University Press.

Strassman, R. (2001) DMT: The spirit molecule: A doctor's revolutionary research into the biology of near-death and mystical experiences. Rochester, VT: Park Street.

Ubisoft Montréal (2012) Far Cry 3. PlayStation 3, Ubisoft.

Weinel, J. (2011) Quake Delirium: Remixing Psychedelic Video Games, Sonic Ideas/Ideas Sónicas, 3 (2): 22–29.

Weinel, J. (2012) "Altered states of consciousness as an adaptive principle for composing electroacoustic music." PhD thesis. Online: http://www.jonweinel.com/media/ASC_Commentary_38.pdf (accessed May 20, 2020).

Weinel, J. (2016) Entoptic Phenomena in Audio: Psychedelic Electroacoustic Compositions, Contemporary Music Review, 35 (2): 202–223.

Weinel, J. (2018) Inner Sound: Altered States of Consciousness in Electronic Music and Audio-Visual Media. New York: Oxford University Press.

Weinel, J. (2019a) Cyberdream VR: Visualizing Rave Music and Vaporwave in Virtual Reality, Proceedings of the 14th International Audio Mostly Conference: A Conference on Interaction with Sound, University of Nottingham, UK. doi: 10.1145/3356590.3356637. Weinel, J. (2019b) Virtual Hallucinations: Projects in VJing, virtual reality and cyberculture, EVA London 2019 (Electronic Visualisation and the Arts). doi: 10.14236/ewic/EVA2019.57.

Weinel, J. (2021, forthcoming) Explosions in the Mind: Composing Psychedelic Visualisations of Sound. Singapore: Palgrave Macmillan.

Weinel, J., and Cunningham, S. (2017) Simulating Auditory Hallucinations in a Video Game: Three Prototype Mechanisms, Proceedings of the 12th International Audio Mostly Conference on Augmented and Participatory Sound and Music Experiences. doi: 10.1145/3123514.3123532.

Weinel, J., Cunningham, S., and Griffiths, D. (2014) Sound Through the Rabbit Hole: Sound design based on reports of auditory hallucination, Proceedings of the 9th Audio Mostly: A Conference on Interaction with Sound. doi: 10.1145/2636879.2636883.

Weinel, J., Cunningham, S., Griffiths, D., Roberts, S., and Picking, R. (2014) Affective Audio, Leonardo Music Journal, 24: 17–20.

Weinel, J., Cunningham, S., Roberts, N., Griffiths, D., and Roberts, S. (2015a) Quake Delirium EEG: A Pilot Study Regarding Biofeedback-Driven Visual Effects in a Computer Game, IEEE Proceedings of the Sixth International Conference on Internet Technologies and Applications 2015, Wrexham Glyndŵr University, Wales.

Weinel, J. Cunningham, S. Roberts, N. Roberts, S. and Griffiths, D. (2015b) EEG as a Controller for Psychedelic Visual Music in an Immersive Dome Environment, Sonic Ideas/Ideas Sónicas, 7 (14): 81–92.

Yakman, G. (2008) STEAM education: An overview of creating a model of integrative education, Pupils' Attitudes Towards Technology (PATT-19) Conference: Research on Technology, Innovation, Design & Engineering Teaching, Salt Lake City, UT.

Zander, T.O., Kothe, C., Jatzey, S., and Gaertner, R. (2010) Enhancing Human- Computer Interaction with input from active and passive Brain-Computer Interfaces, Brain-Computer Interfaces, London: Springer, 181–199.

10

Research-creation as a Generative Approach to Sound Design

Prophecy Sun, Kristin Carlson and Kate Hennessy

1 Introduction

Sound is ever-present and all around us. Some sounds entice, inspire or stimulate us, while others have the power to overwhelm and invade our everyday. Whether they are airborne, structure-borne or liquid-like, sounds continually fill the urban and rural, terrestrial and subterranean, airborne or underwater environment with an overabundance of noise from a myriad of sources, products, infrastructures and electronic devices (Langeveld et al. 2013).[1] This acoustic information obscures and diminishes the human and non-human ability to hear subtle or nuanced tones, frequencies, and variations. The built environment hides almost all of the high or low frequencies ranges which arguably are only discernible when sound, echolocation, or other ultrasonic techniques are employed.[2]

This chapter presents three examples of sound-based installation performance works, each focused on feminist perspectives that foreground and represent emergent experience in research development and production. We suggest that research-creation methodologies are generative engagements with everyday technologies, where the final sound, images and compositions capture the personal immediateness of the performance while evoking embodied aesthetics through their lo-fi properties. These methods employ experimental, embodied and future-focused ways to think *with* and *across* techniques of creative practice (Manning and Massumi 2014). In particular, research-creation is a generative approach for sound design because it supports iterative, propositional experimentation and emergent processes to develop "toward catalyzing an event of emergence" (Manning and Massumi 2014). Research-creation methods encounter aspects of chance and improvisation, and accept meaningful failures along the way as opportunities which can often lead to new perspectives, multidirectional and emergent research and collectively shaped production.

DOI: 10.4324/9780429356360-11

The three artworks we present demonstrate a research-creation approach to the creation of sound works, as the generative nature of artmaking often blends the boundaries of where one project ends and another begins. This is to say that creating in this way provides movement in many new directions. Much like movement, one of the more fluid ways we experience sound is through our bodies and, in turn, tangible connection and interaction with handheld gadgets, laptops and other portable technologies. These accessories connect people to other people, places and experiences. This connection runs deep, deeper than the device itself because data is collected and translated through network cables hidden beneath our feet. These superhighways send data into the skies and back down to the earth again and lay hidden, under buildings, roads and sinkholes sprawled along the ocean floors (Starosielski 2015a). In this way, sound is intrinsically linked to bodies and infrastructure – carrying and connecting us all physically, sonically and virtually along the transoceanic internet traffic-filled superhighway.

Additionally, whether we are conscious of this movement or not, we are implicated in global power exchanges, whether by incidental or purposeful integration of technologies into daily life practices. These phenomena are used as agents, tools, aids, collaborators and weapons to provide fast and reliable communication that extends our relationships beyond fragile human form (Buck-Morss 1992).

Inspired by R. Murray Schafer's ideas on sounds as continuous fields of possibility, Hildegard Westerkamp's environmental installations and, in particular, Henri Lefebvre's writings on lived realms of spatiality (Krause 2016; Schafer 1993; Lefebvre 1991), this chapter introduces three interdisciplinary projects by prOphecy sun and collaborators: *Hunting Self* (2018), *Floating in the in-Between* (2019) and *Nostalgic Geography: Mama and Papa have Trains, Orchards, and Mountains in their Backyard* (2019).[3] The artworks were created between 2018 and 2019 and weave multi-layered moving image, live performance and soundscape composition together with anthropophagic (human), geophonic (non-biological, i.e., wind) and biophonic (natural) sounds as a form of ecofeminist activism.

In the following sections, we contextualize these works in relation to what has been termed the sonic turn (Kelly 2011), the emergence of soundscape studies in the 1960s and 70s; the relationship of sound to site/space/place (de Certeau 1998); and notions of voice as presence (Nancy 2008). The three art projects are presented as case studies to highlight emergent strategies of creating artwork using a myriad of sound design methods of production. Particular focus is on the intersection of the performative and new media works that reflect on current feminist sensibilities. Building on Halberstam's notions of queering time and space, we introduce a research-creation methodology that aims to push sound, bodies and matter into new co-advantageous, experiential moments *with* technology (Halberstam 2005).

2 The Sonic Turn

The twenty-first century has seen the extraordinary rise of new technologies and ways of experimenting *with, across, and through* creative practice in response to complex media environments, sound and the ever-present global hum or auditory glow.[4] Many contemporary artists have utilized the accessibility of new consumer-level technologies to investigate the resonance and/or the stillness of sound by creating new forms of technological, corporeal, cultural and methodological engagements and aesthetic encounters. Author Caleb Kelly describes this phenomenon as the sonic turn, whereby sound is established as a cultural marker or perceptual driver of contemporary reality (Kelly 2011).[5][6]

This turn has opened up avenues for artists, designers, scientists, scholars and theorists to explore auditory stimuli and push experimental research and sound design to the forefront in new emergent ways. Artists use Virtual Reality, drones, cameras, the Web, satellites and smartphones as actants to help them translate, transform or generate new aesthetic fragments and encounters with ethereal visual and auditory landscapes. Take, for instance, Kathryn Brimblecombe-Fox who examines the aesthetics of drone warfare through visual means (Brimblecombe-Fox, website), or Gavin Hood's films (2016). Madam Data (Ada Adhiyatma) uses machine interfaces, computer programs, field recordings and old samplers as tools to explore 1) the embodied ideas of distance and echolocation; 2) how bodies and spaces are defined by separations; 3) the trauma of dislocation; and 4) the impossibility of empathy (Madam Data, website). Gonçalo F. Cardoso and Ruben Pater's LP and booklet offers visual and auditory impressions of flying drones and other strategies to disrupt drone surveillance (A Study into 21st Century Drone Acoustics, website). Writer Thomas Stubblefield explores the insurgence of drones into civilian cultures such as photography, film installation and performance and the intersections between drones, art, technology and power (Stubblefield 2020). These examples show the way that contemporary artists and writers in the context of this sonic turn explore the use of these emerging technologies in sound-based artistic practice, and how this has changed the way we consider everyday sound, art and scholarly research.

2.1 The Birth of Soundscape Studies

Contemporary sound studies have become an important field of research for considering relationships between environment, culture, technology, humans, other animals and other microscopic species. This includes a broad range of interdisciplinary research and creative activity in fields such as environmental humanities, musicology, philosophy, anthropology, architecture and art.

In the 1960–70s, composer R. Murray Schafer popularized the term *soundscape* to describe a combination of sounds made or shaped by humans from the environment. Schafer's research identifies acoustic and ethical best practices of mediating noise pollution (natural and artificial) in cities and urban landscapes. In tandem, Schafer founded the *World Soundscape Project* (WSP) to study the relationship between people and their environments (World Soundscape Project, website).

Schafer and colleagues Barry Truax, Hildegard Westerkamp, Bruce Davis and Peter Huse worked together on the WSP to develop and deepen ecological under-standing perspectives on the impact of sound. Collectively, their research has been published through a variety of formats, including a book and play series called *The Vancouver Soundscape* (1973) and ten one-hour radio programs for CBC Radio called *Soundscapes of Canada* (1974) (Soundscapes of Canada, website).

In the seminal paper "Real-Time Granular Synthesis with a Digital Signal Pro-cessor," Truax discusses how he develops granulation techniques (Truax 1988).[7] Granulation (aka grain) is a micro piece of a sound sample that is isolated, stretched or temporarily changed (Eckel 1995). This and other such synthesis technologies provide artists and composers with ample opportunity to develop a deeper under-standing of spatial compositions and the environment.

Truax's ideas on sound composition as being able to invoke imagination, or rather link audience members to the source and/or original location are useful to help us consider sound as another being or a co-composer of experience (Truax 2002). Under this pretense, sound can be translated in many ways, and much like water or air, it is fluid, open, complex and interconnected, living and undeniably mobile and improvisational.

Contemporary researchers study sound in a multitude of ways, such as examining systematic differences between how aerial and ground technologies operate and how frequencies or levels of disturbance change our perception or experience in specific environments. For example, NASA's project *Design Environment for Novel Vertical Lift Vehicles* (DELIVER) has been developing a series of psychoacoustic tests to determine how much noise is generated by flying machines (Christian and Cabell 2017).[8] Further, institutions such as Bard College's *Centre for the Study of the Drone* have been examining the opportunities and complicated political and social chal-lenges of using drone technologies in society.[9] These examples highlight how sound is portrayed, manipulated and created in other institutional settings and contexts.

Oceanographers and water experts also study how sonic and aerial changes affect how reefs grow and fish populations thrive. For example, Gordon et al. studied the effect of *acoustic enrichment* on the growth of tropical fish communities and the surrounding degraded coral reef habitat (Gordon et al. 2019). Researchers played sounds to a designated section of coral reefs and found that the life of the reef improved with repeated sound playback (Nédélec et al. 2013). Sophie Holles et al. suggest that above water sounds such as boat residue can disrupt the orientation behaviors of coral reef fish (Holles et al. 2013). Or, how sensory-based manipulations can help attract seabirds to newly restored habitats and other conservation sites (Friesan et al. 2017). The above examples show how sound is implemented in other academic capacities outside musical composition.

2.2 Site/Place/Space: Sound Spatiality and Micro-Considerations

Often without notice, our bodies experience sound in fluid, spatial and immersive ways. Sound waves move through our bodies, auditory systems and ear canals in continuous loops. In response, we sound back through language or movement. Indeed, how we experience, embody, perceive or conceptualize it, sound is directly related to our spatial awareness.

Michel de Certeau describes this awareness as instantaneous configurations, or social or practiced spaces, full of known and felt moments of connection (de Certeau 1998).[10] Akin to de Certeau's ideas, Henri Lefebvre and Donald Nicholson-Smith describe human experiences as fluid and full of immersive spatialities – made up of loops or lived realms (Lefebvre and Nicholson-Smith 1991).[11] Sean Street suggests that these auditory and image-based memories can be captured in time, housed in Memory Palaces that are real or imagined image structures used for future recollections (Street 2014, p. 31). These constructed houses, as Paul Ricoeur notes, can be full of memories that can become something else: "halfway between pure memory and memory ... halfway between fiction, and hallucination" (Ricoeur 2012, p. 14). This speaks more to the metaphysical level of experience. Roy Ascott and Edward A. Shanken consider telecommunication systems as spatial tools that extend our hearing, gaze and perception beyond – with many potentialities to amplify our mind and transcend our bodies into unpredictable configurations of creativity (Ascott and Shanken 2007, p. 235–36). This amplification can be a visceral experience, whereby the body and the mind meld. These ideas are explored in Paul Sermon's *Telematic Dreaming* (1992), in which Susan Kozel describes her interactions as a flesh body dancing with projections of live participants in a bedroom environment (Kozel 1994). This is an example of how the body and mind mesh together to create an intimate encounter *with* technology.

Much of sound research today investigates our relationship to site/space/place. For example, the sonic impact humans have on a location, the effects noise has on a person or how the material conditions change the cultural memories of a site. To clarify, we suggest that the term *site* refers to a geographical location, a landscape or an everyday familiar space such as a barn or house. Contemporary Memory discourse scholars and artists are also interested in our cultural history and relationship with a place or site. In turn, some have attempted to define memory according to their particular ways of understanding memory and its relation to various articulations on the body, history and technology.

These ideas can be categorized into Individual, Collective/Globital and Echoic (Sonic) Memory (Reading 2011; Neisser 2014). Each of these ideas tends to exist on a spectrum of cognitive, political and philosophical belief systems that augment in some way events or technological systems. Sound can also trigger memories. These ideas are the premise of Echoic Memory, coined by Ulric Neisser, to describe the sensory register in a person's brain that is responsible for processing acoustic information (Neisser 2014, p. 189–190). In this sense, when auditory and visual information are collected, processed, assimilated and then reconsidered – a memory is born.

2.3 Voice as Presence

In the 1860s, inventor Édouard-Léon Scott de Martinville captured the first recordings of a human voice using a phonautograph (Feaster 2019).[12] The legacy of this ten-second fragment carries an incredible power still today. It imbues a conceptual presence, a vibration or a virtual emotion of a disembodied material, an impression of a location, a space that lives well beyond its years.

According to theorist Patrick Feaster, the phonautograph was originally designed to capture the trace of a voice "onto paper for visual study rather than playback" (Feaster 2019, p. 14–15). Here the aspect of a voice having a pathway and being able to catch an audience's attention through visual impressions on paper versus on air or through a recording is intriguing. He further posits that critics today are obsessed with recordings that can be played back repeatedly to a listener, stating:

> [P]layability has distracted most critics from seeking to understand the phonautograms on their own terms, as visible, archivable documents implicated in motives and uses to which playback was irrelevant, and not a conscious or conspicuous omission.
>
> (Feaster 2019, p. 16)

Scott's invention is still relevant today because it traces and translates aerial sound waves into material form (in this case paper) instead of making recordings solely for continuous playback (Feaster 2019, p. 15–16). In this way, virtual bodies have agency and the power to exist in both material or immaterial forms. Norie Neumark et al. describe this as live or recorded voices finding space to breathe in other forms such as podcasting or *acousmatic* music, which extend beyond the sphere of performance and the origins of where it was first played (Neumark et al. 2010, p. 2).[13] Similarly, Jean-Luc Nancy posits that voices are powerful, disembodied – possessing a presence that moves and breathes beyond a location. Further, he posits how voices are "something-someone-takes distance from the self and lets that distance resonate" (Nancy 2008, p. 20).

Over the last decade, there has been a resurgence of improvisational or *being with* approaches that examine our impact on the environment and how we use technological artifacts to transmit our corporeal voice (Critchley 2008). Gilles Deleuze and Félix Guattari refer to this as acts of *becoming*, or responsiveness in relation to our environments (Deleuze and Guattari 1999). Or what Nance Klehm suggests as best practices on how to appreciate the soil under our feet (Klehm 2019).

This rich history informs our current approach to creative expression, sound design and research in the following ways:

1. Expanding our current approaches on how to activate landscapes. Inspired by Truax's notions of *soundscape composition* (an electroacoustic musical form of composition that incorporates environmental sounds and alludes to the original location) (Truax 1988; 2002).
2. Experiment with new conceptual strategies for composition – which mesh together processed tape, voice, assorted instrumentation, environmental sounds and granulated sampling techniques (Xenakis 1992; Truax 2002; Westerkamp 1994; Bull 1993).
3. Foregrounding eco-feminist and *more-than-human* perspectives (Haraway 2014; Barad 2007).

3 Research-Creation as Methodology

In this section, we discuss the various definitional approaches to the terms research-creation, improvisation, sonic autoethnography and present a feminist research-creation strategy.

3.1 Research-Creation

Contemporary artists, scholars and theorists are continually exploring what artistic research is and how to define, categorize and frame it. In academia and across funding platforms, artistic research is regularly framed as *Research-Creation* in Canada, *Practice as Research* in Australia and Britain, and *Arts-Based Research* in the US (Chapman and Sawchuk 2015).

Research-creation is most commonly understood as a combination of research, art and theory. In this chapter, we describe emergent research that foregrounds research-creation as the generation of new aesthetic approaches of considering the sound design and interdisciplinary research.

Throughout the literature, new sub-categorizations emerge to link how research and creation connect *within* academic practices (Manning and Massumi 2014; Rodgers 2012). For example, Owen Chapman and Kim Sawchuk describe research-creation as a methodology tied to the art of living movement (Chapman and Sawchuk 2015). Erin Manning and Brian Massumi posit it as a speculative, embodied, experimental and future-focused process with much potential for *thinking-with and-across* creative practice (Manning and Massumi 2015). Or, encompassing multidisciplinary, hybrid forms of artistic practice, social science and artistic research (Truman and Springgay 2015).

We present research-creation here as a reflexive, intuitive, experimental and creatively motivated act. It is a practice that is continuous, evolving and built over durations of time – where data and matter mesh together and become insepa-rable. Springgay and Zaliwska describe this as "intra-acting ecologies" or "co-compositions" (Springgay and Zaliwska 2015, p. 140).

This section also articulates three important facets of research-creation for our work in sound: improvisation, autoethnography and feminist practices. Improvisa-tion is the in-the-moment exploration of a process that is dependent on the artist's expertise and creative problem-solving. Autoethnography, on the other hand, is the process through which an artist is reflective to critique and challenge their ideas while documenting the process and practice.

These facets were selected for this chapter because they highlight how integral research-creation is to artists working within time constraints, on the go and in uncontrollable outdoor environments. These ideas are also part of a larger feminist research practice that takes up consumer-grade technologies, chance, improvisation and autoethnographic methods to explore personal and life experiences.

3.2 Improvisation as Practice Space

Improvisation is commonly understood as a combination of present-day moments or a practice space, where things come into being in the moment. Aili Bresnahan describes improvisation as unplanned, free guided and spontaneous; and contrary to popular belief, it is not an ad hoc activity, but rather, "it involves skill, training, plan-ning, limitations and forethought" (Bresnahan 2015, p. 574). Further, how artists working with improvisational techniques influence artists working in other mediums who then re-enact and respond in new experimental ways (Bresnahan, 2015). Andy Hamilton suggests that improvisation requires a connection between preparation and performance (Hamilton 2002).

Artists in domains such as music have enthusiastically explored interactive digital systems, leveraging the pragmatic aspects of automation into their work, whilst also exploring methods for indeterminacy and situated events. Interactive music systems have been designed to react to a live performer's actions (sonic as well as gestural), that enable a wide range of available opportunities to create a work in the moment (Rowe 1992; Winkler 2001). However, Winkler's discussion of improvisation is lim-ited to the fact that the computer is responding and does not go on to discuss the mapping strategies or interactive models that make the magic of improvisation come alive with digital technology: "a performer may play anything, and the computer will respond with something that is unexpected and unrepeatable, but which, hopefully, makes sense to the performer and to the audience" (Winkler 2001, p. 293).

In improvisation, interactions are tied to many constraints, both in the live environment as well as the digital. Lewis suggests: "for humans, the primary constraints upon musical improvisation in general are in the realms of the body, temporality, memory, and the physical environment" (Lewis 1999, p. 4). Lewis highlights memory as a physical attribute that affects improvisation because the musicians are taking into account what has already been created and projecting where the composition could go in the future, all at the same time they are making a choice to play at that moment. Johnson-Laird elaborates on this idea, stating:

> [I]mprovisers must reason about time in order to conceptualize what is to be articulated in light of what they remember has been played ... in comparison with a tune that is being composed as it is being played (i.e., a free tune), the performance of a well-learned tune (such as a jazz standard) may place fewer demands on remembering, since much of what needs to be recalled (e.g., chord changes) is easily accessible from long-term memory.
> (Johnson-Laird 2002 as cited in Mendonca and Wallace 2004, p. 2)

The balancing of past, present and future in our actions is paramount in techniques around improvisation. Fuller and Magerko suggest that it is in the domain of improvisation where cognitive divergence between performers develops, who need to create common ground in order to develop a shared mental model to work within (Fuller and Magerko 2010). Sawyer and DeZutter suggest that emergent creativity is very dynamic and not easily attributed to the specific instances or authors (Sawyer and DeZutter 2009). We use this example to illustrate how improvised performances are constructed through a variety of constants, decision-making, adaptability and mediations with technology. This implies that performers have the ability to make high-level cognitive decisions, in the moment, and this is essential to construct a composition in relation to all past, current and potential future decisions (Lewis 1999).

Currently, technology is involved in every aspect of daily life. Tools are now widely accessible for consumers to create a synchronous telepresent performance with collaborators across the globe. However, the discussion of improvisation and the cognitive processes that are involved continue to be limited in the context of digital technologies and interactive systems. The rise in cognitive science has provided new approaches to understanding whole self-engagement (as opposed to mind-body dualism) and the multiple sensory systems that contribute to an experience of immersion (Corness 2008).

In particular, telematic performance, glitch and improvisation work hand in hand. Ascott and Shanken suggest:

> We begin to understand that chance and change, chaos and indeterminacy, transcendence and transformation, the immaterial and the numinous are terms of the centre of our self-understanding and our new visions of reality ... The very technology of computer telecommunications extends the gaze, transcends the body, amplifies the mind into unpredictable configurations of thought and creativity.
>
> (Ascott and Shanken 2007 p. 235)

3.3 Sonic AutoEthnography

This research methodology builds on standard models of visual and sonic autoethnography. Autoethnography aligns with this research as qualitative studies traditionally are written and recorded by an individual who is also the project's subject matter (Ellis 2004; Muncey 2014). Creswell and Muncey describe this method as portraying multiple layers of consciousness, where the self in research is implicated: "The vulnerable self, the coherent self, critiquing the self in social contexts, the subversion of dominant discourses, and the evocative potential" that it incurs (Creswell 2018).

Traditionally, autoethnography is defined as a qualitative research approach that uses self-reflection and writing to explore personal experience alongside other cultural, social contexts (Creswell 2018, p. 73). David Hayano first coined the term in 1956 to mean a style of critiquing by anthropologists that investigates power structures and incorporates self-representations (Reed-Danahay and Panourgia 2000). For Carolyn Ellis and Art Bochner, it is more radical than that, being both a process and a product, or a writing strategy that "displays multiple layers of consciousness, connecting the personal to the cultural" (Ellis et al. 2011, p. 273–74; Ellis and Bochner 2000, p. 739–740). Ellis and Bochner propose evocative autoethnography as a genre in which "the emotional and embodied experiences of the researcher are drawn upon to allow for a clearer understanding of particular life experience to emerge" (Ellis and Bochner 2000 as cited in McCormack, 2012).

Sonic ethnography refers to a qualitative research approach to audio recordings, sound art, sound walks, interviews and other sonic materials and our experiences (Droumeva 2016). Steven Feld describes ethnography as ethnomusicology, a growing field of study in which the researchers intrinsically have a material, cultural, analytic and creative relationship with sound (Feld and Brennis 2004). In this way, sound studies open up potentialities for a deeper understanding of our everyday ecological environment (Drever 2002).

Contemporary scholars and artists build on these notions and emphasize "experience-near" approaches that value and incorporate hybrid, blended research methods that account for visual and interdisciplinary expressions that convey aspects from everyday experiences (Collins and Gallinat 2010). For some, visual and sonic moments capture the atmosphere of a place, the smallest details, being like "minor-key variations of reality" (Monnet 2014). For others, images are intrinsically linked to action, thought and knowledge production (Conord and Cuny 2014). For others, images communicate about culture and our lived sonic experiences and should be accumulated and exploited as they hold fragmented gateways to emotion (Edwards 2002).

3.4 Feminist Approaches to Research-Creation

Research-creation across its sub-categorizations tends to be multidisciplinary, with a variety of hybrid forms of artistic practice. This research methodology builds on this and the historical lineage of Feminist Media Art – a social movement that emerged in the late 1960s that brought awareness on topics such as women's perspectives on birth or everyday practices and social interactions using various forms of digital media in unique ways.

We present feminist approaches to research-creation here as a strategy that values co-authorship, practice and collaborative ways of making and doing. This strategy is diverse, inclusive and foregrounds emergent experience in research development, acquisition, production and questions body, culture, history, subjectivity and our relationships to material and digital environments (Barad 2007; Alaimo and Hekman 2008; Butler 2009; Behar 2016; Haraway 2016; Mondloch 2018; Braidotti and Hlavajova 2018). Stacy Alaimo and Susan J. Hekman describe this contested space as bodies, and the materiality inhabiting and transforming ideology discourses to include life and lived experience (Alaimo and Hekman 2008). Or, a Feminist Materialist practice space in which researchers critically engage with theoretical foundations and materiality itself, entangled with matter, bodies and space (Mondloch 2018).

The following pages present this feminist approach through a series of artworks that destabilize a privileging of one element over another and mesh lived experience, everyday technology, bodies, matter and processes together.

4 The Projects

In this section, we discuss three projects, *Hunting Self* (2018), *Floating in the in-Between* (2019) and *Nostalgic Geography: Mama and Papa have Trains, Orchards, Mountains in their Backyard* (2019). The interdisciplinary artworks showcase the creative process through a range of formats and approaches to research-creation and sound design.

Figures 10.1–10.3 *Hunting Self*, 2019. Installation view. Photos courtesy of Kirsten Aubrey.

Figures 10.1–10.3 *Continued*

4.1 Hunting Self*: 15-Channel Audio Installation*

Hunting Self (2018) is a looping 60-minute, 15-channel immersive sound installation (see Figures. 1–3 and video link: https://prophecysun.com/Hunting-Self). The work was created by prOphecy sun and installed at the Arts Commons in the 15+ Soundscape Gallery in Calgary, Alberta. The gallery operates as a public corridor with 15 individual spatialized mono speakers positioned throughout the length of the space.

The interdisciplinary investigation consists of processed improvised vocal loops and melodies threaded together with environmental snippets from rural landscapes taken in situ from multiple locations in Nelson, British Columbia, and the Pacific Northwest of Canada. Each channel is unique because it was recorded live and in one take using smartphone technology. The piece weaves, layers, stretches, condenses and meshes together conscious and unconscious moments in time to highlight how systems are everywhere, part of the air we breathe, part of our bodies and part of our urban and wild cultural ecologies.

4.1.1 Sound Design

Over six months, the artist captured field recordings with consumer-grade technologies such as an iPhone, Shure MV88/A iOS Digital Stereo Condenser Microphone, windjammer, and MOTIV app. These technologies are important to the overall process because they are readily accessible, easy to configure, set-up, take down and transfer data forms. For example, the microphone plugs directly into the phone through a lightning connector. The microphone has cardioid (unidirectional) and bi-directional abilities and has a hinge or rotation design that supports various recording angles. The MOTIV app captured and catalogued the raw sound data. The recorded sounds included improvised vocal humming, birds, air, wind, trees and other ground-level impressions taken from urban forests and other remote locations along the highway from Vancouver to Nelson, British Columbia.

4.1.2 Editing Process

The sound editing process started with exporting the data files from the iPhone onto a laptop and then importing the WAV MP3 into Audacity, an open-source, multi-track sound editing program. Essentially, the program is designed as a tool kit for users to record, edit, splice, compose and adjust the speed or pitch of a recording.

4.1.3 Post-Processing

To make the 15-channels, the artist first isolated each track, adjusted the placement, length and volume, then overlaid additional snippets of field recordings onto the original composition structure andprocessed the waveform. Once the structure and alignment were correct, the gain and spatial position of each track were adjusted by moving the left and right meter. After repeatedly listening over the composition, selected sections were granulated across various time codes. After that, delay and reverb were added to the overall composition and final adjustments of the placement of certain sounds such as birds, wind or human breath were addressed. The final composition was listened to again from beginning to end. Once satisfied, the track was exported in 15 separate files in WAV and MP4 Format. Each section was assigned a number that correlated to the discreet ceiling speakers and the subchannel of the Arts Common's 15+ corridor rig.

4.1.4 The Sound Installation

The Arts Common's 15 + corridor rig setup consisted of an Apple Mac Pro computer, amplified through a MOTU 24Ao USB audio interface which drove the 12.3 array with 12 mono channels (twelve ceiling speaker channels and three sub-channels). The ceiling amplification was positioned in the ceiling throughout the Arts Commons +15 corridor into left and right sequences. Each speaker was designated to a specific sound file number and played continually throughout the exhibition (Arts Commons, website).

4.2 Floating in the in-between

Floating in the in-between (2019) is an interdisciplinary performance by prOphecy sun (see Figures 10.4 and 10.5 and video link: https://prophecysun.com/Floating-in-the-in-between). This work is another example of research-creation and attests to the diverse and emergent potentialities of this methodology.

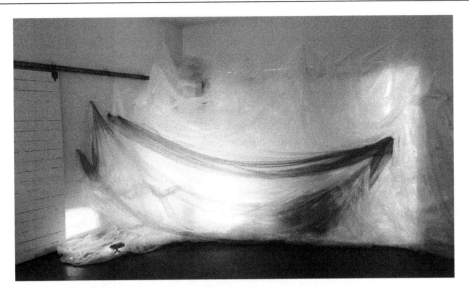

Figures 10.4 and 10.5 *Floating in the in-between*, 2019. Installation and live 15-minute audio/ visual performance. Recycled plastic, fabric, projections, live sound, environmental snippets and processed voice, delay and loop.

The piece was performed in front of a live audience at the Oxygen Art Centre in Nelson, British Columbia. The piece was made in response to Ruth Bieber's play *To See or Not to See* (2019), which employs the quest motif of a visually impaired woman and her subsequent trials and tribulations with her mother and family growing up.

The two-channel immersive installation and 15-minute live performance conjure up images of fragmented, reflective and still waters, woven together through an atmospheric composition consisting of improvised, looping, suspended textures, sounds, processed vocal tones and field recordings.

The performance consists of the vocal artist immersed in the installation, performing a live soundscape through vocalization and technology manipulation. To create the live score, the artist improvised with an assortment of consumer-grade pedals such as a mixer, pitch shift, delay and looping pedals, and a microphone and a monitor. The improvised score was made up of a series of low breathing tones, layered vocal tones and harmonies in 10- to 30-second time increments. The sounds shift in cadence and become granulated and stretched as the artist weaves together the sounds into a tapestry of harmonic and dissonant drones. The resulting sounds were reminiscent of the moving images and an elegiac effort to transcend the darkness.

Complementing the sound is a visual tapestry of two projections, fabric and shimmering plastic. The installation consists of a 20- by 15-foot span of handmade plastic floating mountains and sheer fabric cascading in between multiple sections. Each segment was mounted onto the gallery walls and floor. The section melds into the next mountain, the next one and sets the backdrop for the two projections.

The first projection arrests the viewer with imagery of a female form submersed in deep water. The figure repeatedly tries to swim to the surface, yet is caught and cannot escape being engulfed by the water. Her body repeatedly moves back and forth. The imagery is layered, ghost-like, with the tones that are reminiscent of clouds. The looping imagery eventually shows the figure being carried and held in the water's embrace. The second projection is an enlarged, pixelated iteration of the first projection. The visuals are positioned over the plastic mountains and move in slow motion, unfolding and melding together.

4.2.1 Live Performance and Sound Recording Process

In the opening live performance, the performer placed her body inside the middle of the plastic and fabric cocoon-like installation. The audience would see the silhouette of her form underneath the surface. The sounds created were live using a Sennheiser vocal microphone, Line 6, delay, and pitch-shift pedal. The performance was a channeling experiment of sorts, where the voice emulated the movement, up and down, of the woman falling in the water. The multisensory piece elicits viewers to negotiate reality by transporting them elsewhere, into other places. Ultimately, to witness a woman trapped in her own mind – caught between body, water, and subconscious rhythms.

Figures 10.6 and 10.7   *Floating in the in-between*, 2019. Video stills. Photos courtesy of prOphecy sun.

4.3. Nostalgic Geography: Mama and Papa have Trains, Orchards and Mountains in their Backyard

4.3.1 The Installation

Nostalgic Geography: Mama and Papa have Trains, Orchards and Mountains in their Backyard (2019) is a seven-channel installation created by artists prOphecy sun and Darren Fleet (see Figures 10.8–10.15 and video link: https://prophecysun.com/). The artwork was showcased at Oxygen Art Centre in Nelson, British Columbia.

The installation features an array of digital, corporeal and immersive provocations and embodied ways of situating personal narrative, grief and the process of loss. Looking to the land and its living and non-living materials for direction, the artwork is a meditation on both the presence and the absence of memory embedded

Figures 10.8 and 10.9 *Nostalgic Geography: Mama and Papa have Trains, Orchards and Mountains in their Backyard*, 2019. Process shots. Photos courtesy of prOphecy sun and Darren Fleet.

in the landscape. This is felt through the invocation of sound, text, material and moving images using capture technologies such as drones and smartphones.

Inside the gallery, the remnants of six apple trees are suspended in an oblong circle formation. Found railway ties line the entrance in a strategic formulation, whilst throughout the space, other objects and debris from the rail line are placed at ground level. Two didactic texts line one wall of the gallery – folding together ten years of family stories about picking rocks from the garden. The text describes 43 types of "geography and "nostalgia." Seven projectors sit on the floor display-ing various vignettes, a mix of aerial and ground view perspectives of a female

Figures 10.10a and 10.10b Nostalgic Geography: Mama and Papa have Trains, Orchards and Mountains in their Backyard, 2019. Video stills courtesy of Darren Fleet.

form in red walking through old orchards, railroads – beside water passages and mountains.

Each projection has its own soundtrack which is a blend of voice and other site-specific field recordings. For example, one features the sound of a child laughing while she runs through the tall orchard grass (see Figure 10.10). Another score carries the sound of metal scraping on railroad track ties (see Figure 10.9). While another captures the sound of waves washing ashore on the side of a shady beach (see Figure 10.11). The seventh soundtrack features archival sounds of the family in the kitchen with piano and voice alongside field recordings of the night train. Together, the room sound is an ambient collage that fills the space in waves that reverberate and convey aspects of the visuals and the other assorted ephemera and aged fruit trees lining the space.

4.3.1 Audio and Video Recording Process

The audio and video footage were captured with a DJI Spark drone, iPhone cameras, tripods and the DJI GO 4 app. The raw video and sound files were transferred into Audacity and edited in post-production. The library of sounds features a mix of environmental and personal sounds captured over a two-week residency at Oxygen Art Centre.

The seven compositions play *with* and *against* the moving imagery to create a remediated story, a new, dream-like or blurred timeline of reality. The soundscore is a blend of seven tracks that combine various mixes of processed vocals, piano, field recordings taken in situ, children's laughter and archival audio of prOphecy sun's familial kitchen months before her father's passing. The sounds that were chosen to be in the exhibition embody aspects of the original location. The compositions blend an array of anthropophagic, geophonic and biophonic sounds together such as breathing, voice, wind, walking, air, metal scratching, piano playing, wind chimes, moving water, train horns. The sounds were selected for personal and aesthetic reasons. For example, the sound of the artist's father laughing as he spoke with his granddaughter. Each track was isolated and edited in ways that feature granulated lengths, tempo and tones. The compositions were then amplified throughout the gallery through the projector speakers and through headphones, which were hung beside a large, gnarled and decaying apple tree. The finished sound composition depicts multiple ghost-like personal narratives.

5 Towards a New Sound Methodology

Following a rich legacy of media and audio production, *Hunting Self* (2018), *Floating in the in-Between* (2019) and *Nostalgic Geography: Mama and Papa have Trains, Orchards, Mountains in their Backyard* (2019) offer an iterative, feminist approach to research and creation. These works present emergent aesthetic perspectives and pull the focus both inward and outward in autoethnographic ways to foreground a new sonic sensibility.

Throughout this process, we have been constructing and articulating a sound methodology that embraces improvisational and autoethnographic, research-creation strategies of trial and error, iteration, reflection and further dissemination using consumer-grade technology. This is important because consumer-grade technologies are becoming commonplace options to articulate personal stories and perspectives, whether they are a soundscape, video or an embodied and virtual performance.

Figure 10.11 *Nostalgic Geography: Mama and Papa have Trains, Orchards and Mountains in their Backyard*, 2019. Video still and installation view. Photos courtesy of Darren Fleet and Thomas Nowaczynski.

Figure 10.12 *Nostalgic Geography: Mama and Papa have Trains, Orchards and Mountains in their Backyard*, 2019. Video still and installation view. Photos courtesy of Darren Fleet.

Figures 10.13–10.15 Nostalgic Geography: Mama and Papa have Trains, Orchards and Mountains in their Backyard, 2019. Installation view. Photos courtesy of prOphecy sun.

Figures 10.13–10.15   *Continued*

Figures 10.13–10.15 *Continued*

6 Conclusion

Sound is everywhere and integrated into everything aspect of our lives, experience, infrastructure and systems. It is breathing, living; not fixed or static, and forever in motion. Because sound is part of the basic infrastructure of our world, it directly contributes to what choices we make, what tools we employ that aide us in extending our perception beyond our fingertips and ears.

The use of archival technologies now means that sound can be recorded, taped, played back, shared, sampled, archived and shared time and time again. In this way, technology is more than just an actant – it is a means of communication with other species and non-human environments. We explored three sonic installation works to articulate how the use of everyday technologies can allow for accessible means of sound design, setting up for conversations to happen between a variety of modalities and systems. These works particularly focus on using improvisation and autoethnographic approaches in their creation, leveraging iteration and reflection to refine their salient experiences. Each of these works relied heavily on exploration of sound, the consumer-grade technologies used and a valuation of knowledge that grew, was reflected on and refined to craft these works. We found that considering the knowledge of practice as data helped us to develop these works in ways that rely on concepts of space and memory as infrastructure for research. We have found that the creative process and practice of using research-creation as a methodology are in many ways a logical extension of these emerging technologies, where the final sound, images and composition capture the desired personal immediateness of the initial performance as well as evoking real-life aesthetics through their material properties.

To conclude, we have presented research-creation as an emergent methodology that supports unique creative approaches to the designing of sound composition. The examples presented throughout this chapter demonstrate feminist perspectives on what artistic research is and how it can offer opportunities for creative development and prototyping. Thus, we suggest that hybrid creative processes ensure that artists, scholars, makers and other creatives are able to make salient choices that foreground human experience in addition to contextual considerations.

Definitions

Airborne Sound: Airborne Sound or Airborne Noise is any sound that is transmitted at any frequency through the air and atmosphere.

Anthropophony: First popularized by Stuart Gage and Bernie Krause, the term Anthropophony refers to all sounds produced by humans, including musical composition using technology (Krause 2016).

Arts-Based Research (ABR): Borrowing from James Haywood Rolling (2013), ABR is neither quantitative nor qualitative as it overlaps and borrows from both domains to address questions that cannot be fully measured or generalized with exactitude (Rolling 2013, p. 8).

Biophony: The term Biophony refers to the sound organisms make in their habitat (Krause 2016).

Granular Synthesis: This term refers to a method of sampling or sound synthesis that was found or recorded; samples are broken down into small, micro-level fragments of sound which are called grains (Eckel 1995). This technique allows for precise temporal control, rhythmic and compositional freedom. Iannis Xenakis and Barry Truax are known for creating this style of synthesis.

Geophony: Borrowing from Bernie Krause, the term Geophony refers to non-biological sounds from a habitat (Krause 2016).

Electronic Music: This refers to digital music that uses electronic instruments and circuitry.

Installation Art: Installation Art is a type of art that can be three-dimensional, site-specific, shown indoors or outdoors, in which participants have both a spatial and temporal experience with the work, the exhibition space and other objects (Mondloch 2010, xiii). The work is shown in galleries, museums or in public locations.

New Media: New Media refers to the gadgets, devices and tools we use every day to help us access, organize and communicate information with others (Hansen 2010 p. 172).

Research-Creation: Research-Creation is an emergent category most commonly understood as a combination of research, art and theory. In the literature, artistic research is regularly defined as Research-Creation in Canada, Practice as Research in Australia and Britain and Arts-Based Research in the US (Chapman and Sawchuk 2015).

Screen: A screen refers to any type of moveable or fixed device that filters and/or creates a mediated viewing experience. Screens have existed since the fifteenth century in canvas/screen works by artists such as Leon Battista Alberti or in Camera Obscuras.

Sonic Turn: Jim Drobnick first coined the term the sonic turn. The term refers to our increased theoretical interest in sound in our contemporary everyday culture (Kelly 2011, p. 14).

Sound Art: Sound Art is an interdisciplinary genre of art in which the focus is on the medium of sound.

Soundscape: According to Krause, a soundscape can consist of a combination of Biophonic, Athropophonic and Geophonic sound sources (Krause 2016).

Soundscape Composition: This is a form of electroacoustic music that combines environmental sounds and other sounds that allude or give context to the original location (Truax 1988). Compositions can often invoke the listener's imagination, nostalgic response, or association with a location.

Structure-Borne Sound: Structure-Borne Sound or Structure-Borne Noise is any sound transmitted from an object impacting another surface, such as a person walking on an apartment building floor.

Notes

1 In the paper, *Product sound design: Intentional and consequential sounds*, authors Langeveld et al. describe the sound as manifesting itself in three ways: 1) airborne; 2) structure-borne; and 3) liquid sound. Further, they describe how sound is added to devices to create a feedback loop and how these loops communicate abstract or specific information about daily activities (Langeveld et al. 2013).

2 Lawrence and Simmons describe how the atmosphere filters certain sounds and how different species of bats use varying levels of frequency and echolocation at higher atmospheric levels (Lawrence and Simmons 1982).

3 In the mid-1960s, R. Murray Schafer coined the term "soundscape." Soundscapes are made in response to the landscape, memory and place. In the book *Five Village Soundscape*, he describes how anyone and anything that sounds can offer unique possibilities for composition (Järviluoma et al, 2000). His seminal ideas on composition, visual and analogue sound, environment and humans' role in constructing our aural fields are as prescient and relevant today.

4 We use the phrase "auditory glow" as a way to poetically describe the phenomena of how sound continually is continually all around us and informs our everyday actions and worldview.

5 Since the 1970s people worldwide have reported hearing a hum or a consistent low frequency pitched noise. The *World Hum Map and Database* (WHMD) was created in response to detect and document this phenomenon. Participants pin their experience, location and date and time of hearing the frequency through a public interactive map (Jaekl, 2019).

6 Jim Drobnick first coined the term the *sonic turn* (Kelly 2011).

7 Truax further notes that Gabor (1947), Xenakis (1971), and Curtis Roads developed earlier prototypes of his model of real-time granular synthesis (Truaz 1988).

8 Psychoacoustic tests are an important contribution to sound studies because they highlight how much sound loitering human devices create and describe new methods of evaluation.

9 Bard College's *Center for the Study of the Drone* is an interdisciplinary research space that examines the challenges and potentialities of drone technologies in both the military and civilian sphere (Center for the Study of the Drone, website).

10 Philosopher Michel de Certeau in his essay *Spaces and Places* (2015) discusses the terms *space* and *place*. For example, he defines place as a location in which two or more things coexist beside one another in a relationship, or in an "instantaneous configuration of positions" (Doherty 2015, 118). Space is described as an intersection of mobile elements: unstable, fluid, constantly in motion – composed of a series of movements that are in contractual proximities with one another (Doherty 2015, 119). De Certeau's definitions of place and space are useful when considering sound as a medium that defies boundaries and contends that relationships can be instantaneous and unique.

11 In the book *Production of Space*, Henri Lefebvre describes how humans experience spatiality through embodied, conceptual and perceptual realms (Lefebvre 1991). This idea is important in sound discourse because it acknowledges new approaches, open processes and ways of expressing embodiment.

12 In the article the *Enigmatic Proofs: The Archiving of Édouard-Léon Scott de Martinville's Phonautograms*, Feaster describes how Édouard-Léon Scott's first recording was of a lullaby called "Au clair de la lune" ("In the moonlight") (Feaster, 2019).

13 In the paper, "Sound and Narrative: Acousmatic Composition as Artistic Research," James Andean
    discusses how composer Pierre Schaeffer in the 1940–1950s created a style of concerts that are now
    commonly referred to as *Acousmatic Music* (Andean 2014).

References

A Study Into 21st Century Drone Acoustics. http://droneacoustics.org/. Accessed 27 Jan. 2021.

Alaimo, Stacy, Susan Hekman, and Susan J. Hekman, eds. *Material feminisms*. Indiana University Press,
    2008.Andean, James. "Sound and Narrative: Acousmatic composition as artistic research." Journal
    of Sonic Studies, 7. 2014.

Ascott, Roy, and Edward A. Shanken. *Telematic Embrace: Visionary Theories of Art, Technology, and
    Consciousness*. Berkeley: University of California Press, 2007. pp. 232–246.

Barad, Karen M. *Meeting the Universe Halfway: Quantum Physics and the Entanglement of Matter and
    Meaning*. Durham: Duke University Press, 2007.

Behar, Katherine. "An Introduction to OOF." *Object-oriented Feminism*. University of Minnesota Press,
    2016, pp. 1–38.

Braidotti, Rosi, and Maria Hlavajova. *Posthuman Glossary*. Bloomsbury Publishing, 2018.

Bresnahan, Aili. "Improvisation in the Arts." *Philosophy Compass* 10.9 (2015): 573–582.

Brimblecombe-Fox, Kathryn. "Kathryn Brimblecombe-Fox | Arthives.Com." Accessed February 11, 2020.
    http://www.visualartist.info/visualartist/artist/default2_byname.asp?p=kathrynbrimblecombe-fox.

Buck-Morss, Susan. "Aesthetics and Anaesthetics: Walter Benjamin's Artwork Essay Reconsidered."
    October, vol. 62, 1992, pp. 3–41. doi:10.2307/778700.

Bull, Hank. "Radio Art in a Gallery?" *TDR* (1988-) 37.1 (1993): 161–166.

Butler, Judith P. *Giving an account of oneself*. Fordham University Press, 2009.

Center for the Study of the Drone. https://dronecenter.bard.edu/. Accessed 27 Jan. 2021.

Chapman, Owen, and Kim Sawchuk. "Creation-as-Research: Critical Making in Complex Environ-
    ments." *RACAR: Revue d'art Canadienne / Canadian Art Review*, vol. 40, no. 1, 2015, pp. 49–52.

Chapman, Owen B., and Kim Sawchuk. "Research-Creation: Intervention, Analysis and 'Fam-
    ily Resemblances.'" *Canadian Journal of Communication*, vol. 37, no. 1, Apr. 2012. doi:10.22230/
    cjc.2012v37n1a2489.

Christian, Andrew W., and Randolph Cabell. "Initial investigation into the psychoacoustic properties of
    small unmanned aerial system noise." *23rd AIAA/CEAS Aeroacoustics Conference*. 2017.

Collins, Peter, and Anselma Gallinat, eds. *The ethnographic self as resource: Writing memory and experi-
    ence into ethnography*. Berghahn Books, 2010.

Conord, Sylvaine, and Cécile Cuny. "Towards a 'Visual Turn' in Urban Studies? Photographic Approaches."
    Visual Ethnography, 2014.doi:10.12835/ve2014.1–0028.

Corness, Greg. "The musical experience through the lens of embodiment." *Leonardo Music Journal*
    (2008): 21–24.

Creswell, John W. *Qualitative Inquiry and Research Design: Choosing Among Five Approaches*, Sage Pub-
    lications, Inc., 2018.

Creswell, John W, and Clark V.L. Plano. *Designing and Conducting Mixed Methods Research*. Sage Pub-
    lications, Inc., 2018.

Critchley, Simon and Reiner Schürmann. "On Heidegger's being and time." 2020.

de Certeau, Michel, and Pierre Mayol. *The Practice of Everyday Life: Living and cooking*. Vol. 2. Univer-
    sity of Minnesota Press, 1998.

Deleuze, Gilles, Eugene W. Holland, and Félix Guattari. *Deleuze and Guattari's Anti-Oedipus: introduction
    to schizoanalysis*. Psychology Press, 1999.

Disruptivemedia. *Erin Manning (Concordia University) – Against Method*. 2014. https://www.youtube.
    com/watch?v=ZEUZ6PWzJqU.

Doherty, C., *Situation: Documents of Contemporary Art*. 2015. London/Cambridge: Co-published by
    Whitechapel and MIT Press, 2009.

Drever, John Levack. "Soundscape composition: the convergence of ethnography and acousmatic music."
    Organised Sound 7.1 (2002): 21–27.

Droumeva, Milena. "Curating Aural Experience: A Sonic Ethnography of Everyday Media Practices." *Interference: A Journal of Aural Culture* 5. 2016.

Eckel, Gerhard, Manuel Rocha Iturbide, and B. Becker. "The Development of GiST, a Granular Synthesis Toolkit Based on an Extension of the FOF Generator." *ICMC.* 1995.

Edwards, Elizabeth. "Material beings: objecthood and ethnographic photographs." *Visual studies* 17.1 (2002): 67–75.

Ellis, Carolyn. *The Ethnographic I: A Methodological Novel About Autoethnography.* Walnut Creek, CA: AltaMira Press, 2004.

Ellis, Carolyn, Tony Adams and Arthur Bochner. "Autoethnography: An overview." *Historical Social Research* 36 (4), 2011, pp. 273–290.

Ellis, Carolyn, and Art Bochner. "Autoethnography, Personal Narrative, Reflexivity: Researcher as Subject." In N.K. Denzin & Y.S. Lincoln (eds.), *Handbook of Qualitative Research (2nd ed.).* Sage Publications, p. 733–768

Feaster, Patrick. "Enigmatic Proofs: The Archiving of Édouard-Léon Scott de Martinville's Phonautograms." *Technology and culture* 60.2 (2019): S14–S38.

Feld, Steven, and Donald Brenneis. "Doing anthropology in sound." *American Ethnologist* 31.4 (2004): 461–474.

Friesen, Megan R., Jacqueline R. Beggs and Anne C. Gaskett. "Sensory-based conservation of seabirds: a review of management strategies and animal behaviours that facilitate success." *Biological Reviews* 92.3 (2017): 1769–1784.

Fuller, Daniel, and Brian Magerko. "Shared mental models in improvisational performance." *Proceedings of the intelligent narrative technologies III workshop.* 2010.

Gabor, Dennis. "Acoustical quanta and the theory of hearing." *Nature* 159 (1947): 4044.

Gordon, Timothy A.C., Andrew N. Radford, Isla K. Davidson, Kasey Barnes, Kieran McCloskey, Sophie L. Nedelec, Mark G. Meekan, Mark I. McCormick & Stephen D. Simpson. "Acoustic enrichment can enhance fish community development on degraded coral reef habitat." *Nature Communications* 10.1 (2019): 1–7.

Halberstam, J. Jack. *In a queer time and place: Transgender bodies, subcultural lives.* Vol. 3. NYU press, 2005.

Hamilton, Andy. "15 The art of improvisation and the aesthetics of imperfection." *Teaching music in secondary schools: a reader.* 209. London: Routledge in association with the Open University, 2002.

Hansen, Mark B.N. " New Media." *Critical Terms for Media Studies.* University of Chicago Press, 2010, 172–185.

Haraway, Donna J. *Simians, Cyborgs, and Women: The Reinvention of Nature.* London: Routledge, 2014.

Haraway, Donna J. *Staying with the trouble: Making kin in the Chthulucene.* Duke University Press, 2016.

Holles, Sophie, Stephen D. Simpson, Andrew N. Radford, Laetitia Berten, David Lecchini. "Boat noise disrupts orientation behaviour in a coral reef fish." *Marine Ecology Progress Series* 485 (2013): 295–300.

Hood, Gavin. "Interview: Gavin Hood," April 25, 2016. https://dronecenter.bard.edu/interview-gavin-hood/.

Hunting Self | Arts Commons. https://artscommons.ca/whats-on/2018/hunting-self/. Accessed Jan. 27, 2021.

Jaekl, Philip. "What Is the Mysterious 'global Hum' – and Is It Simply Noise Pollution?" *The Guardian,* March 13, 2019. https://www.theguardian.com/cities/2019/mar/13/what-is-the-mysterious-gl-hum-and-is-it-simply-noise-pollution.

Kelly, Caleb. *Sound.* Whitechapel: Documents of Contemporary Art, MIT Press, 2011.

Klehm, Nance. *The Soil Keepers: Interviews with Practitioners on the Ground Beneath Our Feet.* Terra Fluxus Publishing, 2019.

Kozel, S. "SpaceMaking, experiences of a virtual body." Dance Theatre Journal, 11 (3), (1994).

Krause, Bernie. *Wild soundscapes: discovering the voice of the natural world.* Yale University Press, 2016.

Langeveld, Lau, René van Egmond, Reinier Jansen and Elif Özcan. "Product sound design: Intentional and consequential sounds." *Advances in industrial design engineering* 47.3 (2013).

Lawrence, Beatrice D., and James A. Simmons. "Measurements of atmospheric attenuation at ultrasonic frequencies and the significance for echolocation by bats." *The Journal of the Acoustical Society of America* 71.3 (1982): 585–590.

Lefebvre, Henri, and Donald Nicholson-Smith. *The Production of Space*. Vol. 142. Blackwell: Oxford, 1991.

Lewis, George E. "Interacting with latter-day musical automata." *Contemporary Music Review* 18.3 (1999): 99–112.

"Madam Data: Sounds and Software." Accessed February 11, 2020. http://www.madamdata.net/.

Manning, Erin, and Brian Massumi. *Thought in the Act: Passages in the Ecology of Experience*, University of Minnesota Press. 2014.

McCormack, David. "Book Review: Creating autoethnographies," by T. Muncey, 2010, London, Sage. *British Journal of Guidance & Counselling* 40.2 (2012): 182–184.Mendonça, David, and William A. Wallace. "Studying organizationally-situated improvisation in response to extreme events." *International Journal of Mass Emergencies and Disasters* 22.2 (2004): 5–30.

Mitchell, W.J.T., and Mark B.N. Hansen, *Critical terms for Media Studies*. University of Chicago Press, 2010.

Mondloch, Kate. "Screen Subjects." *Screens: Viewing Media Installation Art*. Electronic Mediations, Volume 30. University of Minnesota Press, 2010.

Mondloch, Kate. "Inhabiting Matter: New Media Art and New Materialisms Informed by Feminism." *A Capsule Aesthetic: Feminist Materialisms in New Media Art*. University of Minnesota Press, 2018, 1–21.

Monnet, Nadja. "Photoethnography of the Urban Space, or How to Describe the Urban World beyond Words: Presentation of a Multimedia Essay." *Visual Ethnography*, 2014.

Morton, Timothy. "Thinking Big." *The Ecological Thought*. Harvard University Press, 2010, 20– 58.

Muncey, Tessa. *Creating Autoethnographies*. London: SAGE, 2014.

Nancy, Jean-Luc. "The being-with of being-there." *Continental Philosophy Review* 41.1 (2008): 1–15.

Nédélec, Jean-Claude. *Acoustic and electromagnetic equations: integral representations for harmonic problems*. Vol. 144. Springer Science & Business Media, 2013.

Neisser, Ulric. *Cognitive Psychology*, 2014, 189–190.

Neumark, Norie, Ross Gibson and Theo Van Leeuwen, eds. *Voice: Vocal aesthetics in digital arts and media*. MIT Press, 2010.

Reading, Anna. "Memory and digital media: Six dynamics of the globital memory field." *On media memory*. Palgrave Macmillan, London, 2011, 241–252.Reed-Danahay and Neni Panourgia. "Book Reviews – Auto/ethnography: Rewriting the Self and the Social." *American Ethnologist*. 27.2 (2000): 551.

Ricoeur, Paul. "Memories and Images." *Memory*. Whitechapel: Documents in Contemporary Art. MIT Press, 2012, 66–70.

Rodgers, Tara. "How Art and Research Inform One Another; or, Choose Your Own Adventure." *Canadian Journal of Communication,* vol. 37, no. 1, March 2012. https://www.cjc-online.ca/index.php/journal/article/view/2521.

Rolling, James Haywood. *Arts-based research primer*. Peter Lang Publishing, 2013.

Rowe, Robert. *Interactive music systems: machine listening and composing*. MIT press, 1992.

Sawyer, R. Keith, and Stacy DeZutter. "Distributed creativity: How collective creations emerge from collaboration." *Psychology of aesthetics, creativity, and the arts* 3.2 (2009): 81.

Schafer, R. Murray. *Five village soundscapes*. No. 4. Vancouver, BC: ARC Publications, 1977.

Schafer, R. Murray. *The soundscape: Our sonic environment and the tuning of the world*. Simon and Schuster, 1993.Stubblefield, Thomas. *Drone Art: The Everywhere War as Medium*. University of California Press, 2020.

"Soundscapes of Canada." Accessed February 5, 2020.

https://www.sfu.ca/sonic-studio-webdav/WSP/canada.html.

Springgay, Stephanie, and Sarah E. Truman. "On the Need for Methods Beyond Proceduralism: Speculative Middles, (In)Tensions, and Response-Ability in Research." *Qualitative Inquiry*, vol. 24, no. 3, Mar. 2018, pp. 203–14. doi:10.1177/1077800417704464.

Springgay, Stephanie, and Zaliwska Zofia. "Diagrams and Cuts: a Materialist Approach to Research-Creation." *Cultural Studies <–> Critical Methodologies*. 15.2 (2015): 136–144.

Starosielski, Nicole. "Fixed Flow: Undersea Cables as Media Infrastructure." *Signal Traffic: Critical Studies of Media Infrastructures*. University of Illinois Press, 2015a, pp. 53–70.

Starosielski, Nicole. "Introduction: Against Flow." *The Undersea Network*. Duke University Press, 2015b, pp. 1–45.

Street, Seán. *The memory of sound: preserving the sonic past.* Routledge, 2014.

"Toward a Process Seed Bank: What Research-Creation Can Do." *NMC Media-N*, Sept. 25, 2015, http://median.newmediacaucus.org/research-creation-explorations/toward-a-process-seed-bank-what-research-creation-can-do/.

Truax, Barry. "Real-time granular synthesis with a digital signal processor." *Computer Music Journal* 12.2 (1988): 14–26.

Truax, Barry. "Genres and Techniques of Soundscape Composition as Developed at Simon Fraser University." *Organised Sound* 7, no. 1 (2002): 5–14.

Truman, Sarah E., and Stephanie Springgay. "The Primacy of Movement in Research-Creation: New Materialist Approaches to Art Research and Pedagogy." *Art's Teachings, Teaching's Art.* Dordrecht: Springer, 2015, 151–62.

Westerkamp, Hildegard. "Bauhaus and Soundscape Studies—Exploring Connections and Differences." *From Bauhaus to Scoundscape Symposium.* 1994.

Winkler, Todd. *Composing interactive music: techniques and ideas using Max.* MIT press, 2001.

"World Soundscape Project." Accessed February 4, 2020. https://www.sfu.ca/~truax/wsp.html.

Xenakis, Iannis. *Formalized music: thought and mathematics in composition.* No. 6. Pendragon Press, 1992.

Xenakis, Iannis. "Formalized Music: Thought and Mathematics in Music." Stuyvesant, 1971.

11

The Soundscape Approach
New Opportunities in Sound Design Practice

Ian Thompson

1 Introduction

Despite our permanent exposure to soundscape, we tend to only notice it when a distant sound grabs our attention (favorably or otherwise) or triggers specific emotional
responses (annoyance, anger, fear, etc.). Sustained exposure to certain sounds which
elicit primal responses to threat (fight or flight) can cause long-term damage to health
and wellbeing if we are unable to control or avoid them (Ising and Kruppa 2004; Goines
and Hagler 2007). Soundscape research seeks to assess relationships between communities and sonic environments, with improvement of quality of life by creating desirable,
sustainable and healthy places to live in and visit being the ultimate goal. Studies which
investigate and describe how a location's soundscape is perceived can inform responsive
modification of the sonic environment, through legislative control, and design interventions by planners, architects, acousticians and, increasingly, sound designers.

Although the word soundscape is relatively self-explanatory and broadly understood in everyday English, the term didn't fully enter the lexicon much before the late
1960s (figure 11.1). Its use has increased almost exponentially since the turn of the
century, as investigation of soundscape's role in affecting the quality of (not exclusively human) life through auditory perception has permeated discourses across the
arts, humanities and sciences (Kang et al. 2016; Kang and Aletta 2018).

The earliest known methodical soundscape study, Michael Southworth's MIT
master's thesis in planning *The Sonic Environment of Cities* in which "the perceptual form of the [Boston] soundscape is investigated by means of a field analysis"
(Southworth 1967, 2), makes no conceptual distinction between sonic environment
and soundscape; the terms are used interchangeably. Subsequent studies by the World
Soundscape Project (WSP 2020) in the 1970s, and more recently by the Soundscape
of European Cities and Landscapes Network (Kang et al. 2013), sustain the understanding of soundscape as an environmental phenomenon, concurring with Barry

DOI: 10.4324/9780429356360-12

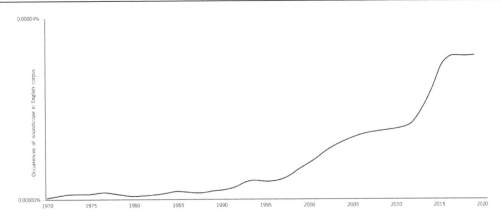

Figure 11.1   Google Ngram search results returned for soundscape based on the Google Books English 2019 corpus, 1970 to 2019.

Michel et al. 2011.

Truax's definition: "an environment of sound (or sonic environment) with emphasis on the way it is perceived and understood by the individual, or by a society" (Truax 1999). More recently, the International Organization for Standardization (ISO) has defined soundscape as an "acoustic environment as perceived or experienced and/or understood by a person or people, in context." (ISO 2014, 2).

2 Studying soundscape

> "old pond
> a frog jumps into
> the sound of water"
>
> (Bashō 1686/2008, 152)

Reflections upon aesthetic properties of soundscape can be found in many poetic and musical works centuries before the invention of electromechanical recording: systematic approaches to measuring, analyzing, and evaluating soundscape are relatively recent.

2.1 Boston, 1967

Masters student Michael Southworth's pioneering study, The Sonic Environment of Cities, attempted to evaluate sight- and hearing-impaired residents' perception of sonic and visual aspects of Boston. Southworth's study combined quantitative and qualitative data collection with detailed and innovative mapping (1967; 1969). Southworth concluded that *delight and acceptance* (ibid.) of an urban environment might be improved by sonic (re)design alone, rather than (generally more costly) visual intervention. He suggested adding features such as sonic signs and sculptures to 'beautify' areas perceived as sonically unappealing. Although this was perhaps naive at the time (Truax 2001, 72–74), the idea of improving urban soundscape through innovative acoustic design and engineering has recently been revisited (Lavia et al. 2015), and soundscape design ('soundscaping') has become a new context for sound design practice (Lacey 2017).

With the benefit of hindsight afforded by more recent discourses in research design (Creswell and Creswell 2017), it appears that a post-positivist, deductive approach constrained the Boston study: Southworth aimed to test hypotheses based on how he expected participants to perceive the soundscape, in a series of experiments which detached them from everyday experience (being blindfolded, wearing ear defenders, being pushed in a wheelchair, etc.). A more social constructivist or pragmatic worldview (ibid.), such as surveying the participants' lived experience of soundscape, would have allowed deeper insights into everyday perception to emerge. Yet despite these shortcomings, Southworth's enthusiastic study is significant in its attempt to understand human perception of an urban soundscape, and also in demonstrating how methodical study of sonic environment might inform new approaches to urban planning and design in the public realm.

2.2 World Soundscape Project, 1973–1975

The conceptual frameworks which underpin contemporary soundscape research are largely derived from work by the World Soundscape Project (WSP), founded by musician and educator R. Murray Schafer at Simon Fraser University, Vancouver, in 1973. This research group of six "highly motivated young composers, activists and students" (WSP 2020) was established with the purpose of documenting and drawing attention to what Schafer regarded as degradation of the urban soundscape by the noises of industry, traffic, communication technology, etc. Although well-intentioned in the context of late-1960s North American countercultural politics, Schafer's definition of noise as "any undesired sound signal" (Schafer 1969, 19; 1970) restritced more detailed investigation into how noise is perceived, and implied there is a subjective consensus as to what it is: more recent discourses in soundscape research and sound studies reveal that the situation is more nuanced.

Two major WSP studies, *The Vancouver Soundscape* (Schafer 1973) and *Five Village Soundscapes* (Schafer 1978), document and analyze sonic environments in detail. *The Vancouver Soundscape* states the aim of the WSP as being "to bring together research on the scientific, sociological and aesthetic aspects of the [sonic] environment" (Schafer 1973, 1). The aim of the Five Villages study two years later was to obtain data for comparison with the Vancouver study, from which a deeper understanding of the nature of soundscape could be derived.

Although no specific research design was described, the WSP's studies deployed a mixed-methods approach (Creswell and Creswell 2017, 14–16) arising from a pragmatic philosophical worldview (Cherryholmes 1992; Given 2012, 672–75; Creswell and Creswell 2017, 10–11) informed by arts and music epistemologies. As such, the group was not restricted by established scientific research conventions, and could draw freely on a variety of methods and data collection techniques. This contrasts with Southworth's Boston study, rooted in more constrained paradigms of late 1960s urban planning. A pragmatic approach (and to be fair to Southworth, more funding and resources) enabled the WSP to liberally draw on quantitative and qualitative methods from:

- acoustics (measuring sound pressure levels)
- sociology (narrative interviews with local people; *earwitness* accounts (Schafer 1994, 272))
- ethnography (audio recording soundscapes and people within them)
- history (investigating how a soundscape may have evolved over time)
- geography (using a variety of mapping techniques to illustrate sounds in space)
- linguistics (asking people to describe their perception of the soundscape)
- music theory (using musical notation to record features of the soundscape)

As published studies, rather than concluded research projects, *The Vancouver Soundscape* and *Five Village Soundscapes* are artifacts in themselves, their presentation style (both in print and as collections of sound recordings) reflecting a prevailing intent to interrogate soundscape in aesthetic terms. The depth of analysis in the WSP's studies led to a comprehensive construct of soundscape as a distinct phenomenon, expressed in definitive publications such as Schafer's *Soundscape: Our Sonic Environment and the Tuning of the World* ([1977] 1994), and Barry Truax's *Handbook for Acoustic Ecology* ([1978] 1999) and *Acoustic Communication* ([1984] 2001). Copies of the original WSP studies, along with further documentation, can be found on Simon Fraser University's *Sonic Research Studio* website – a valuable sound design resource (WSP 2020).

2.2.1 Key Concepts Arising from WSP Studies

Soundscape / Landscape

The WSP's soundscape taxonomy (with which readers may already be familiar) identifies key components in the sonic environment. The suffix 'scape' implies soundscape is the aural equivalent of landscape – a logical analogy insofar that soundscape also represents a sensory boundary. But, unlike sight, hearing is not limited by an anatomical field of vision, or a pictorial frame; it's spatially immersive and temporally constant. "The sense of hearing cannot be closed off at will. There are no earlids"

(Schafer 1994, 11). Visibly, a landscape's composition changes very slowly (meteorology and illumination notwithstanding), and can be accurately represented in a photograph or painting with relative ease. Soundscape is in constant flux, and cannot be captured instantaneously; meaningful investigation calls for longitudinal study. Despite these differences, the landscape comparison is still helpful when describing features in the sonic environment. As landscape can be characterized by specific landmarks (e.g., tall buildings) within a wider panorama, soundscape also contains distinctive features: *soundmarks* (e.g., passing trains) and *sound signals* (e.g., ice-cream van chimes), set against pervading and continuous *keynote* sounds (e.g., road traffic, wind in trees) which together define the acoustic character of a particular place (Schafer 1994, 9). Sounds can traverse categories too; church bells may become so regular and familiar to a local population that they become soundmarks rather than a signals, and so on. This taxonomy is useful in sound design; sounds can be organized into these categories and processed with regard to their function within a composed soundscape.

The landscape analogy notwithstanding, the WSP's studies tended to separate the heard from the seen, and as Jonathon Sterne observes this detachment of sound from vision often happens in cultural studies (invariably at the expense of sound). Sterne refers to the *audio-visual litany* (2012, 9): a clichéd set of assumed differences between looking and listening, combined with a tendency to regard each as a functionally separate sense. Perception (our main concern in soundscape research) is ultimately a function of both – with other senses also in the mix. Awareness of context (rural, urban, park, street, etc.) is essential for any investigation into soundscape is essential; a sense of place is defined as much by vision, smell, touch, and even taste, as it is by hearing.

Acoustic Community

Another important concept to emerge from WSP research is Barry Truax's notion of *acoustic community*[1] (2001, 65–92). Although perception of sound is an individual, personal experience, consensus forms among groups of individuals as sounds become familiar, recognized, expected, understood and accepted within that community, giving a soundscape what might be described as its 'default setting'. Unexpected, unfamiliar sounds instinctively associated with threat or danger, appearing from beyond the community's *acoustic horizon*[2] (ibid., 26), can raise alertness, cause unease, stress and initiate flight responses based on how the community perceives them. From a sound design perspective, a cinema, theater, or similar audience can be considered an acoustic community within Truax's definition.

Hi-fi Soundscape

Schafer (1994, 43) proposes the concept of a 'hi-fi' sonic environment, in which component sounds are clearly distinguishable from one another and easily identifiable, with a sense of clarity and balance (in the sense that sounds don't compete for space within the frequency spectrum). A hi-fi soundscape contributes to a sense of wellbeing among the acoustic community. The opposite, 'lo-fi', is regarded as unpleasant, alienating, distorted, unclear and has a long-term negative impact. The link between sustained exposure to what may be considered lo-fi soundscapes and ill health is self-evident, but in reality sonic environments tend to sit along a spectrum between hi- and lo-fi, with (to continue the home audio metaphor) a signal-to-noise ratio (SNR) that varies throughout the day and during particular events. These temporary fluctuations in SNR may even serve as soundmarks in themselves (e.g., rush hour traffic) and be tolerable in the soundscape for limited durations.

Like Schafer's notion of noise as undesired sound (1969, 19; 1970), the hi-fi/lo-fi dichotomy limits deeper investigation of nuanced and cultural attitudes towards noise. The metaphor (a reference to the relative quality of audio equipment marketed in the 1960s) has become less meaningful in the digital era (Sterne 2003), but proposes a rational foundation for considering how perceived quality of soundscape might be described.

2.3 ISO 12913: A Paradigm Shift

Since the turn of the century, the WSP's work and Michael Southworth's master's thesis have been revisited as part of a desire among planners, architects and acousticians to establish a methodology for evaluating soundscape perception and preference in the built environment (Brown et al. 2011; Kang et al. 2016). To this end, an International Organization for Standardization 'Perceptual assessment of soundscape quality'[3] working group was established in 2008. The Soundscape of European Cities and Landscapes Network (SECLN) research group, funded by the European Union as a European Cooperation in Science and Technology (COST) Action, consisting of several working groups across a spectrum of disciplines, was convened a year later. Between 2009 and 2013, the SECLN explored new approaches to managing environmental noise; noise persistently proving to be more conceptually complex than just unwanted sound.

> Reducing sound level, the focus of EU environmental noise policy, does not necessarily lead to improved quality of life in urban/rural areas, and a new multidisciplinary approach is essential. Soundscape research represents this paradigm shift as it involves not only physical measurements but also the cooperation of human/social sciences (e.g. psychology, sociology, architecture, anthropology, medicine), to account for the diversity of soundscapes across countries and cultures; and it considers environmental sounds as a 'resource' rather than a 'waste.'
> (Kang et al. 2013, 8)

The SECLN's final publication, *Soundscape of European Cities and Landscapes*, contains a set of research papers reviewed as part of the COST action, available online (Kang et al. 2013). Much of the material is also covered in the book *Soundscape and the Built Environment*, edited by action chair Jian Kang and vice-chair Brigitte Schulte-Fortkamp (2015). Both publications are comprehensive soundscape research references, and their scientific format makes an interesting stylistic contrast to Schafer and Truax's earlier texts. Many of the SECLN's recommendations have been incorporated into a new International Organization for Standardization (ISO) specification, *ISO 12913- Acoustics - Soundscape*, outlining a research methodology in three parts:

1. definition and conceptual framework (ISO 2014)
2. data collection and reporting requirements (ISO 2018)
3. data analysis (ISO 2019)

ISO 12913 specifies a methodology through which to evaluate perception of soundscape within the built environment, and its publication in a period when urban populations are increasing is timely. Cities need to adapt by way of new (or re-) development while at the same time mitigating potential damage to the wellbeing of residents from long-term exposure to various forms of noise (not necessarily loud sounds), which has been shown to be detrimental to health (Ising and Kruppa 2004; Goines and Hagler 2007; Lercher et al. 2015; Epstein 2019). The specification aims to describe soundscape in the public realm accurately and meaningfully, to inform subsequent intervention and alteration through whatever means are appropriate. It also presents a consistent a framework for peer review, and evaluation of soundscape design interventions at a location over time.

2.3.1 The Problem with Noise

Since the mid-20th century, urban noise control measures have been implemented in response to higher-than-specified sound pressure level (SPL) readings from fixed locations at particular times of day and/or night, usually without further consideration of the types of sound perceived as noise, or their context. Complaints about noise in urban settings often arise from differences in sound preference and lifestyle between acoustic communities, rather than the presence of generically annoying sound. For example, a vibrant late-night concert in a public square that disturbs a neighborhood's sleep is common in many regenerated city districts where young and socially active populations co-exist with older and more sedentary indigenous residents who desire "peace and quiet" after 10pm. Brandon LeBelle explores these ideas further in his book *Acoustic Territories*: "the street is an acoustical instrument for the propagation and diffusion of multiple sonorities, which the city itself comes to feedback." (LeBelle 2010, 130).

In a similar vein, several writers within sound studies (Lacey 2017, 28–38) have criticized Schafer's tendency to reject the "noisy" urban in favor of the bucolic and natural. This idealization of sounds limited the scope of the WSP's earlier research to understand noise as a cultural construct, rather than just a broad category of undesirable sounds. Since the turn of the century, acknowledgement that noise is perceptively complex, and an inherent part of the urban soundscape, has informed discourses in planning and architectural acoustics (Aletta and Kang 2015; Vogiatzis and Remy 2017). Simultaneously, the nature of noise has characterized critical discourses in sound studies (Voegelin 2010; Goodman 2010; Hainge 2013; Cox 2018). Sound designers can benefit from exploring both of those academic territories.

ISO 12913 defines noise as "sound that is deemed to be unpleasant, unexpected, undesired or harmful" (ISO 2018, 2), and also states that "the term 'noise' is not intended as a value judgement." (ibid., 15): identification of sounds as noise is deferred to research participants. The measurement of psychoacoustic parameters – loudness, sharpness, tonality, roughness, fluctuation strength (ISO 2019, 12–15; Fastl and Zwicker 2007, 203–65) – is also part of ISO 12913, and a far more effective measure of implied noise than SPL alone. When correlated with perception survey data, this affords significantly more insight than Schafer's hi-fi/lo-fi dichotomy.

2.3.2 Perceptual Construct of Soundscape

Part one of ISO 12913 – *Definition and conceptual framework* (ISO 2014), deals almost exclusively with describing a perceptual construct of soundscape (Figure 11.2).

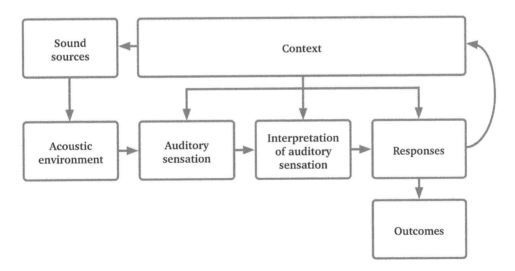

Figure 11.2 The perceptual construct of soundscape used in ISO 12913.
Adapted from ISO 12913-1:2014 (ISO 2014, 2).

The construct recognizes that perception of soundscape is a cognitive process for each listener, with *context* identified as the most significant component: "The context includes the interrelationships between person and activity and place, in space and time." (ISO 2014, 2). Context influences soundscape through:

- *auditory sensation*; hearing (airborne, structure-borne and bone/tissue conduction) – which may be affected by weather, hearing impairment, etc.
- *interpretation of auditory sensation*; listening (recognition of sounds), combined with other sensory information (vision, smell), etc.
- *responses*; may be affected by emotional state, memory association, personal taste, ability to tolerate sounds, etc. Does the listener stay or leave?

(ibid.)

Context also determines the *sound sources* present (birds, people, traffic, etc.) within the *acoustic environment* that affects their propagation (reflection, absorption, or diffusion of sound by buildings, foliage, etc.). Sound sources shaped by the acoustic environment are detected by auditory sensation, then interpreted through listening, leading to a response which contributes to an understanding of the context (e.g., from experience, a listener might come to understand the context as inviting or hostile). The *outcome* of the perceptual process is the "overall, long-term consequence facilitated or enabled by the acoustic environment. Outcomes include attitudes, beliefs, judgments, habits, visitor/user experiences (e.g. activities, actions and mental states), health, wellbeing and quality of life, as well as reduced social costs for society" (ibid., 3).

This perceptual construct is not dissimilar to procedures in studio-based sound design, in which sound sources (audio recordings), acoustic environment (adjusted by signal processing), and context (story, dialogue, etc.) are chosen and modified with the aim of affecting outcome (audience response). Sound design in the public realm also calls for strategies to shape perception (introduction or removal of sound sources, design or modification of the acoustic environment), to affect response, context and outcome.

2.3.3 Research Design and Triangulation

Francesco Aletta and Jieling Xiao observe that soundscape research is "positioned somewhere in the 'intersection' (some might even say 'union') among diverse disciplines, such as sociology, environmental psychology, music, acoustic ecology, urban planning, noise control engineering, architecture and more." (2018, 1). As the WSP demonstrated in the 1970s, soundscape research exemplifies a mixed-method approach, now refined in ISO 12913 as a methodology in its own right and referred to as the 'soundscape approach' (Davies et al. 2013; Aletta and Kang 2015;

ISO 2018). In terms of research design, a social constructivist worldview prevails across ISO 12913 (Given 2012, 817–820; Creswell and Creswell 2017, 6–10), in that the research goal is to describe a location's soundscape by way of investigating the perceptions of individuals who inhabit it. This represents a paradigm shift from the pragmatic theory-building of the WSP, towards an approach based on the concept of triangulation.

Triangulation (a term borrowed from nautical navigation) is the principle that a single accurate position (or a likely conclusion) can be located at the convergence of outcomes measured from three different positions. It is suited to mixed-methods research, where results from one method can be tested against two others to reduce errors and inconsistencies, and strengthen any correlations that may be inconclusive between only two sets of results. It is especially appropriate in situations where a single phenomenon can be understood from different points of view.

Triangulation can be applied to different theories, data sources, methods or investigators (Given 2012, 893–894). Theories arising at each stage of the perceptual construct of soundscape (relating to hearing, semiotics, cognition, etc.) (figure 11.3) suggest a mixed-methods research design that includes:

- data triangulation (quantitative survey, narrative interview, measurement)
- method triangulation (analysis of perception data, grounded theory, psycho-acoustic analysis)
- investigation triangulation (survey leader, local experts, recording devices).
  The ultimate description of soundscape lies at the convergence of:
- people
- acoustic environment
- context

Lercher and Schulte-Fortkamp 2013; Schulte-Fortkamp
and Fiebig 2015, 82; figure 11.3

This concurs with Schafer's earlier observation that "the home territory of soundscape studies will be the middle ground between science, society and the arts" (1994, 4).

2.3.4 Data collection methods

Part two, ISO/TS[4] 12913-2 – *Data collection and reporting requirements* (ISO 2018), specifies a mixture of quantitative and qualitative methods. As the key objective is to evaluate soundscape perception in an open environment, the primary measurement device is the ear and the research setting is usually public space. Thus, overheads are relatively low compared to laboratory-based research, raising the prospect of ISO 12913 (even without total compliance to the specification) being adopted for small-scale local studies as a context for activism, opening debates around soundscape

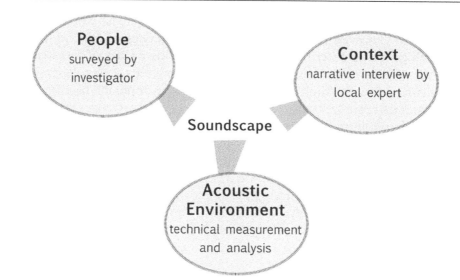

Figure 11.3 Methodological and investigator triangulation in the soundscape approach. (ISO 2014; 2018; 2019). The principle of triangulation applies across the ISO 12913 specification.

Adapted from Lercher and Schulte-Fortkamp 2013; Schulte-Fortkamp and Fiebig 2015, 82.

quality in communities where tranquility and other characteristics of soundscape quality are perceived to be under threat.

Data collection methods in ISO 12913 include activities with which sound designers may already be familiar, such as soundwalking and binaural recording, presenting clear opportunities for them to contribute professional expertise to soundscape research projects.

The data collection methods specified in ISO 12913-2 (ibid.) are:

- soundwalk
- questionnaire
- guided interview
- sound source taxonomy
- binaural measurements

Soundwalk

Soundwalking as a technique for engaging with soundscape through active listening (Oliveros 2005), rather than passive hearing, has gained popularity among writers, artists and musicians in the early twenty-first century (Drever 2009; Walk Listen Create 2020). Sound designers will already be familiar with ear training and listening exercises, and are well advised to investigate soundwalking as part of their discipline if they have not already done so. ISO 12913-2 requires that study participants are led on a soundwalk route, with close involvement of local experts.[5] Reflections on the soundwalk are then recorded in a group discussion / narrative interview (Czarniawska 2001, 732–750), to establish:

- What was your favorite sound on the walk? Why and where was it?
- What sound did you dislike the most? Why and where was it?
- Where would you make improvements to the sounds you heard? Which would be the most important and why?
- Did the sounds you heard match your expectations of the area? Why/why not?

ISO 2018, 11

Participants also complete questionnaires about perception of the soundscape, outlined below. The soundwalk is perhaps the most radical part of the ISO specification, in that a somewhat esoteric arts practice has been adopted in a context generally associated with prosaic activities such as quality control and safety compliance.

Questionnaire

Three different methods of data collection are specified: A, B, and C (ISO 2018). Methods A and B require soundwalk participants to record perceptions of the soundscape on either five-point ordinal scales (Method A), or five-point continuous-category scales (Method B). Method C is a guided interview.

Method A uses a four-part questionnaire to record:

1. Sound source identification: the extent to which different categories of sound dominate the soundscape:
   - sounds of technology
   - sounds of nature
   - sounds of human beings
2. Perceived affective quality: the extent to which the respondent agrees that the soundscape is:
   - pleasant
   - chaotic
   - vibrant
   - uneventful
   - calm
   - annoying
   - eventful
   - monotonous

3. Assessment of the surrounding sound environment: how the respondent describes the sound environment, between *very good* and *very bad*.
4. Appropriateness of the surrounding sound environment: the extent to which a respondent considers the overall sound environment to be appropriate to the present place, between *not at all* and *perfectly*.

(ibid.)

In Method B, participants are asked to listen at specific points along a soundwalk route, where the soundscape is synchronously binaurally recorded (see below).

The questionnaire in Method B is in three parts:

1. Assessment of the sound environment: respondents are asked to rate their perception of the sound environment at locations along the route on a scale between *not at all* and *extremely* to the questions:
   • how loud is it here?
   • how unpleasant is it here?
   • how appropriate is the sound to the surrounding?

   and finally, between *never* and *very often*:

   • how often would you like to visit this place again?

2. Sound source recognition and ranking: the respondent is asked to list all the sound sources they can hear, starting with the most noticeable.
3. Subsequent comments: an open response to the question *what is going through your mind?* to record thoughts and feelings about the location after listening to the environment.

(ibid.)

Guided Interview

Method C is a guided interview (Lichtman 2014, 241–275), completed by local residents or visitors whose views about the soundscape and attitudes to the wider quality of the environment are relevant. This method is more appropriate in certain circumstances, such as where a soundwalk or recording may not be possible, or soundscape is to be evaluated from inside a building, and so on.

Sound Source Taxonomy

As part of discussions with survey participants, identification and categorization of sound sources is suggested, based on hierarchical cluster diagrams often found in medical research, computing, and data mining (Bones et al. 2018) (figure 11.4). This is a useful organizational technique in sound design practice, helpful when archiving and meta-tagging audio libraries, and also in non-linear sound design using procedural audio and object-orientated games engines. Unlike the WSP taxonomy of

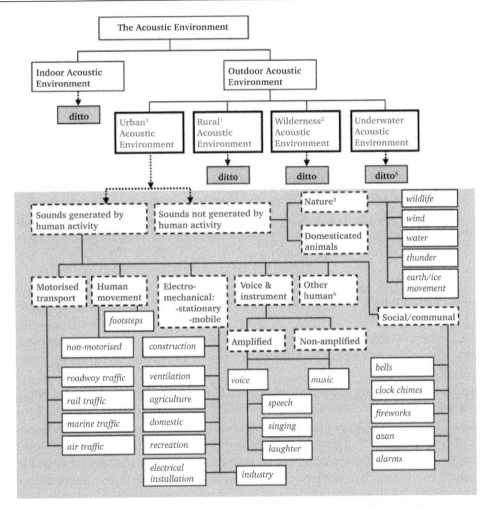

Figure 11.4 A taxonomy of the acoustic environment for soundscape studies showing categories of places (bold boxes), categories of sound sources (dashed boxes), and sound sources (italics).

1 The urban/rural distinction will not always be readily defined, but remains useful.

2 The wilderness category includes national parks, undeveloped natural and coastal zones, large recreation areas, etc., and the wilderness/rural divide will not always be clear cut.

3 While "nature" and "domesticated animals" sources are shown as being "not generated by human activity" there are many areas of overlap – for example the sounds of running water in constructed water features or the sounds of wind on buildings. Domesticated animal sounds will generally be from animals associated with a human activity/facility.

4 Recording, replay and amplification may occur for any type of sound – as for example in installations playing nature/wildlife sounds.

5 Because of the different acoustic impedances in air and water, many of the terrestrial sound sources within the shaded area of the figure would not normally be observed under water, but overall the same classification system is still applicable.

6 Coughing, for example.

From Brown et al. 2011, 390, adopted in ISO 12913-2:2018 (ISO 2018, 13).

keynote, soundmark, sound signal, this grouping isn't concerned with the function of a sound within the soundscape, but merely logging which sounds are present and within each category.

Binaural Measurements

The word "measurements" is significant here; soundscapes are recorded for psycho-acoustic analysis, not for their aesthetic value, though recordings may also be used for later asynchronous playback to survey participants in a studio via headphones, and for reference.

Binaural recordings are usually made with a dummy head microphone array (Nicol 2010) and ISO 12913 specifies that the dimensions of the head comply with ITU-T P.58:2013[6] and ANSI/ASA S 3.36:2012[7] specifications. Several models of head are available commercially at prices ranging from thousands of dollars for a fully compliant recording system, such as that made by HEAD Acoustics,[8] to several hundred for less sophisticated models. Although cheaper models may not fully meet the prescribed specification, they can still produce satisfactory recordings and usable data. It's also possible to use in-ear microphones, however this may compromise consistency of measurements within and between research projects due to variations in head geometry. The main problem to overcome with outdoor binaural recording is wind noise, so a high-quality windshield system is essential.

Ambisonic recording and reproduction are suggested as an alternative to binaural, with no technical or acoustic standard specified. This may be more convenient in the field and more flexible later in the laboratory; recordings can be binaurally decoded, or replayed on a multi-channel loudspeaker system in a listening room. However, context – the main sensory component in the conceptual construct of soundscape (figure 11.2) – is absent in the laboratory or studio. Even with high-quality recording and playback, listeners lack additional sensory information to completely evaluate the soundscape. As an anecdotal example, studio-based listening tests using recordings of a new acoustic vehicle alerting systems (AVAS) for London Buses (TfL 2019) yielded negative responses, compared with field tests in an urban setting where the same system was shown to be effective.[9]

2.3.5 Data Analysis and Presentation

Part 3, ISO/TS 12913-3:2019 (ISO 2019), specifies how survey data should be analyzed and presented. The aim is to build an accurate description of the soundscape under investigation so that informed and appropriate interventions can be made to improve or protect it. Redesigning physical spaces with features such as fountains, foliage, barriers and similar acoustic treatment is costly; the built environment can't be remixed in a studio, so it's essential that those interpreting the research are able to do so effectively before subsequent interventions are commissioned. To that end, presentation standards are specified, supported by graphical presentation of data (ibid.)

Three sets of data collected using the methods outlined in Part 2 (section 2.3.4.) are collated and triangulated:

- survey data from Method A, B or C (quantitative and qualitative)
- interview data from narrative interviews during soundwalks (qualitative)
- psychoacoustic measurements from binaural recordings (quantitative)

Quantitative Data Analysis

Although complete analysis of quantitative data in ISO 12913 calls for an understanding of statistical mathematics beyond the scope of most sound design practice (ibid.), the basic principles are not particularly complex. Ordinal survey data from Method A, and the continuous-category scale data from Method B (section 2.3.4.), are presented graphically.

For Method A, a two-dimensional radar plot of "pleasantness" against "eventfulness" is used (figure 11.5). Pleasantness is a subjective judgement, but descriptively easy to interpret. An eventful environment is defined in ISO 12913 as:

> An ... environment [that] is busy with human activity, for example a city center or other sound events produced by non-human agents, whereas an uneventful environment is completely devoid of human activity, for example a wilderness area or during late evening hours in a residential area without social, commercial and industrial activity.
>
> (ISO 2019, 5)

The components chosen to describe soundscape perception are based on a survey of 100 Swedish students by Östen Axelsson, Mats E. Nilsson and Birgitta Berglund at Stockholm University (2010), and the research paper provides an informative account of how soundscape is understood by psychologists.

The qualitative data from Method B (section 2.3.4.) can be presented as simple histograms or similar graphs, showing the proportion of responses in each category (ibid.).

The quantitative survey data from Methods A and B are linked with psychoacoustic indicators measured from binaural recordings, using statistical analysis such as linear regression (Allen 2017, 865), ANOVA (ibid., 34–36), and Spearman's rank coefficient (ibid., 274–76), to establish if any relationships exist, and measure the extent to which the sets of data correlate.

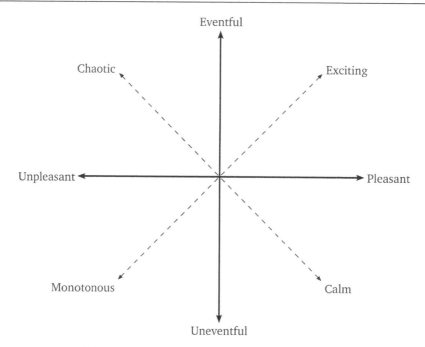

Figure 11.5 Axes used for two-dimensional radar plot of perception data collected in Method A, question 2 (ISO 2018; Axelsson et al. 2010). Co-ordinates (Pleasantness, Eventfulness) for each respondent are calculated with the formulae: P =(p–a)+cos45°(ca–ch)+cos45°(v–m), and E =(e–u)+cos45°(ch–ca)+cos45°(v–m), where: a=annoying; ca=calm; ch=chaotic; e=eventful; m=monotonous; p=pleasant; u=uneventful; v=vibrant. (ISO 2019).

Qualitative Data Analysis

Qualitative data is collected in:

- narrative interviews/discussions with soundwalk participants
- part 3 of Method B
- Method C guided interview

(ISO 2018)

ISO 12913 recommends textual analysis using the Grounded Theory (GT) approach (Corbin and Strauss 2008; Urquhart 2013, 115–130; Bryant and Charmaz 2019). GT analysis systematically abstracts categories, patterns and themes from responses. These are then used to inform and refine theories and impressions which emerge from the interview data through an iterative, cyclical process (figure 11.6). GT is widespread in social science research and usually conducted by someone with a solid

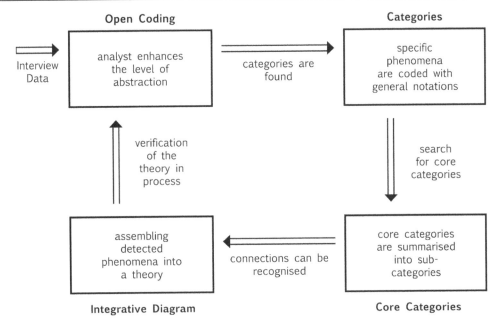

Figure 11.6 Circular systematic analysis used to assess qualitative data from open-ended survey questions and guided narrative interview according to the Grounded Theory.
Adapted from Fiebig and Schulte-Fortkamp 2005; ISO 2019, 11.

understanding of the subject under investigation – such as a professional sound designer in this case. Other suggested qualitative data analysis tools are Qualitative Content Analysis (Flick 2014, 171–183), and Social Network Analysis (Scott and Carrington 2014).

Analysis of Psychoacoustic Components

Binaural recordings made at key points in the soundscape are subjected to psycho-acoustic analysis, using software generally only accessible to industry or research institutions, such as HEAD Acoustics ArtemiS[10] suite and Oros Sound Quality.[11] Some open source analysis tools are available, such as PsySound3 (Cabrera et al. 2007) and PsyAcoustX (Bidelman et al. 2015), although a basic level of expertise with MatLab[12] is required.

The psychoacoustic parameters measured in ISO 122913 are:

- loudness (as perceived by the listener)[13]
- sharpness (relative proportion of high frequency content)

- psychoacoustic tonality (content heard as discrete tones)[14]
- roughness (amplitude fluctuation at ~70Hz)
- fluctuation strength (amplitude fluctuation at ~40Hz)

For further discussion and information on psychoacoustic analysis, see Genuit and Fiebig 2005; Fastl and Zwicker 2007, 203–265.

Once extrapolated from the binaural recordings, values of the psychoacoustic parameters are correlated with survey results and can be presented as part of the final report as noise maps across the survey area (ISO 2019) or as charts (figure 11.7).

2.3.6 ISO 12913 and the Soundscape Approach in Action

At the time of writing, a limited number of studies have been conducted under the ISO 12913 specification. Francesco Aletta et al. (2019) have field tested methods A and B at Roma Tre University, Italy, concluding that although the difference in results between them is negligible, using both is recommend if possible. Tin Oberman et al. (2020) have used it to assess sound art interventions in Croatia. Both of these papers provide a useful overview of the specification in practice. The 2020s will undoubtedly see increased use of ISO 12913, along with revision and reevaluation of the standard.

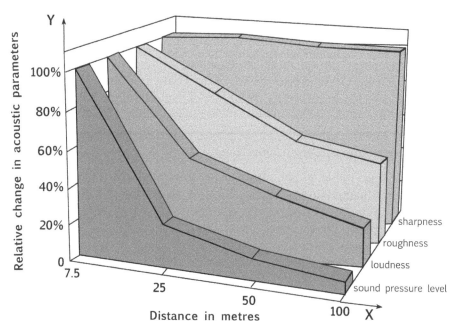

Figure 11.7 Hypothetical percentage change of acoustic parameters of a car passing at the distances 7.5m, 25m, 50m and 100m.

Adapted from ISO/TS 12913:3–2019 (ISO 2019).

At a national level, the soundscape approach and ISO 12913 has been adopted as part of the Welsh government's Noise and soundscape action plan 2018–2023 (Wales 2018), from which (at the time of writing) research and responses are yet to materialize. This is the first national legislation to incorporate to soundscape assessment into a development strategy.

ISO 12913 represents significant progress from the WSP's earlier work in terms of methodological consistency, however it shifts the center of gravity for soundscape research away from the aesthetic and ethnographic towards the often politically contentious realms of urban planning and development. By association with the ISO, there is a risk of soundscape quality becoming considered something only to be measured and certified in the service of limited agendas, rather than appreciated and valued for wider societal benefit. Critical involvement of creative and cultural practitioners – particularly those concerned with listening practice (e.g. sound designers) – will be vital to ensure research integrity. This concern aside, the soundscape approach offers an accessible framework through which interest in the sonic environment can be activated.

New opportunities for sound design practice in the public realm are increasing, and often costly investment in such schemes will need to be justified by studies which employ a soundscape approach as part of the design process. One example is the AVAS electric vehicle alert sound adopted by London Buses (TfL 2019). Although research and evaluation of AVAS were conducted by acousticians, the collaborative input of a sound designer, Matt Wilcock of Zelig Sound, London, was essential (Burgess 2019). Binaural recording of London Buses AVAS in London streets: https://tinyurl.com/AVAS-TfL.

3 Soundscape Research Beyond ISO 12913

While the soundscape approach has arisen from studies of the built environment, soundscape continues to be an essential means of investigating the natural world. Disciplines such as bioacoustics – the study of animal communication through sound (Laiolo 2010), and ecoacoustics, in which soundscape monitoring and recording are used to measure changes to biodiversity (Farina and Gage 2017) – have gained momentum since the turn of the century. The World Forum for Acoustic Ecology (WFAE), established in 1993, continues the work of the WSP in "the study of the social, aesthetic, cultural, and ecological aspects of the sonic environment," and its journal *Soundscape* offers comprehensive coverage across multiple disciplines (WFAE 2020). Also the important work of Bernie Krause, whose soundscape recordings have documented degradation of natural habitats, has raised awareness of acoustic ecology among a growing audience through his *niche* hypothesis of animal communication (1987;1999) and his 2012 book, *The Great Animal Orchestra* (Pijanowski et al. 2011; Krause 2020).

Aesthetic appreciation of soundscape has gathered momentum as part of the broader *sonic turn* across early twenty-first century arts (Kelly 2011; Eng 2017; Herzogenrath 2017). Field recording can be found across contemporary arts practice

(Lane and Carlyle 2011), and websites such as *The London Sound Survey*[15] and *Radio Aporee*[16] have brought awareness of soundscape to a global audience. Soundscape composition, pioneered by WSP member Hildegard Westerkamp, has enjoyed a renaissance among electroacoustic composers such as Brona Martin (2018) and others, too numerous to mention here: for further discussion, see John L. Drever (2002; 2019) and Barry Truax (2002; 2008). Soundscape composition is regularly featured as part of ambient music group Zoviet France's *Duck In A Tree*[17] podcast – recommended listening for any sound designer seeking inspiration.

The cross-disciplinary nature of soundscape research is exemplified by *Locus Sonus*,[18] hosting live open microphone audio streams around the world, and an online platform for projects such as *Soundcamp*:[19] a forum for academic collaboration, community engagement and wider public awareness of acoustic environments (Soundcamp 2019). Tools such as the *AudioMoth*[20] monitoring device facilitate soundscape study through citizen science projects, for example *Silent Cities*[21] which, at the time of writing in 2020, is gathering crowd-sourced audio data to assess the impact of Covid-19 containment measures on soundscapes throughout the world (Challéat et al. 2020).

4 Conclusion

The extensive work of The Soundscape of European Cities and Landscapes Network in formulating ISO 12913 has lead to a paradigm shift in how soundscape is evaluated and understood in the built environment. Despite concern that a standardization of research methodology could shift emphasis from the lived experience of soundscape towards scientific measurement, the evaluation of community perception is foregrounded in ISO 12913, and the incorporation of active listening practice into the research method acknowledges important work begun by the World Soundscape Project in the 1970s.

Actively listening to soundscape is an essential discipline for any sound designer to adopt, and the soundscape approach affords opportunities to develop a deeper understanding of how soundscapes function in different contexts. This has considerable benefit for sound design practices – in the studio, and out into the public realm. ISO 12913 also presents new opportunities for sound designers to become constructively and critically involved in research, collaborating professionally in areas which traditionally have had no need to consider the valuable contribution that sound design can make.

Notes

1 "The acoustic community may be defined as any soundscape in which acoustic information plays a pervasive role in the lives of the inhabitants (no matter how the commonality of such people is understood)." (Truax 2001, 66).
2 The acoustic horizon is "the farthest distance from which sound may be heard." (Truax 2001, 26).

3 Perceptual assessment of soundscape quality working group, ISO/TC 43/SC 1/WG 54: https://www. iso.org/committee/48474.html.
4 "TS" in the title ISO/TS 12913-2:2018, standing for Technical Specification, denotes that the document is iterative and subject to revision pending regular reviews of the methodology's efficacy.
5 Local expert defined as: "person who is familiar with the area under scrutiny either living in the area or having further daily routines related to the area" (ISO 2018, 2).
6 ITU-T P.58 specifications: https://tinyurl.com/ITU-TP-58.
7 ANSI/ASA S 3.36 specifications: https://tinyurl.com/ASA-S-3-36.
8 HEAD Acoustics website: https://www.head-acoustics.com/.
9 Waters, Grant. (Soundscapes Lead, Anderson Acoustics, London UK), in discussion with the author. November 2019.
10 HEAD Acoustics ArtemiS suite: https://tinyurl.com/Head-ArtemiS.
11 Oros Sound Quality software: https://tinyurl.com/Oros-Sound-Quality.
12 MatLab website: https://www.mathworks.com/.
13 HEAD acoustics 'Psychoacoustic Analyses I,' direct PDF link: https://tinyurl.com/HeadPsychoAn1
14 Ecma International Standard ECMA-74 'Measurement of Airborne Noise emitted by Information Technology and Telecommunications Equipment,' available PDF link: https://tinyurl.com/ ECMA-74.
15 London Sound Survey website: https://www.soundsurvey.org.uk/.
16 Radio Aporee website: https://aporee.org.
17 A Duck In A Tree podcast: https://zovietfrance.podbean.com/
18 Locos Sonus website: https://www.locusonus.org.
19 Soundcamp website: http://soundtent.org/.
20 AudioMoth website: https://www.openacousticdevices.info/.
21 Silent Cities website: https://tinyurl.com/SilentCities.

References

Aletta, Francesco, Claudia Guattari, Luca Evangelisti, Francesco Asdrubali, Tin Oberman and Jian Kang. 2019. "Exploring the compatibility of 'Method A' and 'Method B' data collection protocols reported in the ISO/TS 12913-2:2018 for urban soundscape via a soundwalk." *Applied Acoustics* 155 (December): 190–203. https://doi.org/10.1016/j.apacoust.2019.05.024.

Aletta, Francesco, and Jian Kang. 2015. "Soundscape approach integrating noise mapping techniques: a case study in Brighton, UK." *Noise Mapping* 2 (October): 1–12. https://doi.org/10.1515/ noise-2015-0001,

Aletta, Francesco, and Jieling Xiao. 2018. "What are the Current Priorities and Challenges for (Urban) Soundscape Research?" *Challenges* 9, no. 1: 16–27. https://doi.org/10.3390/challe9010016.

Allen, Mike, ed. 2017. *The SAGE Encyclopedia of Communication Research Methods*. London: SAGE. http://dx.doi.org/10.4135/9781483381411.n369.

Axelsson, Östen, Mats E. Nilsson and Birgitta Berglund. 2010. "A principal components model of soundscape perception." *Journal of the Acoustical Society of America* 128, no. 5 (November): 2836–2846. https://doi.org/10.1121/1.3493436.

Bashō, Matsuo, Reichhold, J. (trans). 1686/2008. *Bashō: The Complete Haiku*. New York: Kodansha USA, Inc.

Bidelman, Gavin M., Skyler G. Jennings and Elizabeth A. Strickland. 2015. "PsyAcoustX: A flexible MATLAB® package for psychoacoustics research." *Frontiers in Psychology* 6 (October): 1498. https:// doi.org/10.3389/fpsyg.2015.01498.

Bones, Oliver, Trevor J. Cox and William J. Davies. 2018. "Sound Categories: Category Formation and Evidence-Based Taxonomies" *Frontiers in Psychology* 9 (July): 1277. https://doi.org/10.3389/ fpsyg.2018.01277.

Brown, A.L., Jian Kang and Truls Gjestland. 2011. "Towards standardization in soundscape preference assessment." *Applied Acoustics* 72: 387–92. https://doi.org/10.1016/j.apacoust.2011.01.001.

Bryant, Antony and Kathy Charmaz, eds. 2019. *The SAGE Handbook of Current Developments in Grounded Theory.* London: SAGE. https://dx.doi.org/10.4135/9781526485656.

Burgess, Matt. 2019. "Listen to the mesmerising sound of London's new electric buses." *Wired*, December 20, 2019.

Cabrera, Densil, Sam Ferguson and Emery Schubert. 2007. "'PsySound3': Software for Acoustical and Psychoacoustical Analysis of Sound Recordings." In *Proceedings of the 13th International Conference on Auditory Display, Montréal, Canada, June 26–29, 2007*, 356–363. USA: International Community for Audio Display Available at: https://tinyurl.com/PsySound3. Accessed August 6, 2020

Challéat, Samuel, Nicolas Farrugia, Amandine Gasc, Jeremy Froidevaux, Jennifer Hatlauf, Frank Dziock, Adrien Charbonneau, et al. 2020. "Silent·Cities." *OSF.* March 17, 2020. https://doi.org/10.17605/OSF.IO/H285U.

Cherryholmes, Cleo H. 1992. "Notes on pragmatism and scientific realism." *Educational Researcher* 14 (August-September): 13–17.
https://www.jstor.org/stable/i250199.

Corbin, Juliet and Anselm Strauss. 2008. *Basics of Qualitative Research (3rd ed.): Techniques and Procedures for Developing Grounded Theory.* London: SAGE.
https://dx.doi.org/10.4135/9781452230153.

Cox, Christoph. 2018. *Sonic Flux: Sound, Art, and Metaphysics.* Chicago: University of Chicago Press.

Creswell, John W. and J. David Creswell. 2017. *Research Design: Qualitative, Quantitative, and Mixed Methods Approaches.* (5th ed.) London: SAGE.

Czarniawska, Barbara. 2001. "Narrative, Interviews and Organizations." In *Handbook of Interview Research*, edited by Jaber F. Gubrium and James A. Holstein, 732–750. London: SAGE. https://dx.doi.org/10.4135/9781412973588.n42.

Davies, William J., Mags D. Adams, Neil S. Bruce, Rebecca Cain, Angus Carlyle, Peter Cusack, Deborah A. Hall, Ken I. Hume, Amy Irwin, Paul Jennings, Melissa Marselle, Christopher J. Plack and John Poxon. 2013. "Perception of soundscapes: An interdisciplinary approach." *Applied Acoustics* 74, no. 2 (February): 224–231. https://doi.org/10.1016/j.apacoust.2012.05.010.

Drever, John L. 2002. "Soundscape composition: the convergence of ethnography and acousmatic music." *Organised Sound* 7, no. 1 (April): 21–27.
https://doi.org/10.1017/S1355771802001048.

Drever, John L. 2009. "Soundwalking: Aural Excursions into the Everyday." In *The Ashgate Research Companion to Experimental Music*, edited by James Saunders, 163–192. Aldershot, UK: Ashgate. PDF download at: https://tinyurl.com/DreverSoundwalk. Accessed August 6, 2020.

Drever, John L. 2019. "Soundscape Composition: Listening to Context and Contingency." In *Foundations in Sound Design for Linear Media: A Multidisciplinary Approach,* edited by Michael Filimowicz, 358–377. Oxford: Routledge.

Eng, Michael. 2017. "The Sonic Turn and Theory's Affective Call." *Parallax* 23, no. 3: 316–329. https://doi.org/10.1080/13534645.2017.1339970.

Epstein, Marcia Jenneth. 2019. "Healing the urban soundscape: reflections and reverberations." *Cities & Health.* Online edition.
https://doi.org/10.1080/23748834.2019.1676628.

Farina, Almo and Stuart H. Gage. 2017. *Ecoacoustics: The Ecological Role of Sounds.* Oxford: Wiley.

Fastl, Hugo and Eberhard Zwicker. 2007. *Psychoacoustics Facts and Models.* Heidelberg: Springer.

Fiebig, Andre and Brigitte Schulte-Fortkamp. 2005. "The exploration of the listener's perceptual reality – The potential of explorative methods with respect to community noise research." In *INTER-NOISE and NOISE-CON Congress and Conference Proceedings, InterNoise05, Aug. 7–10, 2005, Rio de Janeiro, Brazil*, 3471–3475. Reston, VA: Institute of Noise Control Engineering. Available at: https://tinyurl.com/GT-reference. Accessed August 6, 2020

Flick, Uwe, ed. 2014. *The SAGE Handbook of Qualitative Data Analysis.* London: SAGE. http://dx.doi.org/10.4135/9781446282243.

Genuit, Klaus and André Fiebig. 2006. "Psychoacoustics and its Benefit for the

Soundscape Approach." *Acta Acustica united with Acustica* 92, no. 6: 952–958. Available at: https://tinyurl.com/Genuit-Fiebig-2006. Accessed August 6, 2020

Given, Lisa M., ed. 2012. *The SAGE Encyclopedia of Qualitative Research Methods*. London: SAGE. https://dx.doi.org/10.4135/9781412963909.

Goines, Lisa and Louis Hagler. 2007. "Noise Pollution: A Modern Plague." *Southern Medical Journal* 100 (March): 287–294. https://doi.org/10.1097/SMJ.0b013e3180318be5. Accessed August 6, 2020

Goodman, Steve. 2010. *Sonic Warfare: Sound, Affect, and the Ecology of Fear*. Cambridge, MA: MIT Press.

Hainge, Greg. 2013. *Noise Matters: Towards an Ontology of Noise*. London: Bloomsbury Academic.

Herzogenrath, Bernd. 2017. *Sonic Thinking: A Media Philosophical Approach*. London: Bloomsbury Academic.

Ising, H. and B. Kruppa. 2004. "Health effects caused by noise: Evidence in the literature from the past 25 years," *Noise and Health* 6, no. 22: 5–13. Available at: https://tinyurl.com/Noise-Health. Accessed August 6, 2020.

ISO. 2014. *ISO 12913-1:2014 Acoustics – Soundscape – Part 1: Definition and conceptual framework*. Geneva: International Organization for Standardization. Available at: https://tinyurl.com/ISO12913-1. Accessed August 6, 2020.

ISO. 2018. *ISO/TS 12913-2:2018 Acoustics – Soundscape – Part 2: Data collection and reporting requirements*. Geneva: International Organization for Standardization. Available at: https://tinyurl.com/ISO12913-2. Accessed August 6, 2020.

ISO. 2019. *ISO/TS 12913-3:2019 Acoustics – Soundscape – Part 3: Data analysis*. Geneva: International Organization for Standardization. Available at: https://tinyurl.com/ISO12913-3. Accessed August 6, 2020.

Kang, Jian, Francesco Aletta, Truls T. Gjestland, Lex A. Brown, Dick Botteldooren, Brigitte Schulte-Fortkamp, et al. 2016. "Ten questions on the soundscapes of the built environment." *Building and Environment* 108 (2016): 284–294. https://doi.org/10.1016/j.buildenv.2016.08.011.

Kang, Jian and Francesco Aletta. 2018. "The Impact and Outreach of Soundscape Research." *Environments* 5, no. 58 (May). https://doi.org/10.3390/environments5050058.

Kang, Jian, Kalliopi Chourmouziadou, Konstantinos Sakantamis, Bo Wang, Yiying Hao, eds. 2013. *Soundscape of European Cities and Landscapes: COST TUD action TD0804*. Oxford: Soundscape-COST. Available at: https://tinyurl.com/COST-TD0804. Accessed August 6, 2020.

Kang, Jian and Brigitte Schulte-Fortkamp, eds. 2015. *Soundscape and the Built Environment*. Boca Raton, FL: CPC Press.

Kelly, Caleb. 2011. *SOUND*. London: Whitechapel Gallery.

Krause, Bernie. 1987. "Bioacoustics: Habitat Ambience & Ecological Balance." *Whole Earth Review* 57 (Winter): 16–21. Available at: https://tinyurl.com/Krause-Niche1. Accessed August 6, 2020.

Krause, Bernie. 1999. *Loss of Natural Soundscapes Within the Americas*. Glen Ellen, CA: Wild Sanctuary. Available at: https://tinyurl.com/Krause-Niche3. Accessed August 6, 2020.

Krause, Bernie. 2020. "Wild Sanctuary." http://www.wildsanctuary.com. Accessed August 6, 2020.

Lacey, Jordan. 2017. *Sonic Rupture: A Practice-led Approach to Urban Soundscape Design*. London: Bloomsbury Academic.

Laiolo, Paola. 2010. "The emerging significance of bioacoustics in animal species conservation." *Biological Conservation* 143, (2010): 1635–1645. https://doi.org/10.1016/j.biocon.2010.03.025.

Lane, Cathy and Angus Carlyle, eds. 2011. *In the Field: The Art of Field Recording*. London: Uniformbooks.

Lavia, Lisa, Max Dixon, Harry J. Witchel and Mike Goldsmith, 2015. "Applied Soundscape Practices." In *Soundscape and the Built Environment*, edited by Jian Kang and Brigitte Schulte-Fortkamp, 246–301. Boca Raton, FL: CPC Press.

LaBelle, Brandon. 2010. *Acoustic Territories: Sound Culture and Everyday Life*. London: Bloomsbury Academic.

Lercher, Peter and Brigitte Schulte-Fortkamp. 2013. "Soundscape of European Cities and Landscapes – Harmonising." In *Soundscape of European Cities and Landscapes: COST TUD action TD0804* edited by Jian Kang, Kalliopi Chourmouziadou, Konstantinos Sakantamis, Bo Wang, Yiying Hao, 120–127. Oxford: Soundscape-COST. https://doi.org/10.13140/2.1.3030.1127.

Lercher, Peter, Irene van Kamp, Eike von Lindern and Dick Botteldooren. 2015. "Perceived Soundscapes and Health-Related Quality of Life, Context, Restoration, and Personal Characteristics." In *Soundscape and the Built Environment*, edited by Jian Kang and Brigitte Schulte-Fortkamp, 90–105. Boca Raton, FL: CPC Press.

Lichtman, Marilyn. 2014. *Qualitative Research for the Social Sciences.* London: SAGE. https://dx.doi.org/10.4135/9781544307756.

Martin, Brona. 2018. "Soundscape Composition: Enhancing our understanding of changing soundscapes." *Organised Sound* 23, no.1 (April): 20–28. https://doi.org/10.1017/S1355771817000243.

Michel, Jean-Baptiste, Yuan Kui Shen, Aviva Presser Aiden, Adrian Veres, Matthew K. Gray, The Google Books Team, Joseph P. Pickett, et al. 2011. "Quantitative Analysis of Culture Using Millions of Digitized Books." *Science* 331, no. 6014: 176–182. https://dx.doi.org/10.1126/science.1199644.

Nicol, Rozenn. 2010. *Binaural Technology.* New York: Audio Engineering Society.

Oberman, Tin, Kristian Jambrošić, Marko Horvat and Bojana Bojanić Obad Šćitaroci. 2020. "Using Virtual Soundwalk Approach for Assessing Sound Art Soundscape Interventions in Public Spaces." *Applied Sciences* 10, no. 6 (March): 2102. https://dx.doi.org/10.3390/app10062102.

Oliveros, Pauline. 2005. *Deep Listening: A Composer's Sound Practice.* Lincoln, NE: iUniverse

Pijanowski, Bryan C., Luis J. Villanueva-Rivera, Sarah L. Dumyahn, Almo Farina, Bernie L. Krause, Brian M. Napoletano, Stuart H. Gage, Nadia Pieretti. 2011. "Soundscape Ecology: The Science of Sound in the Landscape." *BioScience* 61, no. 3 (March): 203–216 https://doi.org/10.1525/bio.2011.61.3.6.

Schafer, R. Murray. 1969. *The New Soundscape: A Handbook for the Modern Music Teacher.* Don Mills, Ontario: BMI Canada PDF download at: https://tinyurl.com/NewSoundscape. Accessed August 6, 2020.

Schafer, R. Murray. 1970. *The Book of Noise.* Vancouver: World Soundscape Project. Available at: https://tinyurl.com/WSP-Projects. Accessed August 6, 2020.

Schafer, R. Murray, ed. 1973. *The Vancouver Soundscape.* Vancouver: World Soundscape Project. Available at: https://tinyurl.com/WSP-Projects. Accessed August 6, 2020.

Schafer, R. Murray, ed. 1978. *Five Village Soundscapes.* Vancouver: World Soundscape Project. Available at: https://tinyurl.com/WSP-Projects. Accessed August 6, 2020

Schafer, R. Murray. 1994. *The Soundscape: Our Sonic Environment and the Tuning of the World.* Rochester, VT: Destiny Books

Schulte-Fortkamp, Brigitte and André Fiebig. 2015. "Impact of Soundscape in Terms of Perception." In *Soundscape and the Built Environment*, edited by Jian Kang and Brigitte Schulte-Fortkamp, 69–88. Boca Raton, FL: CPC Press.

Scott, John and Peter J. Carrington, eds. 2014. *The SAGE Handbook of Social Network Analysis.* London: SAGE. http://dx.doi.org/10.4135/9781446294413.

Soundcamp. (Maria Papadomanolaki, Dawn Scarfe and Grant Smith). 2019. "Biosphere Open Microphones (BIOM) – Towards a network of remote listening points in the UNESCO Biosphere Reserves." *Soundscape – The Journal of Acoustic Ecology* 18: 23–34. https://tinyurl.com/Soundscape18.

Southworth, Michael. 1967. "The Sonic Environment of Cities." Cambridge, MA: MIT. http://hdl.handle.net/1721.1/102214.

Southworth, Michael. 1969. "The Sonic Environment of Cities." *Environment and Behavior* 1, no.1 (June): 49–70. https://doi.org/10.1177/001391656900100104.

Sterne, Jonathan. 2003. *The Audible Past: Cultural Origins of Sound Reproduction.* Durham, NC: Duke University Press.

Sterne, Jonathan. 2012. "Sonic Imaginations." In *The Sound Studies Reader*, edited by Jonathon Sterne, 1–17. London: Routledge.

Sterne, Jonathan. 2013. "Soundscape, Landscape, Escape." In *Soundscapes of the Urban Past: Staged Sound as Mediated Cultural Heritage*, edited by Karin Bijsterveld, 181–194. Bielefeld: Transcript Verlag. https://www.jstor.org/stable/j.ctv1xxsqf.11.

TfL. 2019. "TfL to trial innovative new bus sound to improve road safety. Accessed August 6, 2020. https://tinyurl.com/TFL-AVAS.

Truax, Barry. 1999. *Handbook for Acoustic Ecology.* Vancouver: World Soundscape Project. Available at: https://tinyurl.com/WSP-Projects. Accessed August 6, 2020.

Truax, Barry. 2001. *Acoustic Communication.* Westport, CT: Ablex.

Truax, Barry. 2002. "Genres and techniques of soundscape composition as developed at Simon Fraser University." *Organised Sound* 7, no. 1 (April): 5–14. https://doi.org/10.1017/S1355771802001024.

Truax, Barry. 2008. "Soundscape Composition as Global Music: Electroacoustic music as soundscape." Organised Sound 13, no. 2 (August): 103–109. https://doi.org/10.1017/S1355771808000149.

Urquhart, Cathy. 2013. *Grounded Theory for Qualitative Research: A Practical Guide.* London: SAGE. https://dx.doi.org/10.4135/9781526402196.

Voegelin, Salomé. 2010. *Listening to Noise and Silence: Towards a Philosophy of Sound Art.* London: Bloomsbury Academic.

Vogiatzis, Konstantinos and Nicolas Remy. 2017. "Soundscape design guidelines through noise mapping methodologies: An application to medium urban agglomerations." *Noise Mapping* 4 (March): 1–19. https://doi.org/10.1515/noise-2017-0001.

Wales. 2018. "Noise and soundscape action plan 2018–2023." https://tinyurl.com/Wales-Noise. Accessed August 6, 2020.

Walk Listen Create. 2020. "walk · listen · create." https://walklistencreate.org/. Accessed August 6, 2020.

WFAE. 2020. "World Forum for Acoustic Ecology." https://www.wfae.net/. Accessed August 6, 2020.

WSP. 2020. "World Soundscape Project." *Simon Fraser University.* https://tinyurl.com/SFU-WSP. Accessed August 6, 2020.

12

Developing a Sound Design Creative A.I. Methodology

Miles Thorogood

1 Introduction

Sound design is essential element in video games, animation and virtual reality to connect elements of a story, create identities, arouse moods and give context to scenes. Pioneering film sound designer Walter Murch (2005) discusses sound design as existing between encoded signals used to convey the meaning of a message and the purely embodied signals given by music that is "experienced directly, without any intervening code." Game sound introduces additional challenges to adapt to on-screen actions, player interaction and respond to the game's non-linear properties. Karen Collins (2008) expresses these concerns relating to important player perceptions to: access and interact with the story: a preparatory function affecting the players decision enhancing the overall structure of the game; and identifying characters, moods, environments and objects. Creating these types of experiences is challenging, and many tasks of sound design are repetitive. An example is listening through an extensive database of audio files, segmenting selected files based on salient characteristics and mixing segments into a cohesive sonic experience. Automating such tasks will facilitate a better workflow – enabling a sound designer to investigate innovative ways of interacting with sonic creativity and generate an expanded set of alternative design solutions.

The sound designer's task involves cycles of considering an existing partial design, comparing it with the design goal, deciding on a transformation to get closer to the goal and then applying that transformation to the partial design. Dorst (1995) describes the design process as *rationalizable* process and *reflection-in-action*. Describing design as a process of reflection-in-action works particularly well in the conceptual stage of the design process, where the designer has no standard strategies to follow and is proposing and trying out problem/solution structures. Lopez et al. (2011) study the cognitive process of designers through protocol analysis using an

DOI: 10.4324/9780429356360-13

experiment with different design sessions, then applying a combination of outcome-based and process-based evaluations. In the study of parametric modeling, Shireen et al. (2011) outline the process of analyzing the designers work by videotaping each participant's computer screen during a task experiment. These videos provided authors with insight into designers' intentions through a chain of iterated actions, along with the design completion time. Unfolding the process in sound design provides indications of how machines may do the work.

2 Sound Design as Creative A.I.

Plut and Pasquier (2020) survey the current state of the art of generative music in video games and demonstrate that generative systems for creating interactive sound for video games are becoming more commonplace. These tools model different aspects of creative tasks ranging from rule-based systems to perceptual goals. As described by Pasquier et al. (2017), systems performing such tasks exhibit behavior considered creative if a similar output could be demonstrated by a human – ranging from generating musical scores to interactive tools in performance and sound design. Such creative systems are designed by either modeling creativity as it could be and evaluating the output, or as it is by encoding specific human processes. Evaluating creative systems is essential for elucidating progress in the research, and different frameworks are designed to evaluate aspects of Creative A.I. systems (Pease and Colton, 2011). Creativity support tools (CST) is a branch of human-computer interaction (HCI) studying systems for assisting in human creativity tasks. A fundamental element of HCI is evaluating systems with the appropriate metrics and evaluation methodology. For example, Candy and Edmonds (1997) identify criteria for evaluating interaction in software systems. The motivating factor of these criteria is to understand the interactions of users with the system in creative applications. Cherry and Latulipe (2014) define another set of criteria, named the Creativity Support Index (CSI), for evaluating computer-assisted creativity tools based on the engagement of users, factoring in immersion, enjoyment and collaboration as criteria.

A survey of generative sound design systems (Thorogood, 2019) classifies the different approaches to automating creative tasks. These approaches range from simulating an environment as the sound sources in that space, to highly abstracted representations taking a more artistic approach in terms of perceptual criteria. Developing computational systems for generating rich interactive sound for multimedia, Pietrocini and Lopa (2019) outline the creative approach in sound design as a systematic *listening* and *task execution* pipeline. Hendricks and McPherson (2016) explore observing sound designers in situ in the design of computational audio systems for interactive applications, noting the importance of perceptual and embodied criteria in the work. As evidenced in the literature it is thus necessary to model sound perception for computational systems to synthesize these creative behaviors.

Towards establishing the perceptual concepts to be analyzed and modeled, sensory evaluation techniques (Meilgaard, 2006) study how people perceive the particular properties of a stimulus. Statistical methods modeling the stimulus and participant responses then result in a set of terms that describe the sensory space under investigation. As a specific area of research, sound studies aims to identify subjective descriptions of sound in different environments through survey question-naires and spontaneous descriptions, noting subjects' affective responses (Axelsson et al., 2010). These methods are well suited to sound design research for eliciting concepts to be represented in a computer program simulating human-like behavior.

A corpus of annotated audio files is essential for developing machine learning models with predictive capabilities to make sound design decisions in Creative A.I. systems. In the affective computing community, affective rating instruments are tools for collecting annotations for a stimulus. Such tools find application in video, music, speech, movement and soundscape emotion recognition (Fan et al., 2016). Due to the different contextual situations and cultural backgrounds, each level's meaning on a rating scale may change across annotators. Therefore, both ratings from different annotators and the same annotator may not be consistent. Ranking is an alternative approach for eliciting responses from subjects that circumvents many reliability problems (Yang and Chen, 2010). In either approach, the aim for sound design research is to obtain a set of labelled audio examples for analysis and model-ing to endow machines with human-like listening.

2.1 Audio Features

Audio features common to environmental sound classification are time-domain features such as zero-crossing rate, RMS, loudness, and frequency domain features, including Mel Frequency Cepstral Coefficients (MFCC) and spectral centroid. Loudness and MFCC have proven useful audio features for the classification of environmental sound. Used by Hall et al. (2013), loudness is a measure for rating the vibrancy and pleasantness of urban settings recordings. McKinney and Breebaart (2003) have shown that perceptual features, including loudness, are useful audio features in the classification of non-musical sounds, such as speech, noise and crowd noise. Loudness is a key feature used by Oldoni et al. (2010) to mimic a human psy-chological system for processing soundscapes.

Loudness is the characteristic of a sound associated with the sensation of inten-sity. The human auditory system affects the perception of the power of different fre-quencies. A model of loudness is provided by Zwicker and Fasl (2013), which takes into account the disparity of loudness at different frequencies along the Bark scale, which correspond to the critical bands of hearing. Specific loudness is the loudness associated with each Bark band (Peeters, 2004), with the total loudness being the sum of particular loudnesses in all bands.

Another audio feature that is effective for sound classification is the Mel-frequency Cepstral Coefficients (MFCC). MFCC represents the shape of the sound by a small number of coefficients from a short-time spectrum of an audio signal spaced along the Mel scale – a non-linear scale of pitches that models the response of human auditory system behavior (Peeters, 2004). A comparison of different implementations of MFCC is presented by Zheng (2001). The common representation of the MFCC is of 40 filter bank values, linearly spaced filters at lower frequencies, followed by logarithmically spaced filters at higher frequencies. The logarithm of the magnitude of each filter bank output is computed, and, finally, the discrete cosine transform is calculated for the filter bank values. Using MFCC features, Ma et al. (2006) demonstrated a high level of classification accuracy. Categories in their research are background sound in a fixed set of noise environments, including bar, beach, bus, car, office, street and railway. Classification of background and separation of foreground for environmental monitoring is discussed by Moncrieff, Venkatesh and West (2007). Again, using MFCC, their adaptive background sound model updates over time, allowing for the extraction of foreground sounds that deviate from the background model. Context-dependent modeling of the background using MFCC is also given by Chu et al. (2009). They were computing the audio features of the audio file results in a set of values for each analysis window.

In most cases, a summarization of the features representing the audio file is used. The bag-of-brames (BOF) is a method of collapsing multiple feature vectors by a statistical distribution of the individual overlapping short time frames. It considers structures that represent a signal have possibly different values, and the aggregation of the frames provides a more useful representation than a singular frame. Frames are aggregated using various statistical methods, such as the mean and standard deviation of features, or more complex modeling such as Gaussian Mixture Models. Parameterizing a BOF approach can be done in many ways: frame size and overlap; number of features and feature vector dimensions; statistical reduction methods. The vector containing the summarized features is then used to train a supervised classifier. The bag-of-frames (BOF) approach is employed by Aucouturier et al. (2007) using an MFCC feature vector for the general classification of environmental audio contexts. They see this approach as a holistic similarity between sounds, such as 'sounds like parks' rather than 'sounds like an airport.' Results using the BOF technique showed the classifier had a near-perfect accuracy. The BOF approach considers longer durations of audio data and presents an attractive model for sound design classification tasks.

2.2 Sound Classification

A critical function of a Creative A.I. system is to make decisions on selecting and manipulating audio files. Sound classification and prediction aim to label an audio signal automatically. A range of approaches is suited to model audio signals by testing and ranking various audio features, classifiers and windowing options (Fu, 2010). Classification techniques are parameterized by the number of categories and category labels, number of audio features and the classifier used to model them. Prior work in audio classification has focused on evaluating systems by adapting metrics of established MIR methods, such as precision, recall, accuracy and F-Measure (Downie, 2010). Galliano et al. (2009) describe these measures for the ESTER evaluation criteria as the aggregate duration of inserted class events relative to ground truth segment boundaries, and where they occur (recall) and where they are detected (precision). These measures provide the best-in-class indicators for classification and segmentation algorithms.

One such classification algorithm is Support Vector Machines (SVM). SVM have been widely used in environmental sound classification problems and consistently demonstrated good classification accuracy. An SVM is a non-probabilistic classifier that learns optimal separating hyper-planes in a higher-dimensional space from the input, representing a non-linear decision boundary in the original space, discussed in detail Cortes et al. (1995). The sequential minimal optimization (SMO) algorithm is state of the art for optimally training the SVM and replaces missing values and output coefficients after first normalizing the data, as covered by Witten and Frank (2002). Excellent classification accuracy was achieved by Barkana and Burak (2010) using an SVM classifier for environmental sounds. They found that the SVM outperformed a k-means clustering classifier by a significant amount. Similarly, results by Chen et al. (2006) show that an SVM classifier consistently outperformed other classifiers for environmental sound, speech and mixed sounds. They posit that an SVM classifier is better suited to hybrid and ecological sound because of its characteristic of performing well on non-linearly separable classes. For the application of detecting complex sound scenes in film audio, Moncrieff et al. (2001) employ an SVM classifier trained using the SMO algorithm. From results, they found a high level of classification accuracy achieved using the SVM.

3 A Sound Design Research Case Study

When developing ambiances, the sound designer's role is to produce different audio elements to generate a mix that aligns with the concepts in an environment or script. For example, to create the ambiance for a spooky scene of a cat walking up the stairs of an old house requires a combination of sounds of a haunting nature. To create this mix, the sound designer makes decisions when selecting audio files or segments to situate the viewer. In this selection behavior, there are multiple levels of signification the sound designer is actively listening for in the audio. The first level of the search includes semantic criteria. That is, does the audio file contain sounds of objects represented in the environment. From our example, are there the sounds of a cat walking on a wooden stairs? The second level of search involves the feeling of the scene – is the sound spooky? Finally, the sound designer will listen to the sound to fit their particular artistic sense for the composition.

Typically, a user searches a computer for audio files at the semantic level through the file and folder structures of a standard WIMP interface. In this system, the user will name folders and files to create an indexing system for retrieving files at a later time. The standard means of creating a semantic index is done by adding tags to the recording in a database system. One such way to organize different audio files in using a taxonomy, a survey of which is detailed in Thorogood and Pasquier (2019), where sounds are given categories and the folder structure can be traversed to locate a particular file. For example, using the Urban Sound taxonomy (Salamon et al., 2014), the file "cat walking.wav" would be found in the directory /nature/animals/domestic/cats/cat-walking-inside.wav.

At the second-level criteria, a sound's expression of mood is a subjective perception of what the listener feels when presented with a recording. Often at the sound designer's judgment, an audio file is appended with a particular mood label (e.g., cat-walking-inside-spooky.wav). One question that arises from the self-evaluation of mood is if other listeners would also find the sound to have the same mood. In the sound design creative practice, this may not matter that much – as the artistic sensibilities of the designer makes the sound make sense. On the other hand, the idea of mood is more generalizable to a population of people. One means of making such generalizable subjective judgments is to ask many people to listen to the recording and rate the mood they feel the recording conveys. Such a rating could take on the form of a list of feelings (e.g., happy, neutral, spooky) that the listener would select from, or a range of how much the audio conveys a particular emotion (e.g., not spooky<1–10>spooky) that they would mark. After collecting many such responses, one then looks at the results and applies the mood label that is generally agreed upon (e.g., cat-walking-inside-spooky.wav, or cat-walking-inside-spooky_0.9.wav).

The third level of search criteria is based upon the individual sound designer's aesthetic style. Like the mood-based criteria, the sound designer will apply the label to an audio file related to the quality of the sound. For example, a sound designer wants to classify audio recordings that have the quality of 'sizzle.' We use sizzle here to indicate the character of a sound that has a high level of brightness and presence. Like the mood criteria, aesthetic criteria can be either scaled to a population of listeners or stay at the designer's subjective scope. Appending the particular aesthetic quality to the file name, we now have cat-walking-inside-spooky_0.9-sizzle_0.6.wav (and on it goes).

In the case where there are a more significant number of files and labels, an SQL database is better suited for the search and retrieval of audio files. In an SQL database, files are indexed in a table based on appropriate search terms that the database system will retrieve after given a query containing those terms. In our example, the SQL database table will have a column for semantic, sentiment and aesthetic terms. The sound designer can then make a query in a text-based interface [Semantic: cat walking] [Sentiment: spooky] [Aesthetic: sizzle]. After submitting the query, the database will return with suggestions that match that criteria, if any. Once audio files are retrieved from the file system or database, a sound designer then gets to the work of sequencing and manipulating sounds to align with the script or description of the environment to be represented with the soundtrack. The process of manually tagging recordings provides a strong indication of the semantic, sentiment and aesthetic elements in the recording. However, as collections of recordings become large, the process of tagging becomes long and automating this procedure has the advantage of reducing listening fatigue and speeding up the tagging process.

Altogether, the process of listening to audio recordings, labelling and indexing these into a database is an essential but time-consuming and repetitive task. When the amount of recordings increases, as when curating an existing library of recordings or processing one's own recordings, this task costs an unreasonable amount of time. To speed up this process, we will develop a computer program capable of analyzing, labelling and indexing a sound file. To break this program down, this chapter demonstrates a procedure of developing supervised learning models for predicting high-level audio features used in sound design. In short, the process is to create a corpus of audio files labelled by a human. Then, using audio feature extraction techniques to decompose the complex audio signal into relevant representations, to train a machine learning algorithm with these labels and features. Finally, to use the model to classify or predict new sounds. To show the procedure, we will go step-by-step through an example of a sound designer wanting to index a database of sound files based on an aesthetic criterion.

3.1 Creating the Corpus

A corpus of audio files is a structured collection of files labeled with a set of tags. These tags are typically semantic elements of the sounds present in the recording. Still, any labelling can be applied, such as 'sentiment,' or in our example, the aesthetic indicators called sizzle. It is often the case that a recording will be long and change over time. Ideally, the corpus contains short duration and acoustically consistent examples to feed samples to the machine learning algorithm. The optimal duration will allow a person labelling the sample to adequately perceive the stimulus while at the same time reducing the computational cost of extracting features later on. As demonstrated through empirical studies (Thorogood et al., 2016), audio examples with a consistent acoustic profile of between four and six seconds find a good balance between listening potential and cost.

Another essential property of the corpus is its size: how many audio files it contains. A more significant number of files is superior for the machine learning tasks. If the number too small, then the machine learning algorithm will be good at classifying those files but not others – a conundrum called over-fitting that reduces the search space within the coverage of those examples. Therefore, we aim for a more significant number of audio files so that the learning algorithm will have a higher diversity of examples that cover more of the search space. As such, it will be more robust when presented with a greater range of sounds. The number of examples to feed to the algorithm will depend on what algorithm is chosen. In the case of Artificial Neural Networks, training sets of 1,000 is modest. If samples are diverse, then for linear models, such as multiple linear regression, a sample size of 100 is adequate.

To build the corpus, start by gathering examples from a collection of recordings to label and log in a spreadsheet such as a .csv file. The labelling depends on the type of machine learning model that will be used. For a classification problem, the corpus will contain many examples of what we deem to have sizzle, and many examples with no sizzle. On the other hand, regression gives us the quantity of a property we are looking at. That is, a particular sound expresses a percentage of sizzle. As such, the corpus will contain examples that are labelled with the amount of perceived sizzle between 0.0 for no sizzle and 1.0 for a high level. The filename of the recording and value are stored in a table, such as a.csv file used for later analysis.

To assist in labelling recordings, a simple tool for quickly reviewing and applying a value to sounds is useful – especially if the dataset is large and entering values into a table takes an unreasonable amount of time (figure 12.1). After listening and annotating the audio recordings using a slider, the table with columns for file and value is copied and pasted into a.csv file.[1]

3.2 Audio Feature Extraction

Once the corpus has been created, we need to collect some audio properties for the machine learning algorithm to model the sizzle category. Audio feature extraction is the process of decomposing a complex sound signal into properties that represent particular characteristics that we can present to a machine learning algorithm. Familiar properties such as pitch and loudness are audio features that are calculated from analyzing a sound file. Music Information Retrieval (MIR) research has developed many other features to represent spectral, perceptual, rhythmic and statistical properties of sound. The simplest way of extracting audio features is using one of the standard toolboxes used in MIR research. One such toolbox that is well supported is the MTG software Essentia (Bogdanov et al., 2013). The process of audio feature extraction involves reading an audio file in chunks of a particular number of

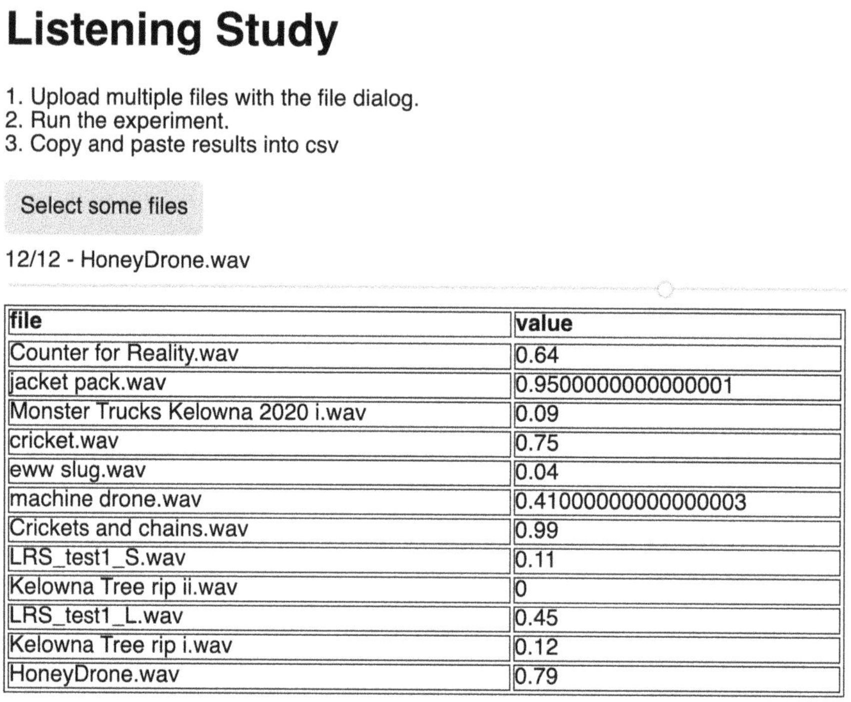

Listening Study

1. Upload multiple files with the file dialog.
2. Run the experiment.
3. Copy and paste results into csv

Select some files

12/12 - HoneyDrone.wav

file	value
Counter for Reality.wav	0.64
jacket pack.wav	0.9500000000000001
Monster Trucks Kelowna 2020 i.wav	0.09
cricket.wav	0.75
eww slug.wav	0.04
machine drone.wav	0.41000000000000003
Crickets and chains.wav	0.99
LRS_test1_S.wav	0.11
Kelowna Tree rip ii.wav	0
LRS_test1_L.wav	0.45
Kelowna Tree rip i.wav	0.12
HoneyDrone.wav	0.79

Figure 12.1   Annotation tool to assist in labelling recordings

samples (typically ~20ms worth), applying a windowing function that smooths out the edges of the audio fragment, calculating the specific features on this chunk, then moving along the length of the file and repeating this process. The chunk size and windowing function have an effect on the feature values and can be modulated. As the algorithm moves along the file's length and the sound changes, each audio chunk will have different values for the audio features. As such, the mean and standard deviation of all the audio features is calculated, a procedure called the bag-of-frames approach (Su et al., 2014). Upon extracting features from items in the corpus, the corresponding set of audio features are appended to the.csv file for further analysis.[2]

3.3 Audio Feature Selection

The selection of features is critical to the ability of the machine learning model effectively to model the corpus. The incorrect choice or overabundance of features will result in a model that makes flawed classifications or erroneous sound predictions. To select the most relevant features to the predictive model, we can either make a best guess on the type of features to fit the model or conduct an automatic feature selection step. While we can make a pretty good guess at features once we are familiar with the flavor and characteristics they represent, we will not necessarily have the best-case scenario for the particular model or data set. Recursive Feature Elimination is one automatic feature selection method that builds and tests numerous machine learning models with different combinations of features. Another approach is Univariate Feature Selection that tests individual features, and the top k scoring features are kept.

3.4 Training and Evaluating Machine Learning Models

Training a machine learning model involves fitting a model using known examples to make predictions on new audio data. The examples we give to the model are the rows of features and labels in the working .csv. New data is shown in the form of audio features extracted from new audio clips. The model will then output a label (a category in the case of a classifier, or a range of values in a regression algorithm).

Classification and regression are two families of machine learning tasks used in sound design research to apply labels to audio files. In the classification case, the algorithm will tell us what category an audio file belongs to (sizzle, or not sizzle). Such a classifier is named a binary classifier (some examples of this type of classifier are neural networks, Support Vector Machines and decision trees). This type of algorithm also works for a broader set of multi-class classification problems. Suppose we have a bunch of categories (sizzle, muddy, rough). In that case, a multi-class binary classifier methodology will decide the category the audio file belongs to. On the other hand, regression gives us the quantity of a property we are looking at. For example, how much a particular sound expresses an amount of the perceived sizzle. Linear regression and Support Vector Regression are two such regression algorithms used in machine learning tasks (for a survey of supervised learning algorithms see Aly, 2005).

When training, different machine learning algorithms have configuration variables required by the model when making predictions. Their values define the behavior of the model when fitting. It is essential to review these variables when implementing a model as they affect its predictive capabilities. A review of different machine learning algorithms reveals the specific configuration steps. One type of machine learning algorithm widely used in music information retrieval is the Support Vector Machines. Support Vector Machines is a binary non-probabilistic linear classifier that learns the optimal separating hyper-plane with the maximum margin of the data from two classes. Non-linear decision boundaries, common with complex sounds, can be represented linearly in a higher dimension space than the input space with a kernel function. In the example code accompanying this chapter, we implement a Support Vector Regression algorithm (SVR) with an RBF kernel. Other kernels will provide different results and should be investigated when experimenting on the type of model you are to use. The configuration variables for the

SVR with RBF kernel are *C* and *gamma*. The *C* value defines the simplicity of the model parameters, with a lower cost making a simpler model at the training samples' misclassification. The *gamma* value determines how much influence a single sample has on the model parameters.

The model parameters are tuned to optimize the model accuracy. When evaluating the model, it is good practice to have a training set and a test set. The training set fits the model's parameters, and the test set is used to evaluate and tune the model. If we were to use the entire corpus to train and validate the model, we would run into overfitting. Overfitting is a situation where the model is only good at making predictions on this corpus and not new data. One approach to dividing the corpus into training and test sets for evaluating the model is *k-fold cross-validation. k-fold* is typically used in conditions where the sample size is limited. The technique involves splitting the corpus into *k* number of equally sized groups. For each group, hold as the test set, train and evaluate the model with the remaining groups and evaluate the test model. The results from the evaluations are then summarized and taken as a result. For generously sized corpora, keeping two-thirds for the training set and leaving the remainder for testing is standard practice. In either approach, the goal, in the case of regression, is to minimize the mean squared error (MSE), which is a calculation of the difference between the estimated values and the actual values. The lower the MSE, the higher the prediction accuracy of the model. Once optimized, the model parameters are saved to file to be used in deployment for predicting new audio examples.[3]

3.4. Deployment

In sound design, automatically predicting sounds significantly reduces the amount of time used for labelling and retrieving audio recordings from a database. When building a sound library, the developer can use the model to insert values into an SQL table. A user queries the database with these search criteria (for a description of audio databases, see Thorogood and Pasquier, 2019). Automatic prediction also gives search capabilities to generative systems. These systems need to decide what sounds to select and manipulate based on sound design criteria (for a breakdown of generative sound design systems, see Thorogood, 2019). In the accompanying code example to this chapter, the trained model analyzes and predicts a directory of audio files. It should be noted that the audio file is summarized in its entirety. Large changes in audio perspective are not captured. Implementation to predict the sizzle at different points involves scanning the recording with the model and noting value at discrete points. As it is, the code will average the values, which is not an issue if the audio signal is relatively stable. The predicted sizzle value and file name are logged to a.csv file. To retrieve the files with the highest amount of sizzle, load the.csv file into a spreadsheet editor (such as LibreOffice) and sort in descending order.

4 Conclusion

Listening to the minutia of detail in large libraries of audio recordings to retrieve those that align with an aesthetic goal is very time consuming. State-of-the-art systems that can generate sound design solutions need to have the ability to represent listening criteria so that they can make human-like decisions. The answer to these challenges is to develop machine learning models to analyze and predict high-level features in recordings that will enrich sound libraries and endow machines' with sound design specific search criteria. This chapter has demonstrated the procedure for developing new high-level aesthetic audio features with a music information retrieval approach. This procedure includes creating a labelled corpus of audio examples, extracting low-level audio features, selecting features that best represent the corpus and training and evaluating a machine learning model. The final step of deploying the model in production environments points toward how this technology finds its place incorporated into audio libraries and generative systems. The future vision for developing new high-level audio features is in sound design production environments. Designers and machines collaborate in dynamic and artistic workflows. Digital audio workstations will make aesthetically driven recommendations. Non-linear environments, such as game engines, will adapt sound designs with real-time player interaction. In the future, machines will do the work while creators move through the artistic space of possibilities.

Notes

1 The annotation tool in figure 12.1 and corpus can be downloaded from https://digitalmedia.ok.ubc.ca/sound_research/ and https://digitalmedia.ok.ubc.ca/sound_research/sizzle_corpus.zip respectively.
2 The code feature Extractor.py extracting audio features and adding to the .csv fie from the listening study can be downloaded from https://github.com/aume/SoundDesignResearch
3 Code for modelBuilder.py incorporating feature selections and model training/evaluation can be downloaded from https://github.com/aume/SoundDesignResearch. The output from this program is a SVR model and feature list to be used in deployment.

References

Aly, Mohamed. "Survey on multiclass classification methods." *Neural Networks* 19 (2005): 1–9.
Aucouturier, Jean-Julien and Boris Defreville. "Sounds like a park: A computational technique to recognize soundscapes holistically, without source identification." 19th International Congress on Acoustics. 2007.
Axelsson, Östen, Mats E. Nilsson and Birgitta Berglund. "A principal components model of soundscape perception." *The Journal of the Acoustical Society of America* 128.5 (2010): 2836–2846.
Barkana, Buket D., and Inci Saricicek. "Environmental noise source classification using neural networks." 2010 Seventh International Conference on Information Technology: New Generations. IEEE, 2010.

Bogdanov, Dmitry, et al. "Essentia: An audio analysis library for music information retrieval." Britto A., Gouyon F., Dixon S., editors. 14th International Society for Music Information Retrieval Conference (ISMIR), Nov. 4–8, 2013, Curitiba, Brazil, p. 493–498.International Society for Music Information Retrieval (ISMIR), 2013.

Candy, Linda, and Ernest A. Edmonds. "Supporting the creative user: a criteria-based approach to inter-action design." *Design Studies* 18.2 (1997): 185–194.

Chen, Lei, Sule Gunduz and M. Tamer Ozsu. "Mixed type audio classification with support vector machine." 2006 IEEE International Conference on Multimedia and Expo. IEEE, 2006.

Cherry, Erin, and Celine Latulipe. "Quantifying the creativity support of digital tools through the creativity support index." *ACM Transactions on Computer-Human Interaction (TOCHI)* 21.4 (2014): 1–25.

Chu, Selina, Shrikanth Narayanan and C-C. Jay Kuo. "A semi-supervised learning approach to online audio background detection." 2009 IEEE International Conference on Acoustics, Speech and Signal Processing. IEEE, 2009.

Collins, Karen. *Game sound: an introduction to the history, theory, and practice of video game music and sound design.* MIT Press, 2008.

Cortes, Corinna, and Vladimir Vapnik. "Support-vector networks." *Machine learning* 20.3 (1995): 273–297.

Cunningham, Stuart, et al. "Supervised machine learning for audio emotion recognition: Enhancing film sound design using audio features, regression models and artificial neural networks." *Personal and Ubiquitous Computing* (2020): 1–14.

Dorst, Kees. "Analysing design activity: new directions in protocol analysis." *Design Studies* 2.16 (1995): 139–142.

Downie, J. Stephen, et al. "The music information retrieval evaluation exchange: Some observations and insights." *Advances in music information retrieval.* Berlin, Heidelberg: Springer, 2010. 93–115.

Fan, Jianyu, Miles Thorogood and Philippe Pasquier. "Automatic soundscape affect recognition using a dimensional approach." *Journal of the Audio Engineering Society* 64.9 (2016): 646–653.

Fu, Zhouyu, et al. "A survey of audio-based music classification and annotation." *IEEE transactions on multimedia* 13.2 (2010): 303–319.

Galliano, Sylvain, Guillaume Gravier and Laura Chaubard. "The ESTER 2 evaluation campaign for the rich transcription of French radio broadcasts." Tenth Annual Conference of the International Speech Communication Association. 2009.

Hall, Deborah A., et al. "An exploratory evaluation of perceptual, psychoacoustic and acoustical properties of urban soundscapes." *Applied Acoustics* 74.2 (2013): 248–254.

Heinrichs, Christian, and Andrew McPherson. "Performance-led design of computationally generated audio for interactive applications." *Proceedings of the TEI '16: Tenth International Conference on Tangible, Embedded, and Embodied Interaction.* 2016.

Lopez-Mesa, Belinda, et al. "Effects of additional stimuli on idea-finding in design teams." *Journal of Engineering Design* 22.1 (2011): 31–54.

Ma, Ling, Ben Milner and Dan Smith. "Acoustic environment classification." *ACM Transactions on Speech and Language Processing* (TSLP) 3.2 (2006): 1–22.

McKinney, Martin, and Jeroen Breebaart. "Features for audio and music classification." *Proceedings of International Symposium on Music Information Retrieval.* (2003).

Meilgaard, Morten C., B. Thomas Carr and Gail Vance Civille. *Sensory evaluation techniques.* CRC press, 2006.

Moncrieff, Simon, Chitra Dorai and Svetha Venkatesh. "Detecting Indexical Signs In Film Audio For Scene Interpretation." *ICME.* 2001.

Moncrieff, Simon, Svetha Venkatesh and Geoff West. "Online audio background determination for complex audio environments." *ACM Transactions on Multimedia Computing, Communications, and Applications* (TOMM) 3.2 (2007): 8ff.

Murch, Walter. "Dense clarity–clear density." *The transom review* 5.1 (2005): 7–23.

Oldoni, Damiano, et al. "Computational soundscape analysis based on a human-like auditory processing model." 1st. European Acoustics Association EAA-EuroRegio 2010: Congress on Sound and Vibration. Slovenian Acoustical Society (SDA), 2010.

Pasquier, Philippe, et al. "An introduction to musical metacreation." *Computers in Entertainment* (CIE) 14.2 (2017): 1–14.

Pease, Alison, and Simon Colton. "On impact and evaluation in computational creativity: A discussion of the Turing test and an alternative proposal." *Proceedings of the AISB symposium on AI and Philosophy*. Vol. 39. 2011.

Peeters, Geoffroy. "A large set of audio features for sound description (similarity and classification) in the CUIDADO project (Cuidado project report)." *Paris, France: Institut de Recherche et de Coordination Acoustique Musique (IRCAM)* (2004).

Pietrocini, Emanuela, and Maurizio Lopa. "Music: Creativity and New Technologies. A Systemic Approach Towards Multimedia Project and Sound Design." *Systemics of Incompleteness and Quasi-Systems*. Cham: Springer, 2019. 117–131.

Plut, Cale, and Philippe Pasquier. "Generative music in video games: State of the art, challenges, and prospects." *Entertainment Computing* 33 (2020).

Salamon, Justin, Christopher Jacoby and Juan Pablo Bello. "A dataset and taxonomy for urban sound research." *Proceedings of the 22nd ACM international conference on Multimedia*. 2014.

Shireen, Naghmi, et al. "Design Space Exploration in Parametric Systems: Analyzing effects of goal specificity and method specificity on design solutions." *Proceedings of the 8th ACM conference on creativity and cognition*. 2011.

Su, Li, et al. "A systematic evaluation of the bag-of-frames representation for music information retrieval." *IEEE Transactions on Multimedia* 16.5 (2014): 1188–1200.

Thorogood, Miles. "Soundscape Generation Systems." *Foundations in Sound Design for Interactive Media: A Multidisciplinary Approach*. Routledge. 2019, 259.

Thorogood, Miles, Jianyu Fan and Philippe Pasquier. "Soundscape audio signal classification and segmentation using listeners perception of background and foreground sound." *Journal of the Audio Engineering Society* 64.7/8 (2016): 484–492.

Thorogood, Miles, and Philippe Pasquier. "Soundscape Online Databases State of the Art and Challenges." *Foundations in Sound Design for Interactive Media: A Multidisciplinary Approach*. Routledge. 2019, 333.

Witten, Ian H., and Eibe Frank. "Data mining: practical machine learning tools and techniques with Java implementations." *Acm Sigmod Record* 31.1 (2002): 76–77.

Yang, Yi-Hsuan, and Homer H. Chen. "Ranking-based emotion recognition for music organization and retrieval." *IEEE Transactions on audio, speech, and language processing* 19.4 (2010): 762–774.

Zheng, Fang, Guoliang Zhang and Zhanjiang Song. "Comparison of different implementations of MFCC." *Journal of Computer science and Technology* 16.6 (2001): 582–589.

Zwicker, Eberhard, and Hugo Fastl. *Psychoacoustics: Facts and models*. Vol. 22. Springer Science & Business Media, 2013.

13

Sonification Research and Emerging Topics

Sandra Pauletto and Roberto Bresin

1 Introduction

The ability to make sense of sounds is generally taken for granted (McAdams and Bigand 1993). Humans can distinguish and make sense of a large set of musical and everyday sounds. Research that contributes to the understanding of how sound can convey information and meaning is interdisciplinary, involving expertise from areas such as cognitive science of human audition (McAdams and Bigand 1993), music, sound and emotions (Peretz and Zatorre 2003; Juslin and Sloboda 2011; Eerola and Vuoskoski 2013), everyday listening (Gaver 1993; Guastavino 2007) and acoustic communication (Truax 2001), to name a few.

Sound design is the process of creating non-speech audio to make an intention audible (Susini 2011). Designed sounds communicate, broadly, two types of intention: function and form. The sound of a car door being shut, for example, tells us simultaneously that the door is now shut (function), and what kind of car (expensive, old) we are listening to (form). This information needs to be clearly heard and correctly interpreted for the design to be considered successful. Creating sound effects for films and games is a sound design process, as much as creating an auditory icon, for example the computer's trash bin sound, or transforming an athlete's movement data into sound (sonification) in order to create an auditory feedback of that performance.

In this chapter, we will be focusing on a specific type of sound design research: sonification research. In contrast to other kinds of sound design, sonification implies the existence of a dataset that someone – a scientist, a teacher, an athlete, a designer, an artist – aims to portray through sound.

Displaying data through sound can satisfy a number of functions (Buxton 1989; Kramer 1994; Walker and Kramer 2004) such as monitoring a dynamic process in real-time, highlighting trends in large datasets, producing a public engagement or educational tool or creating an artistic experience.

DOI: 10.4324/9780429356360-13

Sonification designs can have a variety of forms which adhere to traditional musical structures, are reminiscent of existing everyday sounds, or are highly abstract. Different sonification forms will be appropriate for different functions, and will be considered relevant by different audiences in different contexts.

Reasons for using an auditory display such as sonification include improving accessibility for visually impaired users, providing more channels of information (the auditory channel in addition to the visual one, for example) to facilitate multitasking. We can find applications of sonification in many fields including assistive technologies, health, environmental science, mobile computing, intelligent alarms and many more.

In this chapter we will provide a short overview of sonification research, its historical background and main aim, and then we will look at more recent developments and discussions followed by short sections dedicated to five emergent themes in the field.

2 A Very Brief History of Sonification

The use of sound to convey information predates the modern era (Worrall 2018). But the most iconic tools that have successfully transformed information into sound are the stethoscope, invented in 1816 by French physician René Laennec to be able to listen to bodily sounds (Sterne 2003; Rice 2010), and the Geiger counter, developed in 1928 to hear the presence of radiation (Korff 2013). Both tools, and the sounds that they produce, have entered the imagination of global audiences through the use of these sounds in media such as cinema (Reddell 2018). A less well-known historical auditory display, that was re-discovered more recently in sonification research, was developed in the 1960s by psychoacoustician Sheridan Speeth, who transposed seismic signals into the audible domain in order to distinguish earthquakes from underground nuclear explosions (Volmar 2013). Furthermore, sound has a long tradition of being used as a way to monitor complex temporal processes in some professions, for example, in the automobile (Krebs 2012) and other industries (Mody 2005; Bijsterveld 2006). Car mechanics still use listening as a means of diagnosing faults, and workers in the shop-floor of, for example, textile companies used the sound of the machine to monitor its functionality even if that was at the expense of their hearing.

With the advent of computing, many processes became silent. The transition from noise to silence, welcomed by many, came however with its own pitfalls such as the lack of awareness of computing processes taking place around us and of gestures' feedback from digital interfaces. The need for the creation of computer sounds emerged initially in the form of auditory icons, i.e., sounds for specific discrete computing events, rather than in the form of sonification of datasets. Two important works in this area appear in the 1980s (Gaver 1989; Bly 1982). For a more complete overview of sound in computing see Robare and Forlizzi (2009).

While sonifications are not necessarily digital, there is no doubt that the process of digitization, which, since the late 1990s has affected many industries, aspects of life and society, has contributed to the rapid development of sonification research. The first International Conference on Auditory Display (ICAD) took place in 1992 at the Santa Fe Institute[1] "after nearly two years of searching for the few researchers focused on auditory displays" (Kramer 1996). The first book framing sonification and auditory displays research followed in 1994 (Kramer 1994). In 1999, in the *Sonification Report: Status of the Field and Research Agenda*, Kramer formulated a long-lasting definition of sonification as "transformation of data relations into perceived relations in an acoustic signal for the purposes of facilitating communication or interpretation" (Kramer et al. 1999). This definition, in subsequent years, has been expanded and clarified by a number of authors (Hermann 2008; Worrall 2009). The *Sonification Report* not only further defined the field, it also represented a milestone for sonification research, as an agenda for future work was set for the first time.

In addition to the ICAD conference, other research initiatives developed over the years including the Interactive Sonification Workshop[2] (ISon), which started in 2004; the Sonification Handbook (Hermann et al. 2011); the Conference on Sonification of Health and Environmental Data – SoniHED (Pauletto et al. 2014); the Sonic Interaction Design EU Cost Action (2007–2011), which included work on sonification (Rocchesso 2011), and a number of journal special issues on sonification and music (Schedel and Worrall 2014), data sonification and sound design in interactive systems (Pauletto et al. 2016) and interactive sonification (Yang et al. 2019).

3 The Aim of Sonification

We use sound in our everyday life to monitor our surroundings and to understand both rationally and emotionally what is going on around us. When we cross a road, hear the movements of people in the house or hear the wind and rain, we react and make decisions on what to do next on the basis of what we hear. Everyday sounds are full of information. Humans, for example, are usually able to tell if a person is male or female, young or old, light or heavy, calm or agitated, walking or running from the sound of their footsteps alone. We also "hear" the shoe sole's material as well as the floor's, and whether the person is wearing flat or high heel shoes (Visell et al. 2009; Giordano et al. 2014). Similarly, from a knock on a door we can perceive the emotions of the knocker (Houel et al. 2020) and other characteristics. So even the most mundane sounds can be meaningful. Sonification aims to communicate, through sounds, properties and aspects of a dataset in a similarly meaningful way.

Additionally, since a dataset is usually the numerical representation of something (a process, a system, an action), its sonification is a portrayal of the process, system or action that generated the dataset.

There are two main types of sonifications. Data sonification is the display of a data set in the form of a sound file that can be played back. This kind of sonification is a fixed sonic representation of data. Interactive sonification allows dynamic interaction with the sound and it places "the human user being tightly embedded within an interactive control loop for exploring data sets using sound" (Hunt et al. 2004) at the center of its design. Examples of interactive sonification systems are those that allow the exploration of different aspects of the data through scrolling, looping, zooming (Pauletto and Hunt 2007); modifying the data that is producing the sound in real-time, for example, in movement-related sonifications (e.g., Pauletto and Hunt 2006; Frid et al. 2019); or considering the data as a material that produces sound when excited, a type of sonification called model-based sonification (Hermann 2011).

While sonification techniques are in continuous development, there are three approaches that can be considered highly established. *Audification* is the direct translation of data into sound waveforms, after appropriate scaling, so that the final audio signal is within the human audible range. This technique has been used to portray complex data sets from seismology (Dombois 2001), helicopter flight analysis (Pauletto and Hunt 2004; Pauletto and Hunt 2005), statistical analysis (Vogt et al. 2015), to create artistic experiences (Ikeshiro 2014) and more. This technique has the advantage of displaying data characteristics, such as noise and periodicity, extremely clearly as they are heard immediately as noise, frequencies or repetitions in the sound. An additional advantage is that this technique can display large amounts of data in a very short amount of time as, for example, 44,100 data points can be transformed into one second of sound (at CD quality). While these are advantages for some type of dataset, other smaller datasets are better displayed using parameter mapping.

Parameter mapping refers to the use of data to control sound parameters such as pitch and loudness. This is perhaps the most used sonification technique and many examples of designs using this approach can be found. Applications go from auditory graphs (Bonebright 2005; Nees 2018), sonification of maps (Schito and Fabrikant 2018), sport (Schaffert and Mattes 2015) or network data (Worrall 2015), to name a few. The meaning of the resulting sonification is highly dependent on mapping chosen (Dubus and Bresin 2013), polarity, scale and interdependence of acoustic parameters (Walker 2002) and other aspects such as context.

Finally, *model-based sonification* was first developed by German researcher Thomas Hermann (2011), and it involves viewing a dataset as a virtual sounding object, which will produce sound (the sonification) when excited. While this approach aims to exploit our existing ability to experience and learn about the world and objects within it through the sound they make, the interpretation of model-based sonification results remains highly dependent on the relationship between the dataset and the physical model chosen to describe it.

4 The Development of Sonification Research

Despite the growing volume of research in the field of sonification, scientific and professional fields have so far failed to take up sonification as a mainstream way to communicate information. Recently, researchers have begun to unpack the reasons for this, proposing new ways to advance the field. Supper (2015) has provided an in-depth study on the development of the sonification community and its research from a sociological viewpoint. It emerged that, historically, the community has concentrated on the design of sonifications and relevant technical tools, rather than on developing knowledge on how we can analyze and interpret sonified data, i.e., the application of sonification. This has led to sonification remaining, so far, a contested scientific technique. Neuohff (2019) has argued that, in order to develop, sonification research needs to address a number of challenges including, among others, users' individual differences in audition abilities, and the bias towards musical listening[3] (expected from users) and musical education (of designers) in sonification research. In relation to this last point, Neuhoff and Heller (2005) have argued that sonification could leverage our more commonly used mode of listening: causal listening. Causal listening refers to listening to a sound in order to gather information about its source or cause. They state that "it is possible to find changes in acoustic parameters that map unambiguously to changes in source or event characteristics" (2005 p. 3) such as material (from liquid to solid passing through muddy) or action (from slow walking to running) (Neuhoff and Heller 2005). Additionally, researchers have argued that sonification research should draw upon sound design methods and approaches to meaning making already developed in media production (film, TV, radio) in order to increase the communicative power of sonification designs (Pauletto 2017; MacDonald and Stockman 2013; Walus et al. 2016). These approaches have become more feasible due to the development of physical and procedural-based models of sound synthesis (Farnell 2010) and the development of relevant digital tools such as the Sound Design Toolkit (Baldan et al. 2017).

As an answer to these challenges, Neuhoff (2019) has proposed that design efforts should focus in a perceptual space where audition performs well and individual differences are smallest. He also suggests that empirical researchers should evaluate their designs focusing on audition skills that realistically represent the abilities present in their target users. Furthermore, Neuhoff argues that sonification research could benefit from a clear bifurcation towards *artistic sonification*, where the aim of the sonification is to give a "sense" of the underlying data, and *empirical sonification*, where the precise representation of data is prioritized. Nees (2019) has advanced the discussion about sonification as a design science, and proposed recommendations for theory building. After an analysis of sonification as a discipline through the lens of eight general components of design theory (Gregor and Jones 2007), Nees argues that sonification research has so far adequately articulated some components of design theory, for example purposes and scope, while others are underdeveloped, for example principles of form and function and testing of results (something confirmed also by Dubus and Bresin (2013)), or missing, for example reproducibility of research results. Recommendations include establishing protocols for designs that clearly define criteria for success, which are linked to task- and goal-specific outcomes, developing formal heuristics, and conducting replication studies to confirm or challenge existing results.

Sonification has evolved tremendously in the last 30 years, but despite this it is not yet a mainstream way to display data. The recent analyses and reflections on the state of the discipline have brought much awareness and potential directions for the future.

Perhaps, as Scaletti advanced in her keynote to ICAD (2017), the tipping point for the use of sonification is coming in the near future, facilitated by the ongoing process of technological convergence in which all modes of communication (audio, video, text) and information are possible, interlinked and interactive.

5 Emergent Themes in Sonification

Auditory display research in the 1990s appeared to focus mainly on issues of sound-to-data mapping in sonification, accessibility, as well as the development of systems and interfaces for sonification. In the 2000s, as well as informing research on mapping with studies on perception, psychoacoustic and interaction, researchers furthered discussions on the cultural and philosophical aspects of sonification, and advanced the theoretical and aesthetical basis for sonification design. Since the 2010s, more attention has been given to the nature of data (big data, small data, personal data), to issues related to interactivity (for example in relation to games, learning and sport applications) and to the creation of methodologies for sonification design. In the last few years, a large number of topics, often related to design and digitalization, have entered sonification research more explicitly. We can now frequently find in sonification research terms such as big data, small data, sensor technology, internet of things, maker culture, personal, participatory, speculative, everyday and real-world applications along with specific topics such as cybersecurity, automated systems, health and environmental data sonification.

We will focus now on five emerging themes in sonification research which we believe to be key for the future development of the discipline.

5.1 *Learning Through Sonification*

Since its beginnings sonification research has been concerned with producing displays for visually impaired people. This large section of the population is highly underserved in a world that privileges vision in most aspects of life, particularly education and employment, and it is clearly a primary target audience for sonification research. Since the 1990s, this research ranged from sonifying graphical user interfaces (Mynatt and Edwards 1995) or geographical maps (Krygier 1994) to creating auditory equivalents of visual graphs (Flowers and Hauer 1992).

Through the process of technological convergence, creating multimedia interactive applications online has become increasingly simple and affordable. More and more research has gone into developing, for example, sonification for STEM education for visually impaired students (e.g., Tomlinson et al. 2020; Levy and Lahav 2012). But while research on the connection between sonification and learning was initially developed to address an accessibility issue, its impact on sonification research as a whole could potentially be much wider.

The 2017 ICAD Conference was dedicated to "sound in learning," and in her keynote lecture researcher and composer Carla Scaletti made a key argument for the importance of sonification in learning. She argued that the reason why sonification is not yet used as a legitimate scientific method to present data is related to how we learn about science from kindergarten on. From an early age we learn to understand relationships between objects in the form of diagrams and graphs. The great emphasis on visual representations not only creates a very large obstacle for visual impaired people in education and employment, it also restricts the way we all see the world. On one hand it provides us with a "vocabulary" of signs (e.g., the Cartesian plane) that we will take for granted for the rest of our life, on the other it pushes us towards thinking of processes in terms of static snapshots (images).

Scaletti then argues that the creation of effective sonifications to serve visually impaired groups in education can provide the "catalytic innovation" (Christensen et al. 2006) that could ultimately produce social change for all. When all e-books and webpages will have a sound function which allows the reader to switch on a well-produced and engaging auditory display for visually impaired users, perhaps all people will try it and, if they learn something new through sound, they might never switch the auditory display off. There are good reasons to think that this would happen. Pleasurable music and musical training, for example, have been shown to support learning (Gold et al. 2013) and the development of cognitive abilities (Criscuolo 2019). Additionally, dynamic processes might be much better understood by students through sound (an intrinsically dynamic medium) rather than images. So well designed sonifications could help us all understand science and other disciplines.

Scaletti argues that only when the possibility of experimenting with sound and active listening is available to all children from kindergarten, a new generation of researchers will grow up considering sound a transparent way of displaying data. This means that once the possibility of integrating sound in learning becomes easy for teachers and widely accepted, additional advantages of using sound in learning could emerge for the benefit of all. Ultimately, research on how sonification can support learning might not only address issues of accessibility, but it could represent the path towards the scientific legitimization of sonification for all.

5.2 Sonification of New Digital Things

The objects and artifacts we use in today's digitalized society have changed. If in the past objects – a chair, a telephone, even a TV – could be described as something that we perceive as a whole (even if made by different components) and distinct from their surroundings with relatively stable physical properties, nowadays objects are something different. The networking and personalization capabilities, the gathering and use of data, the ubiquitous presence of sensors and actuators have changed the nature of the "things" we use every day. The smartphone, for example, has a physical form but its use is highly dependent on networked apps and customized characteristics. Smart watches, fridges and security devices and autonomous driving systems are additional examples. Redström and Wiltse (2018) propose a useful term to describe the new everyday "things" that we use: *fluid assemblages. Assemblages* because they combine material and immaterial components, local and external to the physical object that often still encapsulates the "thing," communicating with each other producing characteristics that are unique to the specific assemblage. And *fluid* because properties can change dynamically, at runtime, in a highly responsive manner. Social context, for example, can continuously redesign an assemblage by pushing new apps and services, which highlights how networking and interaction capabilities, rather than physical form, primarily define new digital things. These "living, throbbing confederations" (Bennett 2010, p. 23) have blurred the separation between producers and consumers, design and use. Redström and Wiltse (2018) argue that fluid assemblages create great challenges for established design processes still strongly based on a modernist aesthetic which tends towards solutions that emphasize simplicity and ease of use. In this context, the temporal nature of sound, its potential to portray many characteristics at once through dynamic temporal and spectral properties, make it an excellent candidate for portraying the fluidity and interconnectivity of assemblages. Examples of sonification research in this area include real-time sonification of network traffic and anomaly detection (Lenzi et al. 2020; Ballora et al. 2011; Rimland et al. 2013), sonification of quiet cars, autonomous vehicles and in-vehicle interfaces (Denjean et al. 2013; Larsson et al. 2019; Tardieu et al. 2015), sonification of IoT devices, smart cities and energy consumption (Chernyshov et al. 2016; Sarmento et al. 2020; Lockton et al. 2014). But while sound can give form and visibility to such dynamic systems, its success so far is debatable. The interconnections that characterize assemblages can produce large amounts of data, and Supper (2015) has shown that sonification research so far "does not so much solve as echo the problems of big data" (2015, p. 454).

Reasons for the limited success of this kind of sonification can be found in the assumption, common in sonification research, that data structures will emerge easily when data is sonified, and in the consequent lack of studies on the interpretation of sonified data. Supper shows that the historical sonification examples (the Geiger counter, audification of seismic data) often cited to justify these assumptions are insufficient in size, use or historical success. Additionally, we suggest that these expectations might also be the result of the same widespread modernist design approach that, as Redström and Wiltse (2018) propose, needs shifting towards the design of "forms or processes of becoming" (p. 137) to address today's fluid assemblages. In sonification, such a shift could correspond to ditching the expectation that a simple understandable harmony could emerge from a complex structure, to an approach that values, rather than fights against, the ability to "echo" and bring to our perception the complexities of fluid assemblages. As thunder provides us with both a strongly felt perception and useful information about the dynamic process of the lightning it is a product of (despite being a complex, noisy and instantaneous sound), sonifications of new digital "things" might not need to be "simple" or immediately understood; instead their value might be in allowing us to perceive and feel the real complexity of new "things".

5.3 Personalization

Widely accepted standards in sonification have not yet been established. One reason for this could be the importance of individual differences in our perception of sound and in our musical taste. It might be that the auditory modality requires a more individualized approach, and sonification displays become highly effective only when customized to the individual characteristics of the user. Just like each one of us will adjust the volume of the TV to our preferred level, sonifications might need to be adjusted to the individual, and tuned to the environment we listen in in order to be effective. Personalization might be the answer to making sonification display work for all. Bresin et al. (2020) describe the outcome of a workshop that used a design fiction (Bleecker 2009) methodology to imagine what the soundscape of the future could be like. One theme that emerged strongly in this workshop is that of personalization, i.e., the idea that in the future we will be able, through visible or invisible systems, to personalize our immediate soundscape so that it addresses our needs and moods. A push in this direction can be noticed in recent sonification research. Studies have shown that there are important individual differences in the way people perceive and understand data-to-sound connections (e.g., Mauney and Walker 2007). This, combined with studies showing how the effectiveness of a sonification display depends on the task and the understanding of that task (e.g., Verona and Peres 2017), has motivated the recent adoption of more user-centered and task-centered approaches. However, involving users in the design is not a trivial task. Due to a lack of tools and language for supporting independent design, it can be labor intensive. The drive towards personalized sonification displays has also driven research away from the creation of abstract sonification designs, produced with unfamiliar synthetic timbres for example, towards more familiar sounds. Wolf and Fiebrink (2019), for example, propose a user-centered sonification design of Twitter messaging that use familiar everyday sounds to create a personal and "calm" soundscape. Similarly, Barra et al. (2001) have proposed to sonify the operations of a web server through the manipulation of personal background music. These design approaches exploit knowledge about personal preferences in order to produce more usable displays. This can be a particularly important approach for displays aimed at monitoring tasks when they need to sound for a long period of time.

5.4 Sonification in Health and Environmental Data (SoniHED)

The first conference dedicated to the sonification of health and environmental data, named SoniHED, was conceived and chaired by Pauletto in 2014 (SoniHED 2014). While research on these topics had appeared before (e.g., Polli 2005; Flowers et al. 2001; Pauletto and Hunt 2006; Ballora et al. 2004), this was the first time a conference had brought these themes together and focused on their interconnection.[4] The domains of health and environment are closely related. Pollution, noise and other environmental factors can have a strong impact on our health. In our everyday life we are used to monitoring some health and environmental data: for example, we often monitor the weather (temperature, humidity, wind) and our health (blood pressure, temperature, etc.). Two factors have contributed to the creation of this new theme: the development in 2012 of the UN Sustainable Goals (UN SDGs), which highlighted how global development is dependent on the interconnections between different areas such as health, environment, equality, etc., and the process of digitalization which has allowed the storage of large amounts of health and environmental data both at local and global level. In this context, sound might be a new way to perceive and engage with trends and interconnections in this data.

Sonification research in this area has been concerned with energy consumption (Groß-Vogt et al. 2018), air quality (St Pierre and Droumeva 2016; Teixeira et al. 2005), weather data (Harman et al. 2016), health communication (Walus et al. 2016), blood pressure (Barrass 2014), heart rate (Ettehadi et al. 2020), to name a few examples.

6 Sonification of Movement, Performance and Sport

In nature, there is no sound without movement. Sounds can be generated by both biological movements and physical motion of objects. Movements represent a first-class case for applying sonification, especially interactive sonification, being characterized by changes over time which can be portrayed through real-time audio feedback. Additionally, the wide availability of motion capture systems has boosted the possibilities for the design of applications focusing on the interactive sonification of human movements.

Physical parameters can be summarized in five high-level dimensions: kinematics (e.g., acceleration), kinetics (e.g., pressure), matter (e.g., density), time (e.g., event rate), dimensions (e.g., length). In an overview study by Dubus and Bresin (2013), 179 studies were analyzed and it was found that these five classes have been mapped into five auditory dimensions, namely pitch-related (e.g., pitch range), timbral (e.g., brightness), loudness-related (e.g., loudness), spatial (e.g., head-related transfer function), and temporal (e.g., duration). The three most used mappings were location to spatialization, location to pitch and distance to loudness (Dubus and Bresin 2013).

Furthermore, research has shown that an important aspect to consider in these studies is who is in control of the sonification. When subjects interactively sonify their own movements, they focus more on the quality (regularity, energy) of their movements (Frid et al. 2016), while observers sonifying the same movements focus more on the sonic rendering, making it more expressive and more connected to low-level physical features (Bresin et al. 2020).

Applications of body movements sonification can be found in sport, health, arts, robotics.

Providing real-time sonic feedback to actions and movements in sport activities helps athletes to increase awareness about their own performance and to improve their technique. Examples can be found in rowing (e.g., Schaffert and Mattes 2012; Dubus and Bresin 2015), ice skating (Godbout and Boyd 2010), running (e.g., Bolíbar and Bresin 2012) and gymnastics (Hummel et al. 2010).

In rehabilitation, movement's sonification can contribute to an increased awareness of one's own body movements (e.g., Pauletto and Hunt 2006). For an extensive overview of sonification of movements in sports and rehabilitation, see work by Schaffert and colleagues (2019).

Research on expressive music performance and body movements (Godøy et al. 2016) have been used by researchers as inspiration for the design of sonification studies of other expressive movements such as fluid dance movements (Frid et al. 2019), emotional intentions in humanoid robots (Latupeirissa et al. 2020), and to augment and support circus disciplines such as juggling (Bovermann et al. 2007), Cyr wheel and acrobatics (Elblaus et al. 2014).

7 Conclusion

Sonification research began building a distinct agenda and community in the 1990s. Since then, activity has flourished producing new approaches, a vast number of examples, as well as critical discussions about the field's history, its development and its future. This chapter firmly frames sonification as a design discipline (or more specifically a sound design discipline), and aims to assist even the unfamiliar reader through the main concepts, approaches and discussions that characterize sonification research today. Furthermore, we have identified and discussed five main emerging themes in sonification – learning through sonification, the sonification of new digital things, personalization, sonification of health and environmental data and sonification of movement – where we foresee major potential for expansion in the field.

Notes

1 https://www.santafe.edu/.
2 https://interactive-sonification.org/proceedings.
3 Musical listening refers to the ability to attend to perceptual dimensions and attributes of the sound itself, such as pitch, loudness, etc., when listening to a sound (Gaver 1993).
4 In 2016 the ICAD conference adopted SoniHED as one of their topics.

References

Baldan, S., Delle Monache, S. and Rocchesso, D. (2017). The sound design toolkit. *SoftwareX*, 6, 255–260.

Ballora, M., Giacobe, N.A. and Hall, D.L. (2011) Songs of cyberspace: an update on sonifications of network traffic to support situational awareness, *SPIE Conference 8064, Multisensor, Multisource Information Fusion: Architectures, Algorithms, and Applications*; doi: 10.1117/12.883443.

Ballora, M., Pennycook, B., Ivanov, P.C., Glass, L. and Goldberger, A.L. (2004). Heart rate sonification: A new approach to medical diagnosis. *Leonardo*, 37 (1), pp.41–46.

Barra, M., Cillo, T., De Santis, A., Petrillo, U.F., Negro, A., Scarano, V., Matlock, T. and Maglio, P.P. (2001). Personal webmelody: Customized sonification of web servers. In *International Conference on Auditory Display (ICAD)*, Helsinki, Finland.

Barrass, S. (2014). Acoustic sonification of blood pressure in the form of a singing bowl. In *Conference on Sonification of Health and Environmental Data* (SoniHED).

Bennett, J. (2010). *Vibrant Matter: A Political Ecology of Things*. Duke University Press, Durham, NC.

Bijsterveld, K. (2006). Listening to machines: Industrial noise, hearing loss and the cultural meaning of sound. *Interdisciplinary Science Reviews*, 31 (4), 323–337.

Bleecker, J. (2009) Design fiction: A short essay on design, science, fact and fiction. Near Future Laboratory. https://shop.nearfuturelaboratory.com/products/design-fiction-a-short-essay-on-design-science-fact-and-fiction

Bly, S. (1982) Presenting information in sound. In *Conference on Human Factors in Computing Systems* (CHI). Association for Computing Machinery, New York, 371–375. https://doi.org/10.1145/800049.801814.

Bolíbar, J. and Bresin, R. (2012). Sound feedback for the optimization of performance in running. In *TMH-QPSR special issue: SMC Sweden 2012, Sound and Music Computing*, 52 (1), 39–40.

Bonebright, T. L. (2005) A suggested agenda for auditory graph research. In *International Conference on Auditory Display* (ICAD), 398–402, Limerick, Ireland.

Bovermann, T., Groten, J., De Campo, A., Eckel, G. (2007). Juggling sounds. In *Workshop on Interactive Sonification.*

Bresin, R., Mancini, M., Elblaus, L., Frid, E. (2020). Sonification of the self vs. sonification of the other: differences in the sonification of performed vs. observed simple hand movements. *International Journal of Human-Computer Studies,* Vol 144, 102500.

Buxton, W. (1989) Introduction to this special issue on non-speech audio. *Human-Computer Interaction,* Vol. 4, 1–9.

Chernyshov, G., Chen, J., Lai, Y., Noriyasu, V., Kunze, K. (2016). Ambient rhythm: Melodic sonification of status information for IoT-enabled devices. In *International Conference on the Internet of Things,* 1–6.

Christensen, C.M., Baumann, H., Ruggles, R., Sadtler, T.M. (2006). Disruptive innovation for social change. *Harvard Business Review,* 84 (12), p. 94.

Criscuolo, A., Bonetti, L., Särkämö, T., Kliuchko, M. and Brattico, E. (2019). On the association between musical training, intelligence and executive functions in adulthood. *Frontiers in Psychology,* Vol. 10, p. 1704.

Denjean, S., Roussarie, V., Ystad, S., Kronland Martinet, R. (2013). An innovative method for the sonification of quiet cars. *The Journal of the Acoustical Society of America,* 134 (5), 3979–3979.

Dombois, F. (2001). Using audification in planetary seismology. In *International Conference on Auditory Display (ICAD),* Helsinki, Finland.

Dubus, G. and Bresin, R. (2013). A systematic review of mapping strategies for the sonification of physical quantities. *PLOS ONE,* 8 (12).

Dubus, G. and Bresin, R. (2015). Exploration and evaluation of a system for interactive sonification of elite rowing. *Sports Engineering,* 18 (1), 29–41.

Eerola, T. and Vuoskoski, J.K. (2013) A review of music and emotion studies: approaches, emotion models, and stimuli. *Music Perception: An Interdisciplinary Journal,* 30(3), 307–340.

Elblaus, L., Goina, M., Robitaille, M.A., Bresin, R. (2014). Modes of sonic interaction in circus: Three proofs of concept. In *International Computer Music Conference.*

Ettehadi, O., Jones, L. and Hartman, K. (2020). Heart Waves: A Heart Rate Feedback System Using Water Sounds. In *International Conference on Tangible, Embedded, and Embodied Interaction,* 527–532.

Farnell, A. (2010). *Designing Sound.* MIT Press, Cambridge, MA.

Flowers, J.H. and Hauer, T.A. (1992). The ear's versus the eye's potential to assess characteristics of numeric data: Are we too visuocentric?. *Behavior Research Methods, Instruments, & Computers,* 24 (2), 258–264.

Flowers, J.H., Whitwer, L.E., Grafel, D.C. and Kotan, C.A. (2001). Sonification of daily weather records: Issues of perception, attention and memory in design choices. In *International Conference on Auditory Display (ICAD),* Helsinki, Finland.

Frid, E., Bresin, R., Alborno, P., Elblaus, L. (2016). Interactive Sonification of Spontaneous Movement of Children: Cross-Modal Mapping and the Perception of Body Movement Qualities through Sound. *Frontiers in Neuroscience,* 10.

Frid, E., Elblaus, L. and Bresin, R. (2019). Interactive sonification of a fluid dance movement: an exploratory study. *Journal on Multimodal User Interfaces,* 13 (3), 181–189.

Gaver, W.W. (1989). The SonicFinder: An interface that uses auditory icons. *Human–Computer Interaction,* 4 (1), 67–94.

Gaver, W.W. (1993) What in the world do we hear? An ecological approach to auditory event perception. *Ecological Psychology,* 5 (1), 1–29.

Giordano, B., Egermann, H., Bresin, R. (2014). The production and perception of emotionally expressive walking sounds: Similarities between musical performance and everyday motor activity. *PLOS ONE,* 9 (12), e115587.

Godbout A. and Boyd J.E. (2010). Corrective sonic feedback for speed skating: a case study. In *International Conference on Auditory Display (ICAD),* 23–30, Washington, DC.

Godøy, R.I., Song, M., Nymoen, K., Haugen, M.R., Jensenius, A.R. (2016). Exploring sound-motion similarity in musical experience. *Journal of New Music Research*, 45 (3), 210–222.

Gold, B.P., Frank, M.J., Bogert, B. and Brattico, E. (2013). Pleasurable music affects reinforcement learning according to the listener. *Frontiers in Psychology*, 4, p. 541.

Gregor, S. and Jones, D. (2007) The Anatomy of a Design Theory, *Journal of the Association for Information Systems*, 8 (5) 313–335.

Groß-Vogt, K., Weger, M., Höldrich, R., Hermann, T., Bovermann, T. and Reichmann, S. (2018). Augmentation of an institute's kitchen: An ambient auditory display of electric power consumption, In *International Conference on Auditory Display (ICAD)*, Houghton, MI.

Guastavino, C. (2007). Categorization of environmental sounds. *Canadian Journal of Experimental Psychology*, 61 (1), 54–63. https://doi.org/10.1037/cjep2007006.

Harman, A., Dimitrov, H., Ma, R., Whitehouse, S., Li, Y., Worgan, P., Omirou, T. and Roudaut, A. (2016). NotiFall: Ambient Sonification System Using Water. In *Conference Extended Abstracts on Human Factors in Computing Systems* (CHI EA). Association for Computing Machinery, 2667–2672. https://doi.org/10.1145/2851581.2892443.

Hermann, T. (2008). Taxonomy and definitions for sonification and auditory display. In *International Community on Auditory Display (ICAD)*, Paris.

Hermann, T. (2011) Model-based sonification. In *The Sonification Handbook*, Hermann, T., Hunt, A., and Neuhoff, J.G., (eds.). Logos Verlag, Berlin. https://sonification.de/handbook/.

Hermann, T., Hunt, A., and Neuhoff, J.G., (eds.). (2011) *The Sonification Handbook*, Berlin: Logos Verlag, Berlin. https://sonification.de/handbook/.

Houel, M., Arun, A., Berg, A., Iop, A., Barahona-Ríos, A., Pauletto, S. (2020). Perception of Emotions in Knocking Sounds: An Evaluation Study. In *Sound and Music Computing Conference*.

Hummel, J., Hermann, T., Frauenberger, C., Stockman, T. (2010). Interactive sonification of German wheel sports. In *Interactive Sonification Workshop (ISon)*.

Hunt, A., Hermann, T., Pauletto, S. (2004) Interacting with Sonification Systems: Closing the Loop. In *International Conference Information Visualisation,* (IV '04). IEEE Computer Society, 879–884.

Ikeshiro, R. (2014). Audification and Non-Standard Synthesis in Construction in Self. *Organised Sound,* 19 (1), 78–89. DOI: 10.1017/S1355771813000435. ISon https://interactive-sonification.org/.

Juslin, P.N. and Sloboda, J. (eds.). (2011) *Handbook of Music and Emotion: Theory, Research, Applications.* Oxford University Press, Oxford, UK.

Korff, S. (2013). How the Geiger Counter started to crackle: Electrical counting methods in early radioactivity research. *Annalen der Physik*, 525 (6), A88–A92.

Kramer, G. (1994). An introduction to auditory display. In G. Kramer (ed.), *Auditory Display: Sonification, Audification, and Auditory Interfaces*, 1–78. Addison Wesley, Reading, MA.

Kramer, G. (1996). Introduction to Icad. https://www.icad.org/websiteV2.0/Conferences/ICAD96/proc96/introkramer.htm.

Kramer, G., Walker, B.N., Bonebright, T., Cook, P., Flowers, J., Miner, N. et al. (1999). The Sonification Report: Status of the Field and Research Agenda. In https://digitalcommons.unl.edu/psychfacpub/444/

Krebs, S. (2012) Sobbing, Whining, Rumbling – Listening to Automobiles as Social Practice. In *The Oxford Handbook of Sound Studies*, 79–101.

Krygier, J.B. (1994). Sound and geographic visualization. In *Modern cartography series* 2, 149–166.

Larsson, P., Maculewicz, J., Fagerlönn, J., Lachmann, M. (2019). Auditory displays for automated driving-challenges and opportunities. In *International Conference on Auditory Display (ICAD)*, Newcastle-upon-Tyne, UK.

Latupeirissa, A. B., Panariello, C., Bresin, R. (2020). Exploring emotion perception in sonic HRI. In *Sound and Music Computing Conference*, 434–441.

Lenzi, S., Terenghi, G., Moreno-Fernandez-de-Leceta, A. (2020). A design-driven sonification process for supporting expert users in real-time anomaly detection: Towards applied guidelines. *EAI Endorsed Transactions on Creative Technologies*, 7 (23).

Levy, S.T. and Lahav, O. (2012) Enabling people who are blind to experience science inquiry learning through sound-based mediation. *Journal of Computer Assisted Learning*, 28 (6), 499–513.

Lockton, D., Bowden, F., Brass, C. Gheerawo, R. (2014). Powerchord: Towards ambient appliance-level electricity use feedback through real-time sonification. *UCAmI 2014: International Conference on Ubiquitous Computing & Ambient Intelligence*.

MacDonald, D. and Stockman, T. (2013). Toward a method and toolkit for the design of auditory displays, based on soundtrack composition. In *Extended Abstracts on Human Factors in Computing Systems*, 769–774.

Mauney, L.M. and Walker, B.N. (2007). Individual differences and the field of auditory display: Past research, a present study, and an agenda for the future. In *International Conference on Auditory Display (ICAD)*, Montreal.

McAdams, S., Bigand, E., eds. (1993). *Thinking in sound: The cognitive psychology of human audition.* Oxford science publications, Oxford University Press.

Mody, C.C., 2005. The sounds of science: Listening to laboratory practice. *Science, Technology, & Human Values*, 30 (2), 175–198.

Mynatt, E.D. and Edwards, W.K. (1995) Audio GUIs: interacting with graphical applications in an auditory world. In *Conference Companion on Human Factors in Computing Systems*, pp. 85–86.

Nees, M.A. (2018) Auditory graphs are not the "killer app" of sonification, but they work. *Ergonomics in Design*, 26 (4), 25–28.

Nees, M. A. (2019). Eight components of a design theory of sonification. In *International Conference on Auditory Display (ICAD)*, Newcastle-upon-Tyne, UK.

Neuhoff, J.G. (2019). Is sonification doomed to fail? In *International Conference on Auditory Display (ICAD)*, Newcastle-upon-Tyne, UK.

Neuhoff, J.G. and Heller, M.L. (2005). One small step: sound sources and events as the basis for auditory graphs. In *International Conference on Auditory Display (ICAD)*, Limerick, Ireland.

Pauletto, S. (2017). Embodied Knowledge in Foley Artistry. *The Routledge Companion to Screen Music and Sound*, 338–348, Routledge, New York.

Pauletto, S., Cambridge, H. and Rudnicki, R., (eds.) (2014) *SoniHED Conference*. https://www.york.ac.uk/media/c2d2/media/sonihedconference/SONIHED%20CONFERENCE%20PROCEEDINGS%202014.pdf.

Pauletto, S., Cambridge, H., Susini, P. (2016). Data sonification and sound design in interactive systems. *International Journal of Human-Computer Studies*, 85 (C), https://doi.org/10.1016/j.ijhcs.2015.08.005.

Pauletto, S. and Hunt, A. (2004) Interactive sonification in two domains: helicopter flight analysis and physiotherapy movement analysis. In *Workshop on Interactive Sonification*.

Pauletto, S. and Hunt, A. (2005) A comparison of audio and visual analysis of complex time-series data sets. In *International Conference on Auditory Display* (ICAD), Limerick, Ireland.

Pauletto, S. and Hunt, A. (2006) The sonification of EMG data. In *International Conference on Auditory Display (ICAD)*, London, UK.

Pauletto, S. and Hunt, A. (2007) Interacting with sonifications: An evaluation. In *International Conference on Auditory Display (ICAD)*, Montreal.

Peretz, I. and Zatorre, R.J. (eds.). (2003). *The Cognitive Neuroscience of Music*. Oxford University Press, Oxford, UK.

Polli, A., 2005. Atmospherics/weather works: A spatialized meteorological data sonification project. *Leonardo*, 38 (1), 31–36.

Reddell, T. (2018). *The Sound of Things to Come: An Audible History of the Science Fiction Film*. University of Minnesota Press, Minneapolis.

Redström, J. and Wiltse, H. (2018). *Changing things: The future of objects in a digital world*. Bloomsbury Publishing, London.

Rice, T. (2010). 'The hallmark of a doctor': the stethoscope and the making of medical identity. *Journal of Material Culture*, 15 (3), 287–301.

Rimland, J., Ballora, M., Shumaker, W. (2013). Beyond visualization of big data: a multi-stage data exploration approach using visualization, sonification, and storification. In SPIE Conference, The International Society for Optics and Photonics.

Robare, P. and Forlizzi, J. (2009). TIMELINES Sound in computing: a short history. *Interactions* 16 (1), 62–65. https://doi.org/10.1145/1456202.1456218.

Rocchesso, D. (ed.) (2011). Explorations in Sonic Interaction Design. Logos Verlag, Berlin.

Sarmento, P., Holmqvist, O., Barthet, M. (2020). Musical Smart City: Perspectives on Ubiquitous Sonification. *Ubiquitous Music Workshop 2020*

Scaletti, S. (2017) Keynote to the International Conference on Auditory Display (ICAD), State College, PA. https://www.youtube.com/watch?v=T0qdKXwRsyM.

Schaffert, N., Janzen, T.B., Mattes, K., Thaut, M.H. (2019). A Review on the Relationship Between Sound and Movement in Sports and Rehabilitation. *Frontiers in Psychology*, 10, 244.

Schaffert N. and Mattes K. (2012) Acoustic feedback training in adaptive rowing. In *International Conference on Auditory Display (ICAD)*, 83–88, Atlanta, GA.

Schaffert, N. and Mattes, K. Interactive Sonification in Rowing: Acoustic Feedback for On-Water Training, in *IEEE MultiMedia,* 22 (1) 58–67, 2015, doi: 10.1109/MMUL.2015.9.

Schedel, M. and Worrall, D. (2014). Sonification. *Organised Sound,* 19 (1).

Schito, J. and Fabrikant, S.I. (2018). Exploring maps by sounds: using parameter mapping sonification to make digital elevation models audible. *International Journal of Geographical Information Science*, 32 (5), 874–906.

SoniHED Conference (2014). https://www.york.ac.uk/c2d2/seminars/sonihed/.

St Pierre, M. and Droumeva, M., (2016) Sonifying for public engagements: a context-based model for sonifying air pollution data. In *International Conference on Auditory Display (ICAD),* Canberra, Australia.

Sterne, J. (2003). Medicine's acoustic culture: mediate auscultation, the stethoscope and the 'autopsy of the living.' 191–217 *The Auditory Culture Reader*, Berg, Oxford.

Supper, A. (2015). Sonification in the Age of Complex Data and Digital Audio, *Information & Culture*, Vol. 50, no. 4, 441–464, University of Texas Press, Austin.

Susini, P. (2011) Le design sonore: un cadre expérimental et applicatif pour explorer la perception sonore, Dossier d'habilitation á diriger des recherches, Aix-Marseille II.

Tardieu, J., Misdariis, N., Langlois, S., Gaillard, P., Lemercier, C. (2015). Sonification of in-vehicle interface reduces gaze movements under dual-task condition. *Applied Ergonomics*, 50, 41–49.

Teixeira, L.M.L., Barbosa, A., Cardoso, J., Carvalhos, V., Costa, M., Sousa, I., Franco, I., Fonseca, A., Henriques, D. and Rosa, P. (2005). Online data mining services for dynamic spatial databases II: air quality location based services and sonification, In *International Conference and Exhibition on Geographic Information.*

Tomlinson, B.J., Walker, B.N. and Moore, E.B. (2020) Auditory Display in Interactive Science Simulations: Description and Sonification Support Interaction and Enhance Opportunities for Learning. In *Conference on Human Factors in Computing Systems (CHI)*, 1–12.

Truax, B. (2001). *Acoustic Communication*. Greenwood Publishing Group, Westport, CT.

UN Sustainable Development Goals. https://www.undp.org/content/undp/en/home/sustainable-development-goals/background.html.

Verona, D. and Peres, S.C. (2017). A comparison between the efficacy of task-based vs. data-based sEMG sonification designs. In *International Conference on Auditory Display (ICAD),* State College, PA.

Visell, Y., Fontana, F., Giordano, B.L., Nordahl, R., Serafin, S., Bresin, R. (2009). Sound design and perception in walking interactions. *International Journal of Human-Computer Studies*, 67 (11), 947–959.

Vogt, K., Frank, M. and Höldrich, R. (2015). Effect of augmented audification on perception of higher statistical moments in noise. In *International Conference on Digital Audio Effects (DAFx)*.

Volmar, A. (2013). Listening to the cold war: The nuclear test ban negotiations, seismology, and psychoacoustics, 1958–1963. *Osiris*, 28 (1), 80–102.

Walker, B. N. (2002). Magnitude estimation of conceptual data dimensions for use in sonification. *Journal of Experimental Psychology: Applied*, 8 (4), 211.

Walker, B.N. and Kramer, G. (2004). Ecological psychoacoustics and auditory displays: Hearing, grouping, and meaning making. In J. Neuhoff (ed.), *Ecological Psychoacoustics*, 150–175. Academic Press, New York.

Walus, B.P., Pauletto S. and Mason-Jones, A. (2016). Sonification and music as support to the communication of alcohol-related health risks to young people. Study design and results. *Journal on Multimodal User Interfaces*, 10 (3) 235–246.

Wolf, K. and Fiebrink, R. (2019). Toward Supporting End-User Design of Soundscape Sonifications. In *International Conference on Auditory Display (ICAD), Newcastle-upon-Tyne, UK.*

Worrall D. (2009). An introduction to data sonification. In Dean R.T. (ed.). *The Oxford Handbook of Computer Music.* Oxford University Press, Oxford, UK.

Worrall, D. (2015). Realtime sonification and visualisation of network metadata. In *International Conference on Auditory Display (ICAD),* Graz, Austria.

Worrall, D. (2018). Sonification: A Prehistory. In *International Conference on Auditory Display (ICAD),* Houghton, MI.

Yang, J., Hermann, T., Bresin, R. (2019). Introduction to the special issue on interactive sonification. *Journal on Multimodal User Interfaces,* 13 (3) DOI:10.1007/s12193-019-00312-z.

Sound Design in the Context of Sonification

Visda Goudarzi

1 Introduction

The field of sound design originated in sound for image. Chion (Chion 2019, 63–65) introduced the concept of synchresis and how our auditory and vision work together:

> The forging of an immediate and necessary relationship between something one sees and something one hears at the same time (from synchronism and synthesis).

Today, sound design has emerged into our everyday objects, human-computer interfaces (even human-robot interfaces), architectural and environmental spaces. Our daily interactions and experiences with sound that different objects and products emit varies. One could have an aspiration for an electric car because of its sophisticated and quiet engine sound, or one may dislike a vacuum cleaner due to its loud sound. Sound design affects our reasoning, our purchasing decisions and our expectations regarding a product or application and its functionality. The studies on product sound design have confirmed the complementary role of auditory experience on our perception and respond to objects. In other words, a well-designed sound enhances the product experience on ergonomics, usability, pleasantness and satisfaction (Egmond 2008), (Fenko et al. 2011), (Jansen et al. 2011), (Carron et al. 2017) and an emotional appeal to a product can be strengthened using sound (Carron et al. 2014), (Delle Monache et al. 2010). On the contrary, unpleasant auditory experience influences the user's emotional response to the product negatively (Özcan et al. 2017). In recent years, designed sound suggests sophistication in the engineering of the product and increased its value. (For example, in automotive industry (Özcan et al. 2019) and robotics (Robinson et al. 2020)). Some designers even consider "Product Sound Design" as the central activity of their design process (Erkut et al. 2015). In *auditory display design*, sound design is referred to "the ability to imply the identification of a sound with the process" rather than with an object.

DOI: 10.4324/9780429356360-15

The process itself can turn into an acoustic gestalt (Grond and Hermann 2012) which could be recognized and carry information.

Sound design covers a broad range and includes aspects of science, engineering, arts and humanities (Archer 1979). Adapting sound design into Cross' design-oriented research paradigms (Cross 2007), we should focus on people (status and practices of the sound designers), process (innovative methods and tools in sound design) and products (forms, formats and status of the designed sounds).

Sound design may have a variety of definitions depending on the context of use. In this chapter we look at sound design as a medium to "make data audible" or as Susini calls it: "make an intention audible" (Susini et al. 2014). Susini and colleagues give a new description of the overall sound design process, as a combination of three steps (analyzing, creating and testing), which articulate sound perception and sound design. They make use of information available in our sonic environment to take the analysis step. Using psychoacoustics and auditory cognition methods, they create an inventory of sounds that represent the target sonic identity. In sound design for sonification, the first step is very similar. We collect information from the sonic environment of the user. Then, as sonification researchers we combine the scientific knowledge from psychoacoustic/perception to decide which properties of sound to use in our implementation. The outcome is usually a sonification system with a few mapping options of interactions-to-sound that needs to be tested for functionality and aesthetics. The final sonification system can be obtained by the collaborative work between sonification designers, software developers, composers and domain scientists. Although sonification started as a purely scientific field, practitioners have started to appreciate humanities and aesthetic approaches in auditory display design in recent years (Filimowicz 2014):

> The more ubiquitous design has become the more it is apparent that it is neither free expression nor scientific engineering; it is in-between, that is, a 'third culture', sometimes fluxing more towards one than the other; the application of scientific and other organized knowledge to practical tasks."
>
> (Worrall 2020)

The ambiguity introduced to data analysis using entirely aesthetic approaches has been the central concern for scientists. Introducing human-centered design approaches to improve aesthetics while sonifying data with the functionality at the core has brought interdisciplinary collaborations in the field. Aesthetics play an essential role in the design of successful auditory interfaces and heavily impact the communication of data to the user by creating a more gratifying user experience. To tackle the challenge of creating pleasant sonic interfaces for auditory displays with human-centered design emphasis, user experience and design-thinking are adopted in sonification design. These approaches introduce an adaptive and dynamic process with all stakeholders' active engagement. Barrass introduced user-centered design approaches to sonification and took several steps to create a synergy between scientific and artistic methods in sonification (Barrass 2012). Another approach is "ecological interface design." Ecological interface design presents guidelines for the development of displays where a key component is the mapping of real-world properties to the interface. Ecological interface design differs from user-centered design in that the focus of the analysis is on the work domain rather than the end user or specific task. Ecological interface design fundamentals are not limited to visual displays; however, they have been commonly utilized in visual display design. Gaver used ecological concepts in his work on auditory icons and earcons and his technique has also been used to the sonification of real-time data (Gaver 1993). Gaver et al. (Gaver et al. 1991) used ecological approach in the Arkola simulation of a bottling plant, and Hearst used it in a marine-steam power plant (Hearst 1997). Furthermore, Neuhoff continued advances in ecological investigations of auditory perception and tied his work to findings in more traditional areas of psychoacoustics (Neuhoff 2004).

Pierre Shaeffer's modes of listening (Turi and Tuomas 2012) created the foundations of ecological listening and music perception. As understanding and adapting to the listening experience and auditory perception of the users become the focal point of sonification, Schaeffer's taxonomy of modes of listening becomes more relevant. Another ecological consideration has led to efforts in utilizing auditory gestalt in designing auditory display to understand how it fits to the user's environment.

In a more recent approach, Wolf (Wolf 2015) extends user-centered design by empowering the users, giving them control over tweaking and modifying the mapping of parameters. However, in this process the users don't have any control during the design process. They are end-users with some tweaking power. Another approach is when we tried to shift the control fully to the hands of users by utilizing participatory design workshops (Goudarzi 2016). In these workshops, the users are co-designers communicating and prototyping their desired sonification system with sonification researchers. Sonification experts elucidate the process of sound design in sonification and mapping strategies. This collaborative approach and extended communication between sonification designers and end-users create a platform to design more successful sonifications systems. A step closer to the user, Landry and Jeon (Landry and Jeon 2020) utilize participatory and ecological design by focusing on situated and embodied experience of the users. They contextualize their sonification system within user's environment driving data directly from the user's actions.

2 Parameter Mapping Sonification

Parameter mapping is a technique for data sonification – "the acoustic representation of data to convey or communicate information" (Kramer et al. 1999). In a parameter mapping sonification system, changes in data values are utilized to manipulate sonic parameters which makes the communication of the data possible. Successful applications of sonification in exploratory data analysis must be paired with a systematic procedure of understanding the working environment in which this analysis is conducted, along with the psychoacoustic principles that affect auditory perception. Some earlier data sonification tools did not allow end-users to control mapping. Data sonification tools such as Sonification Sandbox (Walker and Cothran 2003), xSonify (Candy et al. 2006), SonifYer (Shoon and Dombois 2009) require no programming skills by the data analysts. The data is imported to text or Excel/CSV files as database support doesn't exist in these tools. The users have a few options for mapping to choose from but not an extensive control.

With the rise of world wide web and other real-time applications, the need for real-time monitoring of multiple data streams, such as for monitoring tasks, has evolved. Some examples are financial data sonification systems (Janata and Childs 2004), (Worrall 2009), Twitter data sonification (Dahl et al. 2011), (Hermann et al. 2012), network data sonification (Worrall 2015), EEG data sonification (Hermann et al. 2006) and sonification of astrophysics data (Alexander et al. 2014), to name a few. Worrall (Worrall 2020) mentions Larry Zbikowski's three cognitive processes (Zbikowski et al. 2002), namely that all humans use three cognitive processes to make sense of the world around them: categorization, conceptual models and cross-domain mapping. These three processes are analogous to mapping streams of data to acoustic parameters and making sense of them. The wicked problem during the development of parameter mapping sonification is the mapping typology (Buchanan 1992) – the relationship between the data parameters and acoustic parameters such as pitch, amplitude, duration or timbre – in order to communicate the data to the listener. Dubus and Bresin (Dubus and Bresin 2013) analyze 179 scientific works in sonification and create a database of 495 mappings. Their results confirm that pitch is by far the most used auditory dimension in sonification applications. In the *Sonification Handbook* (Grond and Berger 2011), Grond and Berger argue that "effective Parameter Mapping Sonification often involves some compromise between intuitive, pleasant and precise display characteristics." Nevertheless, the most effective acoustic parameter to convey a specific data is very subjective and depends on the context of use. For instance, unpleasant sounding parameter mapping could lead to misleading information (Sorkin 1988). Grond and Hermann describe the reason for this challenge in the lack of any sonic reference for most phenomena that is explored through sonification (Grond and Hermann 2012). The sound material and structures presented in sonification may be perceived as arbitrary in their relation to the data sets (or mapping may even contradict specific dynamics of the data).

Walker and Kramer (Walker and Kramer 2005) conducted studies to examine the mapping choices in parameter mapping sonification in a process monitoring task. They split the mappings into four categories: intuitive, okay, bad and random.

They examined the accuracy and speed of response. Unexpectedly, mappings that sonification designers hypothesized to be the most accurate and fastest response did not work the best. Walker carries on investigating the use of systematic evaluation of the perceived correlation between data and acoustic parameters in the context of sonification. He brings attention to the psychoacoustic aspects of mapping, such as polarity and scale under magnitude estimation (the relationship between a sensory stimulus and its associated perceived intensity) of the sonified measures (Walker 2002).

Polarity is the directional aspect of the mapping (e.g., increasing or decreasing distortion mapped to increasing air pollution). Scale is described as the amount of change in the sonic parameter for a given change in the data parameter (e.g., for 10 degrees temperature increase, doubling the pitch in Hz). Walker assumes a better polarity as a factor of "naturalness" of the mapping. Dewitt and Bresin consider synchronization between the data parameter and acoustic parameter as naturalness: If sound events are not synchronized with the event that generated them, the listener will perceive them as unnatural (Dewitt and Bresin 2007).

Ferguson and Brewster (Ferguson and Brester 2019) experiment with polarity based on previous experiments to find out how discernable it is when used in real sonification tasks. They determine that for simple auditory displays the polarities of data to sound mappings based on magnitude estimation do not have a substantial effect on any objective performance measures compared with arbitrary polarities (e.g., inverse polarity). They suggest that auditory interface designers explore and find out if the magnitude estimation is necessary to evaluate a particular data-to-sound mapping for their particular use before considering it, which brings us to a user-centered approach. Grond and Hermann describe the challenge of sound design in sonification as a negotiation between the given constrains of data and qualities of sound and its context (Grond and Hermann, 2012). (A summary of mapping strategies discussed in this section are presented in Table 14.1)

Suisi and colleagues claim that sound design should be informed by sound perception research and should be used to inform sound perception research (Suisi et al. 2014). Typical psychoacoustic parameters used in experimenting sound perception are loudness, sharpness, roughness, fluctuation strength (Zwicker and Fastl, 2013, 754–764) and timbre.

The process of sound design in sonification is procedural (Farnell 2007) similar to game sound and web audio, where interaction with and adaptation of sound is essential. In exploration of data during data sonification, the user needs to interact with data in real time and go back and forth and investigate it with different pace to understand and analyze it. Advances in interactive arts and media have helped to improve AR/VR experiences. One of the main driving factors for this has been the video game industry, where improved rendering technologies have allowed engaging with interactive virtual worlds. As the need for adaptation of dynamic audio for the

Table 14.1. Mapping research strategies

Publication	Mapping challenge	Findings
Kramer et al. 2010	Parameter mapping	Using changes in data values to manipulate sonic parameters.
Walker and Kramer 2005	Magnitude estimation	Mappings that designers hypothesized to be the most accurate and the fastest were less effective in performance.
Farnell 2007	Procedural mapping	Designing sound as an exploratory process during interaction with data.
Grond and Berger 2011	Effective mapping	Effective Parameter Mapping Sonification is a compromise between intuitive, pleasant and precise display characteristics.
Grond and Hermann 2011	Misleading mapping	Unpleasant mapping could lead to misinformation. The reason for this is the lack of any sonic reference for most phenomena that is explored through sonification.
Ferguson and Brewster 2019	Polarity versus inverse polarity	The polarities of data to sound mappings based on magnitude estimation do not have a substantial effect on any objective performance measures compared with arbitrary polarities.

game engines has grown, most video and computer games don't rely on sample-based sound design anymore, which is troublesome due to two main reasons:

- in order to have a wide variation of sound, a large computer memory is required
- when sound components are supposed to change depending on the interaction and context, real-time sound creation in sample-based audio is very limited

Farnell (2007) introduces the concept of "Procedural Audio" where sounds are generated using synthesis and analysis techniques in a programming language (such as Supercollider,[1] Max/MSP[2] or Pure Data[3]) for interaction and control. Sounds can be generated in real time to map parameters and capture the behavior of an object or environment. Instead of having to make most of the critical sound decisions before execution, procedural audio allows this to be continued during the game (or exploration of data during the sonification process) with a high level of flexibility, efficiency and control by the user. This new way of understanding and applying sound principles should be a serious consideration for any form of media which uses interactive sound. Worrall describes this procedural process by stating that it's critical to understand the difference between computerized design and computational design:

> Computerized design is based on a data model, whereas computational design relies on a procedural model. Computational design involves the processing

of information and interactions between elements that constitute a specific environment. The final appearance of these elements, whether they be game objects, sonified information derived from data, or otherwise-induced compositional gestures, is rendered from the processing of intrinsic properties, such as the specific values of data points, important information beacons and formal morphological constraints.

(Worrall 2020)

3 Experimenting Sound Design for Sonification

3.1 Acoustic Ecology

Inspired by Schafer (1993) and Krause (1993), we explored the "acoustic ecology" theory, which presents the existence of a sonic organization in nature, in order to ensure the audibility of every species. Such structure is audible in several dimensions (frequency, intensity and timbre) or time scales (temporal, seasonal) and prevents masking or grouping phenomena. In *The Soundscape*, Schafer defines soundscape as the sonic environment: technically, any portion of the sonic environment regarded as a field for study. The term may refer to actual environment or to abstract constructions such as musical compositions, particularly when considered as environment. Analogously, examining data sonification as a sound species in a new ecosystem, we addressed the following questions: What is the structure of such soundscape? Do overloaded streams of sound emerge in ecological conditions? From analysis of various soundscapes, we then gathered some basic rules in terms of sound parameters. We were also inspired by vehicle sound designers (Misdariis and Cera 2013), (Nyeste and Wogalter 2008), who use categories of soundscapes that combine natural and synthesized sound. (Typical sound categories considered for vehicle sounds could be music, urban soundscapes and white noise.) Another influential domain was motion picture sound design. Film sound designers tend to move away from reality and gravitate toward more synthesized sounds with timbral qualities. Nonetheless, even if they could also be used for other purposes (Alborno et al. 2016), examples coming from motion pictures must be considered carefully. Their ephemeral nature (conveying clear meaning in few seconds) contrasts heavily with the continuous sonic presence and exploratory nature of sonification.

In sonification, soundscapes have been used to create a sound environment where separate streams of sounds fit together to shape an immersive environment, while transferring information about the data. Studies have shown that natural sounds are more easily recognized in an office environment than artificial tones (Mynatt et al. 1998). Mauney and Walker (Mauney and Walker 2004) found these soundscapes useful for sonification as they can be easily distinguished from the background sounds but can also fade out of attention, and not be tiring or intrusive when not desired (Vickers et al. 2014). Furthermore, studies show soundscape sonification could be soothing, peaceful or relaxing (Wolf 2015). Some recent examples include Hermann et al.'s real-time sonification of Twitter streams (Hermann et al. 2012), and Boren et al.'s (Boren at al. 2014) use of ambisonic sound recordings from urban soundscapes of New York City as a layer in an auditory display in their project "I hear NY4D," in which real-time sonifications are created that utilize the content and geo-location of Twitter messages. Another example is when data sonification has been used as a narrative device in soundscape composition (Pigrem and Barthet 2017).

3.2 Experiment Design

Metaphoric sonification is an attempt to create auditory metaphors that relate to the data the sound originated from. In order to design a sonic display, we decided to explore if there are any specific metaphors in the climate domain; at least not to conflict with them when creating sounds. Therefore, we analyzed the collected words from focus group conversations (conducted in a previous study) to hear how the climate scientists communicate between themselves and how they communicate their data to scientists from other research groups. In general, we found few metaphoric terms in the collected words. The participants used the standard science vocabulary, which are not metaphors per se, but they become metaphoric when carried to the auditory domain. In the following paragraph, the italicized words are the specific terms that could become metaphoric when used in the auditory domain.

Climate data is *dynamic*: climate scientists run a simulation or collect time series data. Therefore, the general direction of reading the data follows from the time axis of the playback and is independent of the further processing, filtering, amplification and so on that depends on specific sonification data. *Periodicities* and any associated type of wave phenomena play an important role in climate science and can be directly linked to sound oscillation and rhythmic phenomena. *Resolution* is another essential topic in climate science when comparing different datasets or trying to find phenomena at a certain range; resolution in audio could be given by the sampling rate and bit depth. *Missing data* plays a large role in climate science; an obvious analogy is making it audible as silent pauses that can be used for a quick scanning of a dataset. *An ensemble* in climate science is a group of datasets resulting from different runs of a simulation. Because a single outcome is always the product of dynamic processes, only the ensemble of many simulations can be regarded as trustworthy. In music, an ensemble is a group of different instruments – a metaphor that can also be used in mapping, such as different climate models to different timbres. Climate scientists who work with measurement data or with simulation data know about the *signal-to-noise ratio.* One participant called the atmosphere noisy when high amount of greenhouse gases was found there, and scientists search for long-term trends in everyday weather's noisy/random behavior. Although noise in climate data has a different meaning than noise in sound, it could be a useful metaphor in sonification. Discussing and brainstorming with the climate scientists, we came up with a few mapping strategies. Some common ones include mapping height dimension in climate data (altitude) to pitch; the geographical spread can be used for spatial rendering in audio. Some weather-related phenomena are linked to typical sounds and these can be used (rain or wind sounds). On a more conceptual level, terms like extreme, dramatic or beautiful will have to be transferred to the sound design and evaluated in listening tests by future users. Furthermore, in any event, the control of the audio interface will involve actions that climate scientists are used to, such as calibrating or filtering data or sound.

In a set of studies, 16 participants from two groups (climate scientists and sound experts) expressed their preference of sound for an auditory display. We collected an inventory of existing similar sounds (Hug 2009) and based on that sound corpus, we conducted experimental research to find out perceptual auditory features that characterize similarities, differences and other specific qualities to define guidelines (conception) for designing sound for the sonification parameters. The first round of evaluations was focused on the aesthetics of sound, and the second round was for mapping climate terms to sounds.

For this study, 24 sound samples, 10 seconds duration each, were used. Each stimulus had an amplitude envelope of 0.3-second linear ramp onset (attack) and offset (release). An amplitude envelope was designed to avoid abrupt and unpleasant start and stop of the sound. All sound samples were created by the author to support natural acoustic soundscapes, especially climate-related sounds using Supercollider programming language. They were created so that each three would constitute a group thematically or metaphorically connected to one of eight climate parameters that were to be sonified for the scientists: temperature, precipitation, air humidity, pressure, geopotential height, refractivity, radiation and wind (three stimuli in each group). The reason for this selection was to provide a broad range of sounds which could be used to elucidate whether the participants would be able to associate these sounds to parameters of the domain, and whether this association was unanimous. Each study was divided into two sections; the purpose of the first stage was mainly to evaluate the sound samples (stimuli) aesthetically, while the second stage was used in mapping the stimuli to the climate parameters. The total time for each survey was between 35 and 45 minutes. All participants were given identical settings, listening to the stimuli via the same type of headphones (AKG K240 MKII). They were presented with eight groups of three stimuli. After listening to each group of three in a dissimilarity rating, they were asked to indicate which of the three they liked the most on a scale from 1 ("not at all") to 7 ("very much"). Furthermore, they were asked to describe the characteristic that they liked about it. In the second round, they heard the same 24 stimuli one after another but in a random order. They were given a list of climate related parameters, and for each sound sample, they were asked to choose which parameter best correlated to the sample they just listened to. Each sound stimulus was approximately 10 seconds, and there was a 10 seconds break between successive stimuli to give the participant time for evaluation or mapping.

In order to compare the effect of auditory experience and music knowledge on evaluating the
aesthetics of sounds, the study was repeated with two different groups of participants. The second group were all sound designers. Each group consisted of eight participants. The climate parameters were briefly explained to the participants with no science background since they did not have the domain knowledge. This study showed that not all eight parameters provided perceived links to the sound samples. Especially challenging were the parameters that are more abstract such as temperature, refractivity and geopotential height. The results for non-abstract parameters such as wind were very clear over both stages of the study and over both participant groups. As a result, we decided to use sounds in the sonification tool that satisfy either one of these criteria:

1. a sound was mapped to the same parameter by both groups of participants
2. a sound was rated highest by both groups and mapped to the hypothesized parameter

In a second set of studies sound designers were asked to take each selected stimulus from the previous study as reference tone, changing one attribute of it and generating three variations of the stimulus and putting these variations on scale. They were also asked to identify the polarity of each stimulus. Results from a study by (Ferguson and Brewester 2019) suggest that it is difficult for participants to interpret ten levels of stimuli. We found three levels a good start to ensure the task would be manageable for the participants to interpret. The process included analysis and resynthesis of the original sounds. Participants tried wave-table synthesis and granular synthesis. All prototyping was developed in Supercollider audio processing and synthesis environment. Five parameters were modified and tested by the participants: grain size, distortion, reverb, spatialization, and attack time. After experimenting with all stimuli, participants found changing *grain size* and *attack time* the most efficient attribute for scaling. Rendering sound in an immersive environment was not possible, and a stereo setup did not add enough variability for the parameters to distinguish the change. An immersive setup could add more flexible and realistic representation of data trajectories within the soundscape. Finally, we asked the participants to classify all 24 sound samples and place them on Gaver's sound map (Gaver 1993). This task gave us a great overview on what aspects of sounds we should work on in the next iterations of the prototype to create a more cohesive and consistent ecology.

4 Reflections

This chapter discussed the current state of sound design practice from an auditory display design perspective. In *New Directions in Third Wave Human-Computer Interaction: Volume 1 – Technologies* (Filimowicz and Tzankova 2018), the emergence of user-centered design as a key concept that has opened up both exciting perspectives and tremendous challenges. The conceptual shift to a more comprehensive and emotional view of human-computer interactions has been accompanied by the development of numerous methods and tools for the design and evaluation of interactive systems, especially in the young field of *auditory display* (Goudarzi 2018). The collection of subjective and emotional data on the user experience is a necessary step to understand the users. We address this challenge by using a mix of methods (e.g., field studies like contextual inquiry, lab studies such as listening sessions and questionnaires) to gather requirements and to evaluate our suggested sound designs.

From everyday listening experience we – as human beings – are well trained in resolving complex sounds. Thus, the sonification of high-dimensional data, where full visualization is not possible, has high potential. Still, sound is a rather new medium of communication in science, and specified auditory tools are largely missing. We experiment with how sound design and sound perception are relevant and essential for designing data sonifications. With systematic design process, we collected requirements for the sonification tool using contextual inquiry to better understand the context of use of sonification in climate science (Goudarzi 2015). These studies revealed how climate scientists work with their data, what kind of visualizations tools they use to analyze data, and how and where is an auditory display applicable in their workflow. Furthermore, we found analogies between climate parameters and sound parameters to better design sound. The purpose of our case studies on sound design was to get an overview on the climate scientists' aesthetic preferences and designing data-to-sound mappings that represented their perception. Furthermore, designing sound in the context of sonification gives the scientists new tools to explore and perceptualize data. There are many hidden features and patterns in data that are certainly more perceivable through sonification due to humans' auditory capabilities. Experimenting with the sounds from case studies reveals that more in-depth studies of sound design are essential in this field.

Notes

1 https://supercollider.github.io.
2 https://cycling74.com.
3 https://puredata.info.

References

Alborno, Paolo, Andrea Cera, Stefano Piana, Maurizio Mancini, Radoslaw Niewiadomski, Corrado Canepa, Gualtiero Volpe and Antonio Camurri. "Interactive sonification of movement qualities–a case study on fluidity." *Proceedings of ISon* (2016).

Alexander, Robert L., Sile O'Modhrain, D. Aaron Roberts, Jason A. Gilbert and Thomas H. Zurbuchen. "The bird's ear view of space physics: Audification as a tool for the spectral analysis of time series data." *Journal of Geophysical Research: Space Physics* 119, no. 7 (2014): 5259–5271.

Archer, Bruce. "Design as a discipline." *Design studies* 1, no. 1 (1979): 17–20.

Barrass, Stephen. "The aesthetic turn in sonification towards a social and cultural medium." *AI & society* 27, no. 2 (2012): 177–181.

Boren B., M. Musick, J. Grossman and A. Roginska. "I hear NY4D: Hybrid acoustic and augmented auditory display for urban soundscapes," in *Proc. ICAD*, 2014.

Buchanan, Richard. "Wicked problems in design thinking." *Design issues* 8, no. 2 (1992): 5–21.

Candey, R.M., Schertenleib, A. M. and Diaz Merced, W.L. (2006). "xSonify Sonification Tool For Space Physics," in *Proceedings of the 12th International Conference on Auditory Display (ICAD)*, London, UK.

Carron, Maxime, et al. (2014). "Designing Sound Identity: Providing new communication tools for building brands" corporate sound" *Proceedings of the 9th Audio Mostly Conference: A Conference on Interaction With Sound.*

Carron, Maxime, Thomas Rotureau, Françoise Dubois, Nicolas Misdariis and Patrick Susini. "Speaking about sounds: a tool for communication on sound features." *Journal of Design Research* 15, no. 2 (2017): 85–109.

Chion, Michel. *Audio-vision: sound on screen*. Columbia University Press, 2019.

Cross, Nigel. "Forty years of design research." *Design studies* 1, no. 28 (2007): 1–4.

Dahl, L., Herrera, J., and Wilkerson, C. (2011). "TweetDreams: Making Music with the Audience and the World Using Real-Time Twitter Data." In *Proceedings of the International Conference on New Interfaces for Musical Expression (NIME)*. Oslo, 2011.

Delle Monache, Stefano, Pietro Polotti and Davide Rocchesso. (2010). "A toolkit for explorations in sonic interaction design." In *Proceedings of the 5th Audio Mostly Conference: A Conference on Interaction With Sound*, 1–7.

Dewitt, Anna, and Roberto Bresin. "Sound design for affective interaction." In *International Conference on Affective Computing and Intelligent Interaction*, 523–533. Berlin, Heidelberg: Springer, 2007.

Dubus, Gaël, and Roberto Bresin. "A systematic review of mapping strategies for the sonification of physical quantities." PLOS ONE 8, no. 12 (2013): e82491.

Erkut, Cumhur, Stefania Serafin, Michael Hoby and Jonniy Sårde. (2015). "Product sound design: Form, function, and experience." In *Proceedings of the Audio Mostly 2015 Conference on Interaction With Sound*, pp. 1–6.

van Egmond, R. van. "Impact of sound on image-evoked emotions." In: Rogowitz, B.E., Pappas, T.N. (eds.) *Human Vision and Electronic Imaging XIII*, vol. 6806 of Proc. SPIE, San Jose, CA, 6806OG–1–6806OG–12, Jan. 2008.

Farnell, Andy. "An introduction to procedural audio and its application in computer games." In *Audio mostly conference*, vol. 23. 2007.

Fenko, Anna, Hendrik N.J. Schifferstein and Paul Hekkert. "Noisy products: Does appearance matter?." *International Journal of Design* 5, no. 3 (2011).

Jamie Ferguson and Stephen Brewster. (2019). "Evaluating the magnitude estimation approach for designing sonification mapping topologies." In *Prooceedings of the 25th International Conference on Auditory Display (ICAD 2019)*. Georgia Institute of Technology.

Filimowicz, Michael. "Peircing Fritz and Snow: An aesthetic field for sonified data." *Organised Sound* 19, no. 1 (2014): 90–99.

Filimowicz, Michael, and Veronika Tzankova, eds. *New Directions in Third Wave Human-Computer Interaction: Volume 1 – Technologies*. Springer International Publishing, 2018.

Gaver, William W. "How do we hear in the world? Explorations in ecological acoustics." *Ecological psychology* 5, no. 4 (1993): 285–313.

Gaver, William W., Randall B. Smith and Tim O'Shea. (1991). "Effective sounds in complex systems: The ARKOLA simulation." In *Proceedings of the SIGCHI Conference on Human factors in Computing Systems*, 85–90.

Goudarzi, Visda. "Designing an interactive audio interface for climate science." *IEEE MultiMedia* (2015).

Goudarzi, Visda. (2016). "Exploration of sonification design process through an interdisciplinary workshop." In *Proceedings of the Audio Mostly Conference 2016*, 147–153.

Goudarzi, Visda. "Sonification and HCI." In *New Directions in Third Wave Human-Computer Interaction: Volume 1- Technologies*, pp. 205–222. Springer, 2018.

Grond, Florian, and Jonathan Berger. "Parameter mapping sonification." In *The sonification handbook*, pp. 363–397, 2011.

Grond, Florian, and Thomas Hermann. "Aesthetic strategies in sonification." *AI & society* 27, no. 2 (2012): 213–222.

Hearst, Marti A. "Dissonance on audio interfaces." *IEEE Expert* 12, no. 5 (1997): 10–16.

Hermann, T., Nehls, A. V., Eitel, F., Barri, T. and Gammel, M. (2012). "Tweetscapes: Real-time sonification of twitter data streams for radio broadcasting." In *Proceedings of International Conference on Auditory Displays, 2012*, 113–120.

Hermann, T., G. Baier, U. Stephani and R. Helge. (2006) "Vocal Sonification of Pathologic EEG Features," in *Proceedings of International Conference on Auditory Displays 2006*, 158–163.

Hug, D. "Investigating narrative and performative sound design strategies for interactive commodities." In S. Ystad, M. Aramaki, R. Kronland-Martinet, and K. Jensen, ed., Auditory Display -6th International Symposium, CMMR/ICAD 2009, Copenhagen, Denmark, May 18–22, 2009, Revised Papers, volume 5954 of Lecture Notes in Computer Science. Springer, 2010.

Janata P., Childs E. (2004) MarketBuzz: A Sonification of Real-Time Financial Data. In: Barrass S. and Vickers P. (eds.) *Proceedings of the Tenth International Conference on Auditory Display*, Sydney, Australia.

Jansen, Reinier J., Elif Özcan and René van Egmond. "Psst! product sound sketching tool." *Journal of the Audio Engineering Society* 59, no. 6 (2011): 396–403.

Kramer, G., Walker, B., Bonebright, T., Cook, P., Flowers, J., Miner, N.; Neuhoff, J., Bargar, R., Barrass, S., Berger, J., Evreinov, G., Fitch, W., Gr hn, M., Handel, S., Kaper, H., Levkowitz, H., Lodha, S., Shinn-Cunningham, B., Simoni, M., Tipei, S. "The Sonification Report: Status of the Field and Research Agenda." Report prepared for the National Science Foundation by members of the International Community for Auditory Display. ICAD, Santa Fe, NM, 1999.

Krause, Bernard L. "The niche hypothesis: a virtual symphony of animal sounds, the origins of musical expression and the health of habitats." *The Soundscape Newsletter* 6 (1993): 6–10.

Landry, Steven, and Myounghoon Jeon. "Interactive sonification strategies for the motion and emotion of dance performances." *Journal on Multimodal User Interfaces* (2020): 1–20.

Mauney, B.S., and B. N. Walker. (2004) "Creating functional and livable soundscapes for peripheral monitoring of dynamic data," in *International Conference on Auditory Display ICAD-04*, Sydney, Australia.

Misdariis, Nicolas, and Andrea Cera. (2013). "Sound signature of Quiet Vehicles: state of the art and experience feedbacks." In *INTER-NOISE and NOISE-CON Congress and Conference Proceedings*, vol. 247, no. 5, 3333–3342. Institute of Noise Control Engineering.

Mynatt, Elizabeth D., Maribeth Back, Roy Want, Michael Baer and Jason B. Ellis. (1998). "Designing audio aura." In *Proceedings of the SIGCHI conference on Human factors in computing systems*, 566–573.

Neuhoff, John. *Ecological psychoacoustics*. Brill, 2004.

Nyeste, Patrick, and Michael S. Wogalter. "On adding sound to quiet vehicles." In *Proceedings of the Human Factors and Ergonomics Society Annual Meeting*, vol. 52, no. 21, pp. 1747–1750. SAGE Publications, 2008.

Özcan, Elif, Gerald C. Cupchik and Hendrik NJ Schifferstein. "Auditory and visual contributions to affective product quality." *International Journal of Design* 11, no. 1 (2017): 35–50.

Özcan, Elif, René van Egmond, Alexandre Gentner and Carole Favart. "Incorporating Brand Identity in the Design of Auditory Displays: The Case of Toyota Motor Europe." In *Foundations in Sound Design for Embedded Media*, pp. 155–193. Routledge, 2019.

Pigrem, Jon, and Mathieu Barthet. (2017). "Datascaping: Data sonification as a narrative device in sound-scape composition." In *Proceedings of the 12th International Audio Mostly Conference on Augmented and Participatory Sound and Music Experiences*, 1–8.

Robinson, Frederic Anthony, Oliver Bown and Mari Velonaki. (2020) "Implicit Communication through Distributed Sound Design: Exploring a New Modality in Human-Robot Interaction." In *Companion of the 2020 ACM/IEEE International Conference on Human-Robot Interaction*, 597–599.

Schafer, R. Murray. *The soundscape: Our sonic environment and the tuning of the world.* Simon and Schuster, 1993.

Schoon, A., and Dombois F. (2009). "Sonification in music." In *Proceedings of the 15th International Conference on Auditory Display*, Copenhagen, Denmark, May 18–22, 76–78.

Sorkin, Robert D. "Why are people turning off our alarms?." *The Journal of the Acoustical Society of America* 84, no. 3 (1988): 1107–1108.

Susini, Patrick, Olivier Houix and Nicolas Misdariis. "Sound design: an applied, experimental framework to study the perception of everyday sounds." *The New Soundtrack* 4, no. 2 (2014): 103–121.

"Tuuri, Kai, and Tuomas Eerola. "Formulating a revised taxonomy for modes of listening." *Journal of New Music Research* 41, no. 2 (2012): 137–152.

Vickers, P., C. Laing, M. Debashi and T. Fairfax. (2014). "Sonification aesthetics and listening for network situational awareness," in *Proceedings of the Conference on Sonification of Health and Environmental Data*.

Walker, Bruce N. "Magnitude estimation of conceptual data dimensions for use in sonification." *Journal of experimental psychology: Applied* 8, no. 4 (2002): 211.

Walker, B. N. Cothran, J. T. (2003). "Sonification Sandbox: A graphical toolkit for auditory graphs." In *Proceedings of the International Conference on Auditory Display* (ICAD 2003). 161–163.

Walker, Bruce N., and Gregory Kramer. "Mappings and metaphors in auditory displays: An experimental assessment." *ACM Transactions on Applied Perception (TAP)* 2, no. 4 (2005): 407–412.

Wolf, KatieAnna E. (2015). "Assisting End Users in the Design of Sonification Systems." In *Proceedings of the 20th International Conference on Intelligent User Interfaces Companion*, 125–128.

Worrall, D. (2009). The use of sonic articulation in identifying correlation in capital market trading data. In *Proceedings of the 15th International Conference on Auditory Display*, May 18–22, Copenhagen, Denmark.

Worrall, D. (2015). "Realtime Sonification and Visualization of Network Metadata (The NetSon Project)," in *The 21st International Conference on Auditory Display (ICAD 2015)*, July 8–10, Graz, Austria, 337–339.

Worrall, David. " Computational Designing of Sonic Morphologies." *Organised Sound* 25, no. 1 (2020): 15–24.

Zbikowski, Lawrence Michael. *Conceptualizing music: Cognitive structure, theory, and analysis.* Oxford University Press (on demand), 2002.

Zwicker, Eberhard, and Hugo Fastl. *Psychoacoustics: Facts and models.* Vol. 22. Springer Science & Business Media, 2013.

15

Creating and Evaluating Aesthetics in Sonification

Núria Bonet

1 The Importance of Aesthetics in Sonification

Sonification transmits information through sound. In order to render data understandable to the listener, the sonification designer must make aesthetic choices that support this transmission. The term aesthetics does not necessarily denote something "beautiful" or "pleasing." Rather, aesthetics refers to what can be perceived by the senses. It describes a set of choices that inform a musical or sonic product; a sonification is necessarily aesthetic as it is perceived by our senses. Aesthetic design choices will determine how the sonification sounds, and what effect it produces.

Aesthetics can have an enormous impact on the success of a sonification within a research context. Sonification with appropriate "aesthetics and good sound design" is more likely to be chosen and used (Supper 2012, 180). The acceptance of a sonification is linked with a number of aesthetic attributes including pleasantness, loudness, noticeability, clarity and integration. However, one of the big challenges facing the field is its interdisciplinary nature. Its difficulty lies in "integrating concepts from human perception, acoustics, design, the arts, and engineering" (Walker and Kramer 2004, 7). Calls for an increased collaboration between experts and researchers, as well as an increased focus on sound aesthetics have been commonplace in the field for decades (Blattner et al. 1989; Kramer 1994; Walker and Kramer 2004; Vickers and Hogg 2006; Barrass and Vickers 2011). Unfortunately, successful collaborations between researchers from scientific and musical disciplines remain too scarce; sonification researchers predominantly stem from the musical field (Neuhoff 2019, 328). Attention-grabbing projects such as the CERN's Collide Program which chooses an Artist-in-Residence have been criticized for producing impressive aesthetic results but little transmission of knowledge. Rioji Ikeda's 2015 *Supersymmetry* was described by a journalist as "a lot of sound and light, signifying nothing" (Jones, 2015). There is still much scope for researchers to investigate the

DOI: 10.4324/9780429356360-16

aesthetics and information transmission within the field of sonification, whether for functional or artistic purposes.

The best aesthetics for a given sonification will depend on its objective. When designing an aural display, we need to consider a number of factors, including the original data and what meaning we are seeking to extract. For sonifications that run over a long period of time (for example, a heart rate monitor), "improved aesthetics will likely reduce display fatigue" (Kramer 1994, 53). We should reiterate that "aesthetics" does not necessarily refer to "pleasant" music and sound; it can include unpleasant auditory experiences too. Alarms are loud and unsettling, and unsuitable to prolonged listening. Their aesthetic is highly appropriate to their function as they should raise the listener's attention and prompt them to act. The sonic environment of a hospital provides an insight into the link between a sonification's purpose and its aesthetics. A hospital environment demands quick and precise sonic information. A heart rate monitor should be clearly discernible within a dense sonic environment, without becoming irritating over a long period of time, when there is no emergency. It should be precise and correlated to the severity of an emergency, and raise immediate concern in the listener (Sinclair 2012). The aesthetics of a sonification must be linked to its meaning and purpose.

There is a strong case to suggest that aesthetics are important in research, as they are in music and sound art. Paul Dirac proposed the idea of "beauty" in his *Principle of Mathematical Beauty*. He stated that a mathematician, for example, should strive for beauty in their work because the "simple, that is, the beautiful, brings understanding more readily" (Dirac 1939). This a useful consideration in sonification design too, as simpler displays can bring a greater understanding of the data. Furthermore, aesthetics can support the concept of "usefulness": the "function of the task it is being used to support" (Barrass 1997, 21). The same dataset could be approached with different aesthetics depending on its usefulness (i.e., the purpose it is trying to fulfil) (Barrass 2012, 178). An appropriate aesthetic approach helps to cross the boundary between data and information transmission (Barrass and Kramer 1999). Without context, data does not provide information (Roddy and Furlong 2014, 75); the context is created through aesthetics.

2 The Importance of Evaluation in Sonification

Research is carried out according to established methods of inquiry. The scientific method stipulates that a question is formulated and, in order to reach an answer, a hypothesis is derived and tested, and experiments are analyzed and evaluated. Assuming that a sonification is designed for a purpose, it should also be tested and evaluated to ascertain whether it is fit for the intended purpose. Unfortunately, evaluation is not sufficiently carried out in the field of sonification. At the International

Conference on Auditory Display (ICAD) in 2012, for example, only 1 out of 53 papers included an evaluation (Degara et al. 2013, 167). The 2019 ICAD featured a higher number of papers that included a robust evaluation or testing aspect, but it was still not a ubiquitous practice. One of the challenges that researchers face in this regard is that there are no established or prescribed methods for evaluating sonifications. There are no "specific guidelines" on developing sonifications either (Ibrahim et al. 2011, 77). This gap in the field can result in a lack of rigor and accountability in the research methods of sonification designers. Kramer proposes adopting a "methodical research approach" with a "benchmarking framework that allows for the comparison of sonification algorithms" (Kramer 2004; Degara et al. 2013), as this will contribute to raising the quality and profile of the field.

The question of evaluation becomes more complex because of sonification's interdisciplinary nature, which requires a mixed method approach that borrows testing methods from scientific and artistic research. Functional aims of an auditory display are mostly evaluated through task-based testing and statistical evaluation. Musical evaluation is a far more divided question because of its subjective qualities, as well as a reticence to evaluate the "musical." This argument might stem from a romanticized idea of music as an art that requires inspiration and cannot be fully described. However, composition is largely determined by musical craft and skills that can be taught and learned. Genres are based on common conceptions and rules. It is, therefore, possible to evaluate music according to certain guidelines while accepting that there is an inherently subjective element to musicality that cannot be fully described in words or numbers. The concept of the evaluation of music is important because it introduces accountability in musical research. How can we understand whether we have achieved our research aims and objectives if we have no criteria to answer research questions? To sum up, the evaluation of the aesthetic aspects of sonification is not only possible, but necessary.

Some researchers have proposed evaluation frameworks for sonification. The 2004 Listening to the Mind Listening competition asked different composers to produce a sonification on the same EEG dataset (Barrass et al. 2006). It included an "aesthetics" category rated on a scale from 1 to 5. Degara et al.'s Sonification Evaluation eXchange proposes a community-based platform that enables the definition and evaluation of standardized tasks for the formal comparison of sonification methods, supporting open science standards and reproducible research in the context of ICAD (Degara et al. 2013). While this framework recognizes the urgent need for a community effort to create a standardized process and criteria for evaluation, no specific aesthetic criteria are defined either. Vogt's evaluation criteria includes interesting elements such as "Gain" (what is gained from using sonification over other representation methods) and "(Sound) Amenity," which asks whether the sonification is sonically pleasing (Vogt 2011). Finally, Williams proposes a set of questions to evaluate musification (a sonification with the data subjected to a "set of musical constraints"). These also ask whether the display is "audibly pleasant" (Williams 2016). While some existing frameworks do take aesthetics into consideration, they do not query their correct application sufficiently.

3 Creating Aesthetics

Creating suitable aesthetics for a sonification requires a framework that accounts for every step in the sonification process. Previous research by Barrass and Vickers (2011) and Vickers and Hogg (2008) provides further discussion around the importance of aesthetics and useful design concepts in sonification. The *Data-Mappings-Language-Meaning* framework (Bonet et al. 2016a) incorporates these discourses and divides the process into four parts, which influence aesthetic decisions. The process is not always chronologically linear and the individual factors affect the rest of the process. While the framework was first developed for "artistic sonifications," its principles are applicable to auditory displays with a variety of purposes. It is intended for Parameter-Mapping Sonification (PMSon) (see Goudarzi's Chapter in this volume),[1] but could be adapted to other forms of auditory display such as audification or Model-Based Sonification.

3.1 Data

An excellent understanding of data is required to choose a dataset, apply transformations to it, and choose appropriate mappings and musical context. This refers to an understanding of the dataset's content, but also its meaning. Once we understand both the content and the meaning of the dataset, we can make informed choices regarding the next steps in the sonification process: mappings, aesthetic language and meaning.

The data needs to be suitable for aural display as not all datasets benefit from aural, rather than visual, representation. Thus, the choice of data is an aesthetic choice. At this point we should warn of the danger of the "big data fetish," which refers to using complex data in the belief that bigger data implies scientific credibility (Bjørnsten 2015). Often, the opposite is true because complex data is more difficult to handle and transmit, resulting in an incomprehensible aural display. Furthermore, humans tend to prefer music with average complexity (this is heavily influenced by factors such as age and musical experience, for example); excessive complexity or randomness is commonly disliked by listeners (Güçlütürk and van Lier 2019).

Sound is inherently temporal. Datasets that have a chronology or timeline can therefore offer the designer an obvious starting point for mapping. However, not all types of data have an order that can create a timeline. There are different types of data. Barrass (1997, 43) describes qualitative and quantitative data types that can

Table 15. 1 Data types describing different phenomena

Data types	Definition	Examples
Nominal	Difference without order	Banana, apple, orange
Ordinal	Difference, order and metric	Green, crisp, ripe
Interval	Difference, order and metric	Temperature
Ratio	Difference, order, metric, and natural zero	Rainfall

Barrass, 1997, 43.

be defined as "nominal," "ordinal," "interval" and "ratio" (see Table 15.1). Where there is no time vector, a time dimension needs to be extrapolated from correlations between other vectors to create a chronology; temporal relationships need to be found in the dataset (Rhoades 2014). Adderley and Young's sonification of Sahelian soil samples, *Ground-Breaking*, creates a "timeline" in the non-temporal data. This data manipulation introduces an element of decision-making on the outcome of the sonification by its designers (Adderley and Young 2009, 408). The handling of data before its mapping is paramount to the outcome, but we must be aware of the possibility of skewing its meaning. Referring back to Barrass' (1997, 43) description of data types, those with an "order" describe a temporality, even where it might not appear obvious at first (e.g., an apple – "green, crisp, ripe").

Handling the data before mapping it to sound is integral to the sonification process as we choose what information we want to transmit. Common data handling choices include filtering and compressing. Huge datasets (of dark matter data, for example) require filtering in order to bring forward salient features in the data (Rhoades 2014; Bonet et al. 2016b). Preliminary work with the data can in fact reveal interesting patterns that the designer might choose to highlight in the sonification.

An understanding of music perception can be useful to choose a suitable dataset. Data that presents some sort of musical structure might be inherently more appealing to our ears because we can make use of proven aesthetic constructs. Sonifications that include a form of repetition mirror musical conventions (e.g., chorus in a song). Being able to recognize shapes or patterns in a sonification can encourage the listener to choose and use a sonification because of the "something to hold on factor ... useful musical devices that support the listening experience" (Landy 2007, 23). This is shown, for instance, by the multitude of existing sonifications of DNA (Ohno and Ohno 1986) and EEG waveforms (Wu et al. 2009). Finally, data that has salient features can be sonified in a manner similar to a melody and an accompaniment. A complex dataset might be filtered so that salient data features are assigned more salient sound mapping, while other data points might be sounded as an accompaniment to the main patterns of the sonification (Bonet et al. 2016b).

3.2 Mappings

The mapping of data to sound is the process whereby data becomes audible by mapping data parameters to sound parameters. This process might be considered the most creative aspect of the sonification because it is "as arbitrary as it is decisive" (Adderley and Young 2009, 408). The designer has an enormous range of mapping possibilities which will strongly determine the outcome of the sonification process. PMSon offers a lot of flexibility because mappings can be chosen specifically to suit the dataset (Hermann and Ritter 2004, 734). However, the near infinite possible combinations of data to sound mappings are a challenge in themselves as there "is a very large set of possible mappings but a notoriously small subset of perceptually cognitively, valid mappings" (Roddy and Furlong 2014, 70). Because there is a limited set of effective mappings, they need to be carefully considered and chosen according to the purpose of the sonification. The aesthetic strategies described below can be used by designers to do so.

The concept of primary and secondary data and sound parameters is useful when thinking about effective mappings. Some sound parameters are easier to differentiate and understand; the ear perceives more detail in certain sound characteristics. Pitch and rhythm are parameters that humans can understand precisely as they provoke a neurological response (Large 2010), while differences in other sound parameters, such as panning or timbre, are more difficult to hear. Most sonification designers are empirically aware of it: a 2013 survey shows that pitch is the most-used parameter (23.8 percent) (Dubus and Bresin 2013, 15). In complex datasets, some data parameters are also more cognitively important than others. Some of the features or patterns in the data are more salient or relevant to the information and meaning in the data. Thus, the data can be divided into *primary* and *secondary data parameters*.

We can also define *primary* and *secondary sound parameters*. Respective data and sound parameters can be mapped to each other. "Primary cues" are attributed to sound parameters to which we are particularly sensitive and capable of perceiving even small changes. These sound parameters could include "pitch, tremolo rate, rhythm and attack time." Secondary cues can include "volume, pan position, number of harmonics or envelope shape" and are more difficult to perceive accurately (Ballora 2014, 31). However, supporting auditory cues can help differentiate primary cues. Ballora gives an example: "data points in an increasing pattern may produce a primary cue of ascending pitches, but these pitch changes may be complemented with corresponding changes in harmonic content and/or pan position" (Ballora 2014, 31). The concept of primary and secondary cues has clear parallels with the musical concepts of melody and accompaniment, where an accompaniment can help shape and elevate the perception of a melody. Designers often use primary parameters intuitively (Bonet 2019a, 91); however, a purposeful approach that accounts for human perception can help to produce more successful sonification quicker.

An emotional connection between data and sound can help a sonification to be more successful. Arbitrary data-to-sound mappings and a lack of emotional connection between data and sound often fail to engage the listener (Smith 1990). Intuitively, we feel that medical images, for example, ought to sound different from demographic data or satellite imagery (Smith 1990). Two different datasets could be mapped in a similar manner and produce near identical results, yet the emotional connection between data and sound might be lacking. How this connection might be achieved depends on the choice of mappings, but also on the aesthetic language. Another aesthetic mapping spectrum is called *indexicality*, which denotes how strongly a sound "sounds like the thing that made it" (Vickers and Hogg 2006, 213). High indexicality describes a literal mapping while a low degree of indexicality might denote a more metaphorical mapping, using, for example, musical motifs to transmit data. Gresham-Lancaster uses the term "second order sonification" to describe a mapping that takes advantage of cultural connotations, musical structures and the connection between frequency, timing, timbre, etc. (Gresham-Lancaster 2012, 210). He argues that second order sonification affords more flexibility because it is not a literal mapping; more abstract sonic relations and concepts can be used to effectively transmit information. The challenge of this approach is to retain an emotional connection between data and sound. Once again, the designer must carefully consider the best approach for a given dataset in order to convey the intended meaning.

3.3 Aesthetic Language

The aesthetic language used in any sonification must be chosen to serve the purpose of the sonification. We are referring to aesthetic, rather than musical, language to avoid limiting the scope of the framework solely to artistic applications; although the aesthetic conventions of sound display are closely linked to musical languages. The vast majority of sonification follow the aesthetic conventions of Western classical music, including a chromatic range of pitches, major and minor scales, regular key signatures, etc. The reasons for this are straightforward. Most sonification researchers are familiar with these musical conventions. Also, popular tools and technical standards, such as MIDI, are based on the Western scale.

Shared knowledge of an aesthetic language can be used to transmit information effectively between sonification designer and listener. Major and minor scales often have connotations of sounding "happy" and "sad," while dissonance can be used to signify an error or an alarm. Furthermore, pre-existing sounds or music can be used as sonic material in mapping to create meaning. Auditory icons, for example, are aural metaphors that "provide an intuitive linkage" (Brazil and Fernström 2011, 325) between the world represented and the sound heard. They require "an existing relationship between the sound and its meaning" (McGookin and Brewster 2011,

339), such as the paper crunching noise of the Recycling Bin in Windows. Similarly, in PMSon an existing piece of music can be used as sonic material and modified through mappings to offer new meaning to the sound and the data (Bonet 2019b). However, converting any data into "Mozartian pastiches" is inappropriate as it disregards the needs of the individual dataset (Bjørnsten 2015). Different data types benefit from different aesthetic conventions in sonification. The designer needs to determine the best aesthetics in order to choose suitable mappings.

An important aesthetic choice must be made when considering sounds on either concrete or continuous scales. Most sonification use concrete frequency scales, for example diatonic pitches. The challenge with a concrete scale is that data might need to be arranged in concrete bands before mapping it to specific pitches. An infamous example of this procedure is the Higgs-Boson particle sonification by the LHC Symphony Orchestra (Vicinanza 2012) that mapped particle mass to diatonic notes. Pesic and Volmar (2014) describe this mapping as "pure artefact" because it misrepresents the actual difference in particle mass. A continuous, or even microtonal, scale would have been preferable as it would have translated the data more accurately. Discarding any scales using intervals smaller than a semitone is a waste of sonic resources because the human ear can only discriminate around a quarter of a semitone (varying depending on the frequency and musical experience) (Zarate et al. 2012, 987).

A sound-based aesthetic approach can also produce effective sonification results and create an emotional linkage between data and sound. Earcons use sounds with an existing meaning to describe the data and are particularly useful in functional applications. However, sound manipulation methods from the field of electroacoustic and acousmatic music can be appropriated for data-to sound mappings, and offer new ways of communicating information.[2] Possibilities of sound mappings when working with sounds include frequency filters, dynamics, granulation synthesis parameters and spatialization. The spatial aspect of pre-recorded sounds is particularly interesting for creating immersive sonifications that provide the listener with a new experience of data. Natasha Barrett's work is of particular interest in this regard. Her piece *Viva la Selva* (1999) used spatial data collected from animal vocalization in a jungle over the course of 24 hours. The vocalizations were also recorded and mapped spatially (x-dimension by panning, y-dimension by pitch, z-dimension by filtering and reverberation) to create an accurate and immersive experience of the jungle environment (Barrett 2000, 22). Her work demonstrates potential uses of a sound-based approach to sonification, where the aesthetic language supports the transmission of information and meaning.

3.4 Meaning

The aim of a sonification should be to transmit the meaning of the data. As discussed previously, aesthetics provide the context to make the meaning apparent to the listener. It can be useful for designers to think about the ways that a message

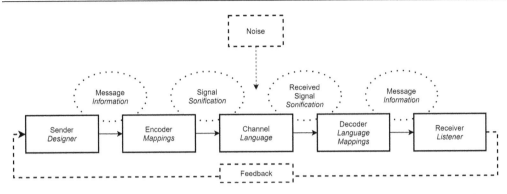

Figure 15.1 Shannon-Weaver Mathematical Theory of Communication applied to the sonification process (in italics). The terms in boxes describe the five parts of the communication process, while the annotated arrows describe the transmission between parts. Noise is an external influence on the channel which can distort the signal and the feedback describes the message returned from the Receiver to the Sender.

gets transmitted to a listener, as described, for example, by the Shannon-Weaver Mathematical Theory of Communication. The model includes an *information source*, a *transmitter*, a *channel*, a *decoder* and a *destination* (Shannon and Weaver 1949, 33–34). The system can also contain *noise*, which is any interference to the signal which creates "distortions" or "errors in transmission" (Shannon and Weaver 1949, 8). *Feedback* denotes the return message from the destination (receiver) to the information source (sender), although it was introduced by later theorists (Chandler 1994). The Shannon-Weaver model can be applied to sonification: the information source is the *sonification designer*, the transmitter (or encoder) is the *mappings*, the channel is the *aesthetic language*, the decoder is the *listener's understanding* of the mappings and language used, and the receiver is the *listener* (see Figure 15.1).

Visualizing the sonification process within the Shannon-Weaver model highlights the importance of aesthetics at every step, but also shows that every step is woven together. Decisions at either stage are not chronological or linear; the designer might need to reconsider the handling of the data, for example, once they have considered the language and mappings. Let us reconsider Roddy and Furlong's statement that information is "data coupled with context" (2014, 75). The message is sent to the listener through the means of sonification. In order to understand this message, the listener needs to be aware of the medium (sonification), method of sonification (mappings) and aesthetics language to be able to decode it and access the message. Thus, the listener requires a context to understand the sonification (see Figure 15.2). This context is provided by a shared understanding between the designer and the listener of certain steps in the transmission process. The understanding stems from a shared sense of aesthetics.

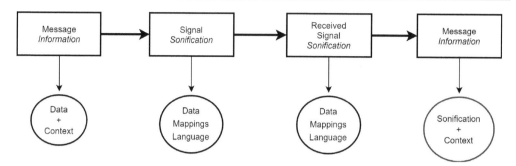

Figure 15.2   The Shannon-Weaver Mathematical Model of Communication describes the Message, Signal and Received Signal as different types of forms that the information takes in the process. The vertical arrows indicate the composition of each within the context of sonification.

4 Evaluating Aesthetics

A mixed method approach is required for the evaluation of the aesthetics of sonification. The framework must combine functional and aesthetic criteria to determine the success of a sonification's sound.

The subjective nature of the perception of aesthetic parameters in sonification renders a definitive evaluation improbable. It is useful, however, to consider some of the ways in which musical works – sonic aesthetic objects – are evaluated; school and conservatory music curricula contain evaluation criteria for performance and composition skills, for example. A musical work can be considered through a technical, cultural, contextual or emotional lens, amongst others. It is possible to assess whether a piece of music conforms to a required technical skill, an assigned genre, an intended meaning, etc. – whether it meets the specific assessment criterion. Such a criterion remains to some extent subjective, but can help towards a general consensus of the quality of a musical work. Similarly, the aesthetics and success of a sonification can be evaluated and statistical significance can be achieved with an appropriate framework and sufficient testing.

The evaluation framework proposed combines existing criteria proposed by Barrass et al. (2006), Vogt (2011) and Williams (2016) and adapts them to focus on the question of aesthetics. Table 15.2 shows the criteria organized according to overarching criteria used in the framework.

Gain describes what is achieved through the aural display of data. It asks about the purpose of a sonification – whether purely functional or with artistic intentions – and whether it has been achieved. It is also linked to the evaluation of the aesthetics, as these will contribute to this purpose. It asks whether the aesthetics serve its purpose. Finally, it might also consider the 'congruency' of a sonification, and whether it accompanies a visualization in a purposeful manner, (see for example Williams 2016). The criterion of *Intuitivity, clarity and learning effort* asks whether

Table 15.2 Sonification evaluation criteria combining criteria from Barrass et al. (2006), Vogt (2011) and Williams (2016).

Barrass et al.	Vogt	Williams	Proposed criteria
/	Gain	/	*Gain*
	Clarity	Intuitivity, Efficiency	*Intuitivity, clarity and*
Accessibility	Learning effort	/	*learning effort*
Mapping	/	/	
Aesthetics	Amenity	Amenity	*Sonification method*
/	Contextability	Congruency	
/		Immersion	
Overall Impression	/	/	*Listener feedback*

the information transmission process is efficient. It also asks which learning effort the sonification requires and whether it is accessible to a wider audience. Within the context of aesthetics, it asks whether these provide the simplest learning effort possible for the listener. This criterion is concerned with the choice of mappings; are the chosen data-to-sound mappings effective? Assessing the *Sonification method* requires an examination of the aesthetic decisions taken during the process; it evaluates the application of the *Data-Mappings-Language-Meaning* framework. *Listener feedback* asks about the participant's experience of the sonification.

The evaluation criteria can be used for testing with listeners. The questions proposed below (see Table 15.3) show how each criterion can be examined. This a mixed method approach that allows for quantitative and qualitative enquiry. The questions can be answered by the listener with a rating (e.g., on a scale from 1 to 5) or a written answer. The combination of both will provide the designer with the nuanced feedback necessary to determine the success of the sonification and its aesthetics.

5 Conclusion

The field of sonification faces recurrent and persistent issues that have frustrated researchers for decades. Its challenges are paradoxical at times: while music researchers are in the majority, sonification aesthetics and success rates often remain unsatisfactory (Neuhoff 2019, 328). We have argued that a stronger focus on aesthetics and systematic evaluation are key to solving some of these long-standing issues. While calls for collaboration between scientists and composers are not new, sound designers might provide the expertise required to complement existing ideas and research in the field (Barrass and Vickers 2011, 164).

Sonification is often carried out on an empirical basis, as the vast differences in individual auditory perception (particularly between musicians and non-musicians) makes it difficult to generalize sonification methods. However, we can certainly synthesize guidelines that tend to be accepted as correct (Neuhoff 2019, 328); this chapter

Table 15. 3 Proposed evaluation questions for sonification.

Questions for evaluation of a sonification		
Gain	Does the data display derive gain from an auditory display? Does the sonification serve its purpose?	A(esthetic) Rating: F(unctional) Rating:
Intuitivity, clarity and learning effort	Are the data and the information clearly heard? Is the sonification accessible? What does the listener need to know or learn to understand the data? Does the sonification create an emotional connection between the data and the sound?	A Rating: F Rating:
Sonification method	<u>Data:</u> Is the data suitable for sonification? <u>Mappings:</u> Do the mappings best transmit the information (clearly, precisely, aesthetically)? <u>Aesthetic Language:</u> Is the aesthetic language suited to this dataset? <u>Meaning:</u> Does it transmit the intended meaning to the listener?	A Rating: F Rating:
Listener feedback	What does the listener hear, understand and feel? How they rate the sonification overall?	A Rating: F Rating:

in fact collates empirical findings from the sonification community. Continued work on the aesthetic perception of auditory display can only advance this endeavor.

The evaluation framework presented here has been developed from empirical experiences of sonification design, as well as listener feedback, in order to offer a method to assess and compare the success of sonifications. It will hopefully be applied and further refined as researchers turn towards a mentality that involves more thorough evaluation. There is no reason that sonification designers and researchers should not be held accountable through robust and universal evaluation methods. A solid research method that includes design and evaluation guidelines which take every step of the sonification process through an aesthetic lens will hopefully be a step in the right direction.

Notes

1 Parameter-Mapping Sonification is the most common sonification method. The sonifications are created by mapping data points to sound events, such as frequency, amplitude, duration, etc. (Hermann and Ritter 2004, 734).
2 See, for example, the suggestion to use Curtis Roads' "aesthetic premises and aesthetic opposition" in sonification (Roads 2004 in Barrass and Vickers 2011, 163–64).

References

Adderley, W. Paul and Michael Young. 2009. "Ground-breaking: Scientific and Sonic Perceptions of Environmental Change in the African Sahel." *Leonardo* 42, no. 5: 404–411.

Angliss, Sarah. 2011. "Euler's Number and the price of fish." Mad Art Lab. http://madartlab.com/eulernumberfish/ (accessed August 18, 2020).

Ballora, M. 2014. "Sonification, Science and Popular Music: In search of the 'wow'." *Organised Sound* 19, no. 1: 30–40.

Barrass, Stephen. 1997. "Auditory Information Design." PhD Thesis, Australian National University.

Barrass, Stephen. 2012. "The aesthetic turn in sonification towards a social and cultural medium." *AI & Society* 27: 177–181.

Barrass, Stephen and Gregory Kramer. 1999. "Using sonification." *Multimedia Systems*, 7: 23–31

Barrass, Stephen and Paul Vickers. 2011." Sonification Design and Aesthetics." In *The Sonification Handbook*, edited by Thomas Hermann, Andy Hunt and John G. Neuhoff, 167–95. Berlin: Logos Publishing House.

Barrass, Stephen, Mitchell Whitelaw and Freya Bailes. 2006. "Listening to the Mind Listening: An Analysis of Sonification Reviews, Designs and Correspondences." *Leonardo Music Journal* 16: 13–19.

Barrett, Natasha. 2000. "A compositional methodology based on data extracted from natural phenomena." *Proceedings of the International Computer Music Conference*: 20–23.

Bjørnsten, Thomas. 2015. "From Particle Data to Particular Sounds: Reflections on The Affordances of Contemporary Sonification Practices." *Journal of Sonic Studies* 10.

Blattner, Meera M., Denise A. Sumikawa and Robert M. Greenberg. 1989. "Earcons and Icons: Their Structure and Common Design Principles." *Human-Computer Interaction* 4: 11–44.

Bonet, Núria. 2019a. "Data Sonification in Creative Practice." PhD Thesis, University of Plymouth.

Bonet, Núria. 2019b. "Musical Borrowing in Sonification." *Organised Sound* 24, no. 2: 184–194.

Bonet, Núria, Alexis Kirke and Eduardo Miranda. 2016a. "*Blyth-Eastbourne-Wembury*: Sonification as a compositional tool in electroacoustic music." *Proceedings of the 2nd International Conference on New Musical Concepts*.

Bonet, Núria, Alexis Kirke and Eduardo Miranda. 2016b. "Sonification of Dark Matter: Challenges and Opportunities." *Proceedings of the Sound and Music Computing Conference*.

Brazil, Eoin and Mikael Fernström. 2011. "Auditory Icons." In *The Sonification Handbook*, edited by Thomas Hermann, Andy Hunt and John G. Neuhoff, 325–338. Berlin: Logos Publishing House.

Chandler, Daniel. 1994. "The Transmission Model of Communication." http://visual-memory.co.uk/daniel/Documents/short/trans.html (accessed January 28, 2021).

Degara, Norberto, Frederik Nagel and Thomas Hermann. 2013. "Sonex: An Evaluation Exchange Framework for Reproducible Sonification." *Proceedings of the 19th International Conference on Auditory Displays*: 167–174.

Dirac, Paul A.M. 1939. "The Relation between Mathematics and Physics." *Proceedings of the Royal Society (Edinburgh)* 59, no. 2: 122–129.

Dubus, Gaël and Roberto Bresin. 2013. "A Systematic Review of Mapping Strategies for the Sonification of Physical Quantities." *PLoS ONE* 8, no. 12: 1–28.

Gresham-Lancaster, Scot. (2012). "Relationships of sonification to music and sound art." *AI & Society* 27: 207–212.

Güçlütürk, Yağmur and Rob van Lier. 2019. "Decomposing Complexity Preferences for Music." *Frontiers in Psychology* (April). https://doi.org/10.3389/fpsyg.2019.00674.

Hermann, Thomas. 2011. "Model-based Sonification." In *The Sonification Handbook*, edited by Thomas Hermann, Andy Hunt and John G. Neuhoff, 399–427. Berlin: Logos Publishing House.

Hermann, Thomas and Helge Ritter. 2004. "Sound and Meaning in Auditory Data Display." *Proceedings of the IEEE* 92, no. 4: 730–741.

Ibrahim, Ag Asri Ag, Fouziah Md Yassin, Suaini Sura and Ryan McDonnell Andrias. 2011. "Overview of Design Issues and Evaluation of Sonification Applications." *2011 International Conference on User Science and Engineering (i-USEr)*: 77–82.

Jones, Jonathan. "Should art respond to science? On this evidence, the answer is simple: no way." *The Guardian*, April 23, 2015.

Joyce, A.E. 2006. "The coastal temperature network and ferry route programme: long-term temperature and salinity observations." *Science Series Data Report*, 43.

Kramer, Gregory. 1994. "An introduction to auditory display." *Studies in the Sciences of Complexity Proceedings* 18: 1–78.

Kramer, Gregory. 2004. "A letter from Greg Kramer: founder of ICAD." *Proceedings of the International Workshop on Interactive Sonification*: 1–2.

Landy, Leigh. 2007. *Understanding the Art of Sound Organization*. Cambridge, MA: The MIT Press.

Large, Edward W. 2010. "Neurodynamics of Music." In *Music Perception*, Springer Handbook of Auditory Research, 36, edited by Mari Riess Jones, Richard R. Fay and Arthur N. Popper, 201–31. New York: Springer.

McGookin, David and Stephen Brewster. 2011. "Earcons." In *The Sonification Handbook*, edited by Thomas Hermann, Andy Hunt and John G. Neuhoff, 399–427. Berlin: Logos Publishing House.

Neuhoff, John G., ed. 2004. *Ecological psychoacoustics*. New York: Academic Press.

Neuhoff, John G. 2019. "Is Sonification Doomed to Fail?" *Proceedings of the 25th International Conference on Auditory Display*, 327–330.

Ohno, Susumu and Midori Ohno. 1986. "The All Pervasive Principle of Repetitious Recurrence Governs Not Only Coding Sequence Construction But Also Human Endeavor in Musical Composition." *Immunogenetics* 24: 71–78.

Pesic, Peter and Alex Volmar. 2014. "Pythagorean Longings and Cosmic Symphonies: The Musical Rhetoric of String Theory and the Sonification of Particle Physics." *Journal of Sonic Studies* 8.

Rhoades, Michael. 2014. "Hadronized Spectra (The LHC Sonifications): Sonification of proton collisions." *eContact!* 16, no. 3.

Roads, Curtis. 2004. *Microsound*. Cambridge, MA: The MIT Press.

Roddy, Stephen and Dermot Furlong. 2014. "Embodied Aesthetics in Auditory Display." *Organised Sound* 19, no.1: 70–77.

Shannon, Claude E. and Warren Weaver. 1949. *The Mathematical Theory of Communication*. Urbana: The University of Illinois Press.

Sinclair, Peter. 2012. "Living with alarms: the audio environment in an intensive care unit." *AI & Society* 27: 269–276.

Smith, Stuart. 1990. "Representing Data with Sound." *Proceedings of the IEEE Visualization 1990*. Piscataway, NJ: IEEE Computer Society Press.

Supper, A. 2012. "Lobbying for the Ear: The Fascination with and Academic Legitimacy of the Sonification of Scientific Data." PhD Thesis, Universitaire Pers Maastricht.

Vicinanza, Dominico. 2012. "LHC Open Symphony." http://lhcopensymphony.wordpress.com (accessed August 21, 2020).

Vickers, Paul and Bennett Hogg. 2006. "Sonification Abstraite/Sonification Concrète: An 'Æsthetic Perspective Space' for Classifying Auditory Displays in the Ars Musica Domain." *Proceedings of the 12th International Conference on Auditory Displays*: 210–216.

Vogt, Katharina. 2011. "A Quantitative Evaluation Approach to Sonifications." *Proceedings of the 17th International Conference on Auditory Display*.

Walker, Bruce and Gregory Kramer. 2004. "Ecological Psychoacoustics and Auditory Displays: Hearing, Grouping and Meaning Making." In *Ecological Psychoacoustics*, edited by John G. Neuhoff, 150–175. New York: Academic Press.

Williams, D. 2016. "Utility Versus Creativity in Biomedical Musification" *Journal of Creative Music Systems* 1, no. 1.Wu, Dan, Chao-Yi Li and De-Zhong Yao. 2009. "Scale-Free Music of the Brain." *PLoS ONE* 4, no. 6, e5915.

Zarate, Jean Mary, Caroline R. Ritson and David Poeppel. 2012. "Pitch-interval discrimination and musical expertise: Is the semitone a perceptual boundary?" *The Journal of the Acoustical Society of America* 132, no. 2: 984–93.

16

Revealing Industry Culture
A Cultural Ethnographic Approach to Postproduction Sound

Vanessa Theme Ament

1 Cars, Computers and Cinema

The 2017 Volvo S60 has a GPS system that is time consuming and takes up to two minutes to log in all of the information for your destination. You must dial in a city, a road and an address letter by letter or number by number, until enough information is accessed; then you turn the dial to the arrow that leaps over to the assemblage of choices that best match what you want. You then dial down the list to the city name, street name and number out of a myriad of possibilities. If you make a decision to alter your trajectory and get off the designated route – most often utilizing the nearest freeway or highway – you will find the GPS routing you back to the previous route. If you are in difficult traffic, you will get a message every few minutes asking you if you want to save a minute by altering your route. Normally, this means getting off your freeway at an upcoming exit, driving on the frontage road for a mile or two, then getting back onto the freeway at a new spot. The messages that pop up must be manually exited or they occlude your GPS map. The voice assistant is very polite, always saying the word, "please" prior to every verbal direction. There are newer systems in more recent Volvo models that allow touch screen access, but the process is the same.

The 2017 Honda Accord has a GPS system that is fast to load. You can enter a business name which is easily entered and filled in, and choices pop up. You click and go. A similar ease occurs when you begin to type in an address. Cities and street names fill in with minimal data. The GPS will give you alternate route information quickly, regardless of how you may have deviated from the original trajectory and does not occlude the map. The system is intuitive and simple to understand and operate. The voice assist interrupts more frequently, and does not say "please" ever.

Both GPS systems will record a previous destination, however the Volvo GPS will often record by zip code if you enter a street name several times, but with different

DOI: 10.4324/9780429356360-17

address numbers. Trying to recall which zip code your desired address includes is obstructive. The Honda GPS keeps previous destinations by address and name.

Why such a different experience on two excellent cars? The Volvo has its own proprietary system,[1] and the Honda uses Garmin, originally designed and manufactured in the US, as its navigation system.[2] The execution of each system seems to be reflective of the culture from where it is derived. The Volvo is designed in Sweden and reflects a more linear and precise method for interaction, relying on the sensibilities of a fairly homogenous society with a shared philosophy of process. Similar to the recognizable functional design of Swedish furniture, which was imbued with a sense of a society built on social democracy constructed to equalize access to taste and design,[3] the Volvo's navigation system incorporates a step by step approach that is non-discriminatory. The Honda's GPS reflects the more intuitive approach necessary for mass democratization of a more heterogenous society that relies less on a common cultural approach, and more on a simplicity of use required by a fast-paced, "snackable" American consumer.

There is a similarity to the comparison of the PC, a Windows-based interface, with the Mac, with its own operating system. The PC will most often be considerably less expensive and requires a more labor-intensive navigation through many steps of utilization, while the Mac is intended to be more fluid, intuitive and to appeal to the more creative clientele in the computer ecosystems. As with the PC and Mac comparison, the intended client and the design of the platform coincides with the culture from where it came: PC from the business oriented culture utilizing documents and spreadsheets, and Mac from the more out-of-the-box model intended for the intuitive approach of creative users and those preferring the fluid and proprietary Apple product.[4]

It is with this cultural sensitivity applied to the observation of a work product that causes me to argue for my chosen method – industrial ethnography – to research and interpret the work practices, cultural environs and aesthetic choices of film sound professionals as I compare the various methods utilized to research the area of sound studies in film, television and other media. Media cultures are as different as the Swedish GPS is from the American designed Garmin, and the cultural influences surrounding the design or production of any creation are deeply relevant to the purpose of and execution by the workers involved with the process. Much as a navigation system or a computer represents the cultures from where they derive, so too do the soundtracks designed by film professionals reveal their cultural roots.

2 From Positivism to Materialism

The accepted approaches for promoting scholarship in sound studies typically involve codified methods for research and reportage of the art, analysis, technology and historical contexts of the discipline. With Rick Altman's call to investigate the "dark corners"[5] of sound studies, there has been an expansion that is a welcome and enriching addition to scholarship and has introduced new perspectives on what should be studied and how. Whether responding to others' research or forging entirely fresh territory based on their own investigations, scholars are developing an amplified body of literature about the history and development of film sound and the progression to modern sound design.

The technological advances that alter the execution and exhibition of sound in film, television, games, as well as online, is an approach that allows the reader to more deeply comprehend the tools and the workflows of the "makers" of film sound. The historical contextualizing of film sound design explains the filmmakers' contributions that led to the progression in film sound and the growth and innovations in the industry itself. A star studies approach, that of revealing more about the most notable celebrities in the field while filtering through the objective lens of the scholar, includes their philosophical approaches to sound, an accounting of the important contributions and qualities of sound design and anecdotes about working on various projects. Analysis of aesthetic choices and their meaning within the media example, lend a more theoretical and interpretive approach, to discern the artistic nature of the sound design, and its meta support for the film narrative. These examples of film and sound studies methods derive from the traditional practices of objective research and analysis.

The nineteenth century practice of positivism, as expressed by August Comte, the founder of present day sociology, requires three principles for objective research: empirical science is the only source of positive knowledge of the world, minds must be cleansed of mysticism and superstition, and scientific knowledge and technical control should be extended to society to make technology "primarily political and moral."[6] These tenets developed into a goal of "applying the achievements of science and technology to the well-being of mankind,"[7] that can be related to the concept of "Technical Rationality," which credits technology with progressing what is considered rational in society, and has some connection to technological determinism.[8]

As a response to the acknowledged need to ground observations in the more practical knowledge, these constructs became a method of observing experiential phenomenon.

As scholars research the production and exhibition of film sound, the recent addition of the perspectives of those who work in the field has allowed the lines between theory and practice to be more fluid and has provided the reader to gain valuable insights into the creative expression of those professionals. To that end, a more ethnographic based research has evolved to include film sound professionals as part of the media industry's subgroup of production studies. Often this approach involves observing the execution of the work and the surrounding environs, interviewing sound professionals themselves and deriving observations and conclusions that support a thesis.

Scholars often utilize what would be considered a respected positivistic method of research; one that includes an external and social scientific perspective of reality. This *Etic* perspective is the more traditional method for research and provides an objective view which offers an analysis based on distance and non-involvement. Ethnographers, however, utilize a more phenomenological method, or *Emic* perspective; one that utilizes the insider's view of reality. Ethnographic research "embraces a multicultural perspective because it accepts multiple realities."[9]

Ideational research theories – cognitive, which "assumes that we can describe what people think by listening to what they say," and ethnosemantic, which argues "we can create taxonomies of how people view the world" from linguistic practices – focus on mental activity as the motivation for fundamental change. In contrast, materialistic theories – those that focus on material conditions such as natural resources, money and modes of operation – focus on cultural change through observable human patterns.[10] Ethnographers use one or both of these categories of theories in research, and tend to use theoretical models that are indirectly related to grand theories, rather than relying on grand theories themselves.[11] The use of fieldwork – going to the environments related to the anthropological research – is "the most characteristic element of any ethnographic research design."[12]

3 Reflecting as Practitioner and Researcher

Traditionally, scholarship has been enculturated into a system of silos: one for objective observation and one for subjective practice. As a consequence, "[t]he hierarchical separation of research and practice is also reflected in the normative curriculum of the professional school." As a reaction to this separation, the concept of "tacit knowing" or "tacit knowledge," first recognized by Michael Polanyi, refers to an innate knowledge that is not easily described or communicated, and has more recently come into favor for ethnographic researchers.[13] A common example of this kind of knowing is the understanding of riding a bike. While the rider is adept at the skill, and has learned it from others and from experience, the actual method of learning was not from being told how to do it as much as the continued practice of doing, which then translates to competence, and eventually expertise. In *The Reflective Practitioner*, by Donald A. Schon, this concept is elaborated upon and expanded to further explore that knowledge in experts of practice regarding how they do their work, how they learn it and how they communicate it to others.[14] The difficulty in communicating how one does her work presents complications for observers and scholars who wish to understand the work process and culture. While observing will provide some useful information, truly understanding how the work is approached and executed requires an understanding of process that is not easily ascertained. "Our knowing is ordinarily tacit, implicit in our patterns of action and in our feel for the stuff with which we are dealing. It seems right to say that our knowing is *in* our action."[15] The practitioner gains expertise by doing, not studying, and reflects upon her work while in the middle of it, in addition to reflecting while thinking about or discussing the work with others.

How then does one who researches the work of the practitioner ensure that the observation and gathering of information is as complete, contextual and comprehensive as is possible? How does that researcher get into the headspace of the practitioner in a way that will reveal more knowledge than previously shared? One method that seems reasonable is that she must become a reflective researcher, one who "cannot maintain distance from, much less superiority to, the experience of practice... [s]he must somehow gain an inside view of the experience of practice."[16] By utilizing a methodology that allows more insight and tacit knowledge of the researcher, the practitioner reveals depth and reflection more readily revealed and understood. The practitioner "reveals to the reflective researcher the ways of thinking that he brings to his practice, and draws on reflective research as an aid to his own reflection-in-action."[17]

As ethnographic techniques have evolved, the writer's personal experiences may be incorporated into the research to add context and depth to the observations and interviews.[18] This approach is intended for both academic and public audiences, as it adds the stories of those interviewed and allows an interpretation utilizing the reflective researcher's experiences and tacit knowledge of the culture and work world of the storyteller. This "New Ethnography," as proposed by H.L. Goodall, Jr., "is constructed out of a writer's ability to hold an interesting *conversation* with readers."[19]

4 Cultural Ethnography Methodology

Over the past two decades, business sociology has led to an industrial ethnography that typically immerses the observer within a labor practice to better understand various aspects of the work product. Often this approach is utilized to observe the practices of those in factories and "making" settings to assess economic applications, labor relations and the thinking behind the design of the product. The observer is most often an objective expert or scholar who interviews, experiences and interprets the ethnography to assert conclusions that effect the business practices themselves,[20] or the relationship between the practices and education,[21] the political philosophy which underpins the practices,[22] in addition to the creative choices effecting the practices.[23] The mechanical and technological tools involved are included in the ethnography as well. A more recent trend is the industrial ethnography within a business or profession that encourages reflection on industrial practices of those who share a particular expertise.[24] These conferences, blogs and panels allow shared experiences to be explored for the betterment of the group and common good.[25]

The area of production studies as a subset of media industries allows this ethnographic observation to expand as a method of gleaning the individual and collaborative artistic decisions within television, film and other media industries.[26] Using a cultural studies approach, writers, costume designers, directors, producers, agents, actors and even audiences are the focus of this method of research.[27] Most often, scholarship in this area is from the viewpoint of those who are familiar with the discipline, but not from the discipline itself. By practicing a more egalitarian social anthropology technique of interviewing as a peer, or from a place of equal status, valuable insights are derived from such methodology, as work not usually observed by the outsider is examined and interpreted by those trained in solid scholarship and media theories.[28]

The exclusive experience of a sound designer is not typically understood by one outside the culture without interviews to better comprehend the thinking behind artistic choices and business decisions. These interviews necessitate reflections of what to inquire, discern the coding of "shop talk" and terms of art and conclude meanings that are accurate in their representation of the work. Considering the added layer of deep texts that only those from the field utilize and comprehend entirely,[29] there are many obstacles to properly and accurately interpreting the creative interactions involved in designing film sound. Even the most esteemed scholar with a corpus of exceptional publications can find herself in a morass of missteps in decoding the peculiar culture of sound design, as it has evolved within a closed system of filmmakers and sound professionals. Add to this the innate nature of promoting one's work as is required to stay employed in film and television, and it is not mysterious why some scholarship might continue the perception that various notables in the field properly represent the true nature of the profession, and that those most visible are valid barometers of the less visible aspects that exist in film sound.

5 From Doing to Observing

As a long-time professional in postproduction sound, I find myself in an unusual situation. I can appreciate and applaud the excellent research published about sound design and postproduction sound. The exposure afforded film through the rigorous scholarship of dedicated academics is a boon to the visibility and understanding of the more recently appreciated contribution to the film studies corpus. Additionally, there are trade publications that lend insight from various sound professionals' stories about their work on specific projects. While there have been articles written about sound professionals that share perspectives not otherwise in print, and various DVDs that include narration by the more renowned sound designers explaining approaches to their art in the film, there is a need for the industrial ethnography that only one who understands the complicated encoding and decoding[30] involved in the enigmatic aspects of film sound can proffer. What is often missing are scholarly articles, chapters and books written with the objectivity and academic rigor of accepted scholarship but also with the added value of the "inside baseball" knowledge of the culture, and the habitus,[31] or "feel for the game," that comes with having been a practitioner in the field. What complicates this kind of contribution is the rarity of scholars who have been, themselves, professionals in the culture of which they write. This very conundrum was the purpose for my perceived exit from the film industry and pursuing a Ph.D. in film.

As a film industry Foley artist and sound editor from 1980 until 2004, when I moved into academia teaching Foley, ADR and sound editing at DePaul University, I noticed a strong divide between those who are "makers" and those who are "scholars" in the navigation of film courses for students. It seemed a rather unnatural chasm to me, as we who have worked on high profile films and television shows are often categorized as non-theorists, when in reality, there is a great deal of industrial and practical theory to every step of filmmaking. While the discussions surrounding the work do not utilize the same lexicon as in film studies, many of the aesthetic concerns and creative decisions involve theorizing how to approach the sound narrative and the meaning conveyed with the artistic choices. What is different is these conversations are encoded with the understanding that they are decoded within a particular knowledge base. Thus, only the trained ear, with a clear understanding of the film practitioner's language derived from the culture, will truly comprehend the depth of the seemingly obvious conversations. The old adage, "it takes one to know one," is particularly apt when referring to the process of creating and executing the industrial business of filmmaking. The culture of the film and television industry is a closed system and an industrial milieu that can be hard to understand if you have never been an inhabitant of it. Thus, the conversations, workflows, aesthetic choices and day-to-day challenges of working in postproduction sound contain nuances best communicated with a colleague who has experienced the work and culture.

Good scholarship requires analysis derived from the education, training and mentorship, provided in graduate studies, as well as the act of writing, editing and publishing. The component of peer review by colleagues provides essential feedback for the scholar, so that publications are as relevant and professional as possible. The desired objectivity of the scholar, which is invaluable, can also be an obstacle to access the coding inherent in a field that has been observed, but not inhabited.

When first learning the language of film studies as my discipline, it was a daunting task for this film industry veteran to adjust to the approach of studying the "object" of media as opposed to involving myself in the practice. Yet, my goal was always to close the gap between those who make the films and those who analyze what is made. There are areas of disconnect between both professions, and the *lingua obscura* of film theory and visual studies bears little resemblance to the typically onomatopoetic communication style of sound experts. There are, however, areas of crossover, much like a Venn diagram. These areas are accessible for a disciplined scholar to discern and evoke when interviewing or observing film and television professionals. There are, however, numerous coded conversations that will not be shared or understood unless observed by a colleague "in the know," who is trusted and has the experiences to interpret subtleties properly, contextualize information and understand the invaluable "follow-up" questions and comments. These crucial conversations require the intimate relationships of those who have shared experiences and know the players. The added value of a practitioner from postproduction sound is the access to a myriad of film professionals who know and trust the interviewer to understand the work and culture. This allows for a comfortability and casualness in the conversations and stories that are shared and understood only as one who has walked the talk can evoke. An industrial ethnographic approach can reveal stories from practitioners in the industry that are essential additions, for they not only amplify the artistic decisions employed and the nuances of the workplace, but "stories are commonly used to make sense out of ambiguous situations or to represent sense-making in earlier events."[32] They also add "detail and color," and can "represent the way the world is... or the way the world should be."[33] Additionally, "what is omitted from their accounts is at least as revealing as what is included."[34] These omissions are intuited by a colleague from the culture and can afford essential opportunities for reflexive follow-up questions that reveal more depth than might be obvious to a more "objective" interviewer.

It has been my goal to relate the cultural and artistic influences involved in postproduction sound from the perspective of the practitioner turned scholar. This endeavor is a balancing act that requires retaining the connection to the workers while including the detached analysis of the academic. This task is not easy if the purpose is to reveal new perceptions, add the voices of the less well-known professionals, and share these observations through the lens of one who has worked in the field, while employing the objective clarity of an observer. So, too, will the practitioner/researcher more easily understand the economics of the industry, and the impact of budget constraints. The workflows of sound editors, sound mixers, Foley artists, sound "designers" who add custom sounds to a soundtrack, or the "sound designer," and supervising sound editor who oversees the entire postproduction soundscape, are familiar terrain for one who has collaborated with professionals in all of these creative crafts. It is with this dual set of qualifications that I pursue my work in industrial ethnography of postproduction sound.

6 From Ideating to Executing

In my first iteration of *The Foley Grail* (2009), my view was that of the Foley artist who was revealing stories and experiences of colleagues in the United States, primarily Los Angeles. My intention was to document the historical background and development of the craft, while illustrating the art and craft of Foley, the execution of the work, and some insights into the culture of working in the field. Additionally, after being part of the faculty at DePaul University in the Digital Cinema program and contributing to Foley and ADR classes at Columbia College in Chicago, I included my perspective on film schools as one who was making the transition from film professional to college instructor. The chapter "The Ivory Tower" was a comprehensive view of the state of film programs at the time.[35]

During my dissertating phase of my Ph.D. program at Georgia State University, I was tasked with writing a second edition of *The Foley Grail* to be released in 2014. My lens had changed considerably as I had participated in panels, designed curriculum and had presented demonstrations and lectures about Foley and postproduction sound. My writing voice had changed as well, as I had been exposed to excellent scholarly writing, had been mentored by experienced academics and was more concerned with revealing the world of Foley through a more objective and distant observer. I altered my introduction to conform to more scholarly books on film and discarded the arguably egocentric voice of my first edition. Chapters were revisited to incorporate more distance between my own experiences and those of my interviewees. With the advances in technology and the inclusion of the adoption of "long form" narrative productions, I included updated trends in the field, and more interviews with professionals from New York and the San Francisco Bay Area. "When society's mode of production changes, so does the nature of work."[36] This awareness required I reengage with mixers and editors who had altered their workflows to accommodate the tightening of schedules and budgets. I also added two chapters about sound for games and animation, and an additional chapter with an often requested "recipe" section of common strategies and practices for the neophyte Foley artist or film student. I altered the chapter on film schools to be more media industries based, as I had been influenced by the theories and methods in production studies and codified media history.

My reach across the oceans to include histories and cultural influences imbued in sound professionals from Russia, France, England, Serbia, Spain and Italy, in addition to interviews with Canadian sound professionals, resulted in observations of national differences and their effects on aesthetics and labor practices. Including this heteroglossia of experiences was a vital addition, since "language [the verbal exchange] is not a neutral medium that passes freely or easily into the private property of the speaker's intention... it is populated – overpopulated – with the intentions of others." As one who has a similar industrial portfolio as those I interviewed, I was able to reveal a more fluid interpretation of their observations and stories, which added a multinational consciousness.

In addition to the updates and deeper dives into theories and methods of postproduction sound, my focus was more on "they" rather than "we," which involved a clear and objective codified approach to the writing. I was more mindful to adhere to the integrity of cultural studies and social anthropology in this edition. My approach to

interviews was similar to previous endeavors; however, I found I was more inquisi-
tive about the broader picture of sound design and included interviews with more
editors and mixers to enhance the comprehensive nature of the book.[37] Both the first
and second editions of *The Foley Grail* have international translations, and thus my
profile as a documenter of film sound professionals has expanded, which allows me
access to my film colleagues in more obscure markets, which I am now including in
the third edition, with a planned release in 2021.

One of the more technologically driven aspects of film sound revolves around
the tools utilized in sound editing and mixing. The ubiquitous use of Pro Tools
as the contemporary industry standard has been well established. However, what
has heretofore been unclear is how film professionals transitioned from analogue
editing and mixing, working with both magnetic film and 24-track tape, to the
digital workstation as is now the norm for sound professionals. The journey was
not monolithic, nor was it smooth. There were several competing technologies in
use simultaneously, and opinions varied as to which would be the standard bearer,
and if analogue was truly going to disappear as a method of production. Much as
the initial introduction of sound to the motion picture was in the 1920s, the introduction of the digital workstation was disruptive and uneven for motion picture
sound professionals in the 1990s. At this time, I was the publisher of Moviesound
Newsletter,[38] a publication focused on film sound, with contributions by various
industry professionals during this technologically innovative era. Included in
this unique contemporaneous publication were several prescient jewels about the
restorations of classic film soundtracks, reviews of laserdisc sound and debates
about whether digital would revolutionize film sound much in the same way as
the CD had replaced vinyl records.

The transition of postproduction sound from analogue to digital technology as
we experienced in it in Hollywood, was that of a single film culture. At the time, New
York and the San Francisco Bay Area, were also highly visible in film production.
Those of us who experienced the qualities of this transition, in any of these three
regions, had some similar issues and experiences. However, the three film cultures
operated differently in the aesthetic philosophy and execution of postproduction
film sound. While this was apparent to film sound professionals at the time in the
three cities, it might not have been as obvious to others. My decision to write about
this transition from the lens of cultural geography as an essential aspect of the era,
was to highlight the specific needs of particular sound professionals, who chose different technologies through their concerns about aesthetics and workflows, as well as
the economics. As one who was immersed in that transition I was aware of the contentious attitudes of some sound professionals when the film industry was grappling
with these new technologies. This important transition was imperative to situate in
recent film history, as a cultural phenomenon, rather than simply a technological
one. With access to colleagues involved with this transition, I was able to interview

some of the key individuals who pioneered the use of digital technologies in New York and the San Francisco Bay Area, in addition to Los Angeles. My upcoming book, *Divergent Tracks: How Three Film Communities Revolutionized Digital Film Sound*, which will be published by Bloomsbury in 2021, is the result of this industrial ethnographic method of research.[39]

During my tenure at Ball State University as the Endowed Chair in Telecommunications (2014–2019), I received a fellowship at the Virginia Ball Center for Creative Inquiry to produce a film about women in film sound. After interviewing student applicants, I assembled 13 students from various disciplines, who enrolled for one semester in my immersive learning course. I undertook the complicated task of coordinating course work relating to ethnography, film sound, gendered labor and advanced film techniques to prepare the students for a professional "Hollywood style" film shoot focused on interviews with women working in production and post-production sound, or who owned sound facilities.

Sound supervisor and director Victoria Sampson led a week-long workshop for the students to further enhance their industrial style of film production, and sociology professor Dr. Rachel Kraus provided classes about gender in the workplace. After five weeks of course work, location scouting, storyboarding, workshops in camera and sound, assigning jobs for each student and arranging interviews and trips to New York and Los Angeles, the students immersed themselves in the task of filming interviews with over 20 women who worked in film sound and subsequently edited and screened the film at the Muncie, Indiana, AMC theatres. The result was the 30-minute documentary, *Amplified: A Conversation with Women in American Film Sound* (2018),[40] which appeared in several film festivals and was awarded two nominations.

The film was a compilation of Foley artists, sound editors, production and post-production sound mixers and facility owners. Two of the interviewees were transgender and revealed details about life before and after transitioning while working as sound professionals. Women discussed juggling child rearing while working, relationships or the lack of them, sexism, ageism and the evolution of working side by side with men during cultural changes in our society and the industry. Understanding some of the struggles and circumstances confronted by the women working in a male-dominated aspect of filmmaking provided me a unique lens through which to consider the content, prepare my students to work with equal status as the interviewees and guide the students to research, prepare and interact with these film professionals with poise. Most of the interview questions were designed by the students with my guidance on how to speak with sound experts as peers rather than as students. I coached the two students chosen to be on-camera interviewers regarding how to bring out their own conversational style, and to listen with an ear for where the narrative might go, as the women told their stories. The students learned how to interview the sound professionals through the lens of industrial ethnography, and since this was a documentary project, to translate the ethnographic approach into the skills of active listening and evocative questioning so that the interviewees could elaborate at will while on camera.

As part of this immersive learning experience, the students edited, scored and titled the film, as I oversaw the production from beginning to end. My role was director and producer, but with the continued mission of empowering the students with exposure to my industrial guidance. This immersive learning experience, which is a fundamental value at Ball State University, was the result of my hybridization as a practitioner/scholar, and my cultural studies approach, which informs all of my scholarship.

7 The Practicing Researcher: Final Thoughts

Whether through newsletters, books, chapters, articles, dissertations, or film projects, the pursuit of an industrial ethnography model in both scholarship and filmmaking about the cultures within film sound reaps multiple rewards for the researcher, the reader and the scholarly discipline. This newer ethnographic method can help film scholars better reveal the nuanced and distinct cultures of the film industry. This approach embraces bringing the culturally distinct and often unheard voices of active sound professionals into the film studies literature. Rather than seeing a divide between the work of industry professionals and academic scholars, we can see and celebrate their points of commonality: their shared love of cinema, their desire to explore and expand the process of filmmaking, as well as their desire to create, shape and interpret meaning. By including more research from the perspective of a reflective researcher who hails from reflective practice, aesthetic choices may be more deeply examined, cultural practices may be more completely revealed and professional conventions may be more accurately decoded. It is an aspirational goal that film studies continues to embrace industrial ethnography as a valuable method to increase cultural insights in our interpretive and theoretical approaches to film studies.

Notes

1 "Navigation Services | Support and Legal Articles | Volvo Support." According to the Volvo Service Center in Ontario, California (which consulted Volvo North America for the purpose of my research) the navigation system is made in Sweden, and uses parts from other companies.
2 "Honda Navigation Updates."
3 Murphy, *Swedish Design*.
4 Between 2006 and 2009, Apple marketed their Mac computer with the "Get a Mac" ad campaign starring John Hodgman as a businessman in a suit and tie ("Hey, I'm a PC"), and Justin Long as a

young creative with an untucked shirt ("Hey, I'm a Mac"), to personify the distinctions between the two computers.

5 Altman, *Sound Theory, Sound Practice* p.71–78.
6 Schon, *The Reflective Practitioner*. p. 31–32.
7 Schon. p. 30.
8 Schon.
9 Fetterman, *Ethnography*.
10 Fetterman. p. 5.
11 Fetterman.
12 Fetterman. p. 7.
13 Schon. p. 26
14 Schon.
15 Schon. p. 49.
16 Schon. p. 323.
17 Schon. p. 323.
18 Goodall, *Writing the New Ethnography*. Italics appear within the quote.
19 Goodall. p. 13.
20 Orr, *Talking about Machines*.
21 Darrah, *Learning and Work*.
22 Murphy, *Swedish Design*.
23 Murphy.
24 Anderson, "Ethnographic Research."
25 "About EPIC."
26 Holt and Perren, *Media Industries*.
27 Mayer, *Production Studies*.
28 Ortner, Sherry B., "Studying Sideways: Ethnographic Access in Hollywood."
29 Caldwell, *Production Culture*.
30 Hall, Stuart, "Encoding and Decoding in the Television Discourse."
31 Bourdieu, *The Field of Cultural Production*.
32 Orr, *Talking About Machines*, p.12.
33 Orr, p. 12–20.
34 Darrah, *Learning and Work*, p.63.
35 Ament, *The Foley Grail* (2009).
36 Orr, *Talking About Machines*, Kindle location 28.
37 Ament, *The Foley Grail*, (2014).
38 Stone, *Hollywood Sound Design and Moviesound Newsletter*.
39 Ament, *Divergent Tracks: How Three Film Communities Revolutionized Digital Film Sound*.
40 Ament, *Amplified: A Conversation with Women in Film Sound*.

References

EPIC. "About EPIC." Accessed June 13, 2020. https://www.epicpeople.org/about-epic/.
Altman, Rick, ed. *Sound Theory, Sound Practice*. New York: Routledge, 1992.
Ament, Vanessa. *Amplified: A Conversation with Women in Film Sound*. Documentary, 2018.
———. *Divergent Tracks: How Three Film Communities Revolutionized Digital Film Sound*. Bloomsbury, 2021.

_____. *The Foley Grail: The Art of Performing Sound for Film, Games, and Animation*. 2nd edition. New York: Routledge, 2014.

_____. *The Foley Grail: The Art of Performing Sound for Film, Games, and Animation*. 1st edition. Focal Press, 2009Anderson, Ken. "Ethnographic Research: A Key to Strategy." *Harvard Business Review*, March 1, 2009.

Bourdieu, Pierre. *The Field of Cultural Production*. Edited by Randal Johnson. 1st edition. New York: Columbia University Press, 1993.

Caldwell, John Thornton. *Production Culture: Industrial Reflexivity and Critical Practice in Film and Television*. Durham, NC: Duke University Press Books, 2008.

Chang, Heewon. *Autoethnography as Method*. Walnut Creek, CA: Routledge, 2009.

Darrah, Charles N. *Learning and Work: An Exploration in Industrial Ethnography*. 1st edition. Routledge, 2013.

Emerson, Robert M., Rachel I. Fretz and Linda L. Shaw. *Writing Ethnographic Fieldnotes*. 2nd edition. Chicago: University of Chicago Press, 2011.

Fetterman, David. *Ethnography: Step-by-Step*. 4th edition. Los Angeles: SAGE Publications, Inc, 2019.

Goodall H. L., Jr. *Writing the New Ethnography*. Writing In Book edition. Walnut Creek, CA: AltaMira Press, 2000.

Hall, Stuart. *Essential Essays, Volume 1: Foundations of Cultural Studies*. Edited by David Morley. Duke University Press Books, 2018.

Hobart, Mark. "The Profanity of the Media." In *Media Anthropology*, 26–35. Thousand Oaks, CA: SAGE Publications, Inc., 2005.

Holt, Jennifer, and Alisa Perren, eds. *Media Industries: History, Theory, and Method*. 1st edition. Malden, MA: Wiley-Blackwell, 2009.

"Honda Navigation Updates." Accessed June 16, 2020. https://honda.garmin.com/honda/site.

Mayer, Vicki. *Below the Line: Producers and Production Studies in the New Television Economy*. Durham, NC: Duke University Press Books, 2011.

Mayer, Vicki, Miranda J. Banks and John T. Caldwell, eds. *Production Studies: Cultural Studies of Media Industries*. 1st edition. Routledge, 2009.

Muncey, Tessa. *Creating Autoethnographies*. 1st edition London: SAGE Publications Ltd, 2010.

Murphy, Keith M. *Swedish Design: An Ethnography*. Cornell University Press, 2019.

"Navigation Services | Support and Legal Articles | Volvo Support." Accessed June 16, 2020. https://www.volvocars.com/uk/support/topics/maps-navigation/navigation-services.

Orr, Julian E. *Talking about Machines: An Ethnography of a Modern Job*. 1st edition. ILR Press, 2016.

Ortner, Sherry B. "Studying Sideways: Ethnographic Access in Hollywood." In *Production Studies: Cultural Studies of Media Industries*, 175–190. Routledge, 2009

Powdermaker, Hortense. *Hollywood, the Dream Factory: An Anthropologist Looks at the Movie-Makers*. Mansfield Center, CT: Martino Fine Books, 2013.

Rosten, Leo. *Hollywood: The Movie Colony, the Movie Makers*. Harcourt Brace, 1941

Rothenbuhler, Eric, and Mihai Coman. *Media Anthropology*. Thousand Oaks, CA: SAGE Publications, Inc., 2005.

Schon, Donald A. *The Reflective Practitioner: How Professionals Think In Action*. 1st edition. New York: Basic Books, 1984.

Stone, David. *Hollywood Sound Design and Moviesound Newsletter: A Case Study of the End of the Analog Age*. 1st edition: New York: Routledge, 2016.

17

Sound Design as Viewed by Sound Designers
A Questionnaire About People, Practice and Definitions

Laura Zattra, Nicolas Donin, Nicolas Misdariis,
Frank Pecquet, David Fierro

1 What is Sound Design? Ask Sound Designers!

What is sound design? Those two words were tied together as early as 1959, when David Collison described audio engineering activity in theatre together with stage management (Collison 2008). It became widespread in the film industry in the 1980s, when Walter Murch was the first to be qualified as a sound designer in the credits of *Apocalypse Now* (1979) (Whittington 2007). During the same period, Raymond Murray Schafer devised his own concept of "acoustic design" (Schafer 1977), opening up a broad field of applications in subsequent sound design works. As a result, the pairing of "sound" and "design" points toward a diversity of meanings depending on the professional context, and there is a lack of agreement about the very definition of "sound design" and its general acceptance compared to the common understanding of "design" (even if the definition of this meta-discipline still remains under discussion), or, with regards to "music composition."

Some recurring components of a definition can be found in either academic or professional explanations of sound design, including the following: sound design is both an artistic and technical discipline; it implies multiple related disciplines and fields of application; it intrinsically fulfils a functionality (Misdariis 2018, 13–19). However, sound design remains a protean discipline under which a great number of artistic, hand-crafted, technical or scientific actions and productions can be consigned.

This being said, the real question is not whether there are common interpretations of this polymorphic and polysemic term, neither about an ultimate and potentially constraining definition to be stated. To this extent, we can rely on what can may be the largest acceptation of "sound design," given by the renowned sound designer, Louis Dandrel: "Sound design should not try to find any definition of itself other than within design" (cited in Rodriguez 2003) – in other words, making sound design is *designing with sounds.* In fact, today one may simply recognize that design concerns better quality (Vial 2015), and designing sound means working with artefacts of multiple kinds, be they physical, digital or even environmental, basically considering that sound has increased power in society, and sound design is a matter of improvement (Rocchesso et al. 2008), either about objects, services or communication (Pecquet 2017).

Furthermore, the profession of "sound designer" has barely been institutionalized and therefore does not benefit from the stabilizing forces of a codified profession. In Europe as well as in North America, sound design is characterized by a fluid and changing distinction between occupations and full professions – a typical feature of emerging professions according to the sociology of work (Durkheim 1893, cited in Demazière & Gadéa 2009; Bourdieu 1979; Menger 2003; McEwen & Trede 2014; Dent et al. 2016; Rogers et al. 2016; Susskind & Susskind 2016). In Europe, not every National Classification of Economic Activity has specific codes to define the productive activity of sound designers (as regards in France cf. Misdariis 2018, 14; Pecquet 2018, 7).

Consequently, rather than trying to offer a top-down definition of sound design, one would better construe it as an umbrella term regrouping different, distinct or intertwined practices whose genealogies must be deconstructed and related to pre-existing disciplines. Professional practices in sound design are carried out in a number of fields, well beyond those where the term was originally coined, including the following: architecture and urban environment, visual and digital arts, cinema and TV, advertising, ecology and acoustic regulation, industry, communication and marketing. Such professional practices are themselves characterized by interdisciplinarity. However, this does not mean that trained sound designers would not be able to coalesce around a set of core components of a definition. By providing the communities of sound design with appropriate instruments of self-understanding, the Analysis of Sound Design Practices Project (henceforth, ASDP) precisely aims at charting this complexity and advancing both knowledge and definition of sound design.

2 Investigating People, Processes and Products: the Analysis of Sound Design Practices Project

Previous research on design has developed methods for understanding how designers think and work. Our paradigm for acquiring knowledge of the discipline and its protagonists is inspired by various texts, among which those from Bruce Archer and Nigel Cross (Archer 1979; Cross 2001), and more recently (Vial 2015). On that basis, by analogy with some considerations of design, sound design may be considered as a proper "discipline," a research subject "on its own," and a specific, or "third" culture being halfway "between arts and humanities" (Archer 1979). The transposition of original formalisms established by Cross regarding the science of design (Cross 2001) is at the root of our framework for exploring our own discipline. Thus, similarly (Cross 2007), there should be three distinct researches *loci* for sound design: (1) "people" and stakeholders, the WHO, namely the status and practices of the sound designers; (2) "processes" and methods, the HOW, in other words the status of the sound design itself, including innovative strategies and tools with regards to sound prototypes (mock-ups, sketches, intermediary objects) or creativity/fixation mechanisms; and finally (3) "products" and sound artifacts, the WHAT, which is to say the status of the designed sounds, their new forms, formats and the listening situations or author/listener relationships they imply.

In a global view, ASDP aimed at exploring those three "P"s – people, processes and products – with, for now, a strong emphasis upon the first two. The project drew from previous research and data-gathering activities conducted amongst sound designers with an ethnographic approach. Amongst others, Hug & Misdariis (2011) was based on grounded theory; Carron et al. (2017) investigated communication tools and strategies (building a sound lexicon) when speaking about sounds with sound designers; Rocchesso et al. (2015), Boussard et al. (2016) or Houix et al. (2016) investigated vocal sketching as a tool for communication and creation; Misdariis & Cera (2017) examined, in a *post-hoc* approach, the role of the sound designer in a large-scale and long-term industrial project; Misdariis et al. (2015) presented an attempt of exploration through sound design by a comparative analysis of two emblematic industrial projects and a focus on the relationships between creativity and artistic/scientific approaches.

Our theoretical approach is also based upon bibliographical references devoted to historical, practical and epistemological issues: examples include amongst others (Farnell 2010; Wyatt & Amyes 2005; Gibbs 2007; Collison 2008; Kaye & LeBrecht 2009; Brown 2009; Dal Palù et al. 2018). The project also relied on written works dedicated to Sonic Interaction Design (SID), an emerging field at the intersection of interaction design, sound and music computing, auditory display, sonic arts and acoustics. Since the mid-2000s, SID researchers have been working actively to formalize tools, methodologies and conceptual frameworks in this scientific domain. A breakthrough advance to auditory interfaces and displays research has been the SOb project (Sounding Object 2001–2003; Rocchesso & Fontana 2003), followed amongst others by the SID COST Action (Sonic Interaction Design, 2007–2011; Rocchesso 2011) and the SkAT-VG project (Sketching Audio Technologies using Vocalizations and Gestures, 2014–2017; Rochesso et al. 2015). Those research and publishing projects (see also Pauletto 2014; Franinović & Serafin 2013; and for a large overview, Rochesso 2014) have helped a community of scientists and practitioners to federate.

With the intention to address the above-mentioned objectives, the ASDP project has been firmly grounded within an interdisciplinary team and is based on its members' previous investigations in domains such as sound design, sound design practices, semiotic of art and design, methodologies and the science of sound design (culminated in Misdariis 2018); sound design theoretical issues and sound anthropology (elaborated in Pecquet & Dupouey 2021); science, technology, society studies and collaboration in music (Zattra 2018); the historical and sociological study of emerging professions such as the computer music designer (Zattra & Donin 2016), as well as software development, automatic analysis and data visualization made by David Fierro. With these multiple shared skills, and by adding to this first round of research, ASDP aims to explore Cross's 3rd 'P' (the 'products' of sound design) and will do so alongside alternative methodological approaches, which are still in progress and therefore will not be addressed in this chapter.

Understanding the two "P"s – people and processes – means studying the presence of a variety of actors in the world, each of them having their own areas of activities, skills, profiles, training, day-to-day business and skills. Such diversity, and such a sparse population, needs to be mapped, geographically and socially. The prosopography (study of the common characteristics of a group of actors by means of a collective study of their profiles, careers, common background, practices) has been acquired from a census and from the analysis of an online questionnaire among different European sound designers. Additional interviews, not presented here, have also been conducted with four senior sound designers. In the following sections we will detail the theoretical framework of our project, the research foundation embodied in a previous similar study (section 3.1), the construction of the questionnaire including its postulates, reasons and contents (section 3.2), our findings in terminology issues, why and how we constructed the census in order to reach our participants (the cartography of this profession, section 3.3), and the inevitable topographical and conceptual limitations we are aware this tactic involves which we've been dealing with since the beginning (3.4). Section 4 will present the main findings of our analysis. While it is impossible to present the entire data set here, we invite anyone interested in further information to visit the ASDP website (https://thesounddesignproject.net/), to directly contact the authors, or to refer to past communications (Zattra et al. 2018 & 2019a, b; Misdariis et al. 2019).

3 Underpinnings, Methodology and Limitations

As dubious as any definition might seem (particularly in this field; see section 2 for further arguments), we sketched a list of general properties of sound design which shape our questions and subsequent findings. The literature and research scholarship along with our personal scholarship experience served as the basis to establish the following working definition of sound design: a collective practice; an answer to a multiple request/specification problem (that is often defined as "wicked," according to the proper design terminology (Rittel & Webber, 1973); a quest for the 'best' or the 'beautiful' sound (targeting a goal grounded on functional and/or aesthetic criteria); a practice where one's identity (musical/artistic) is offered in the service of the project (non "authoriality"); a constant compromise between personal tendencies and production constraints; a scientific attitude inspired by some academic fields like psychoacoustics, psychology, etc.

Together with the integration of a previous similar study (section 3.1), these agreements have shaped the methodology of our project: searching for professionals in this field (section 3.3) and elaborating a questionnaire grid (section 3.2).

3.1 Engaging an Emerging Professional Community Through an Online Questionnaire: Takeaways From a Previous Study

We derived the methodology for this research from a previous study (authored by two authors of this chapter) aimed at illuminating another emerging professional skill in sound and technology, namely 'computer music design.' An offspring of electronic music studios in the 1970s and 1980s, computer music design encompasses managing the technical setup of a new musical piece; helping the composer to cope with current technologies, sound effects and scientific knowledge; translating the composer's artistic ideas into programming languages; taking part in the performance of the piece (Zattra and Donin 2016, 437). Due to the small size of the population of computer music designers (also identified as "musical assistants," "tutors," etc.) at the time of the study, as well as the limited institutional recognition of this profession, even basic information was missing, and the study aimed to "describe this profession" by asking "how the [computer music designers] perceive their profession" as well as analyzing "their age, training and tasks" (438). The questionnaire had to address four main domains: time(s) of activity; legal status and recognition; tasks, skills and training; heritage and technological migration.

The questionnaire balanced quantitative and qualitative data collection. The former included "statistical repartition of ages, periods of activity, number of works and collaborations, training, type of contract, dissemination and archiving." The latter allowed "systematic comparison between assertions from the pre-existing literature that had informed the design of the questionnaire, and every corresponding answer and comment by respondents." The analysis focused on everything that "added to or corrected [those] pre-existing assumptions" (443).

While some computer music designers have been regularly working in the field over a long-time span and expect to enjoy unambiguous recognition of their accomplishments in this capacity, others might be reluctant to take part in such a survey. As a consequence, "the questionnaire should enable the respondents to speak anonymously of problematic aspects of their profession as well as overtly comment on peculiar artistic achievements as desired" (442), which led to anonymity by default, with an optional field allowing respondents to be identified when they agreed.

This study succeeded in sketching the profile of a majority of the estimated 40–60 people under consideration. The fine tuning of the computer music designer questionnaire, with respect to its structure, wording and administration, could be transferred to the Sound Design questionnaire due to many similarities between the objects under study, i.e., uncharted professional population with no formal training before the late 2000s, diversity in labelling the profession, multidisciplinary skills quickly evolving over time, inequalities in institutional support and social recognition.

3.2 The Questionnaire

Our team wrote the questionnaire in order to inventory sound design practices with regard to existing professional activities. The aim was to highlight the social, cultural and economic profile of a sound designer, seeking different types of information in education, methods, processes, project managing, marketing issues together with scientific and artistic research. Our questionnaire contained yes/no questions, multiple-choice, open-ended questions and numerical rating (Likert scale), resulting in a mix of qualitative and quantitative data.

We proposed 39 questions divided into 3 major parts, with both open- and closed-ended questions, with the aim of performing statistical and text analysis. The first part of the questionnaire consisted in a quantitative data collection, with the aim of creating a panorama of the profession. It had 19 questions, divided into 2 sub-parts. Part A (Questions 1–14) concerned the participants' personal profile (age, place of work, years and period of activity, background and training, areas and skills involved in each participant's profession, work status). Part B (Questions 15–19) analyzed the participant's teaching activity (if this is the case): environment, teaching levels, teaching methods, time dedicated to teaching. The second part of the questionnaire was optional. It contained 17 questions, divided into 2 subparts that referred to Methodology (Modus operandi) (Part C – Questions 20–30) and Technical Practice (Part D – Questions 31–36), and that jointly took into account different criteria: working position/framework; professional role and recognition, time frame and communication with stakeholders (duration of projects/phases); typical communication strategy and working methods (brainstorming, development, testing, revisions); personal archival of a project; hardware and software environments; technical features in collecting sound. The third part of the questionnaire (3 questions, optional – Questions 37–39) investigated general thoughts about sound design, especially investigating the sound designer's self-perception together with institutional/legal issues like copyright policy or professional associations.

3.3 Looking for Sound Designers: Language and Geography Issues

In order to send out our questionnaire, the first step was to find out whom to contact, hence to find practitioners in Europe, the first-chosen geographical region. In addition to the professionals we already knew from our respective communities, mapping sound designers has led to searching for names over the Internet. While English is the obvious language (the mother tongue, and etymologically, of the original terms to define this activity), we also systematically sought a proper translation for the words 'sound designer' and 'sound design' in every single country being part of the European Community, given that no attempt so far has been fully realized to standardize terminology. Such a strategy in research remains necessary to refine the investigation, integrating cultural differences on the topic and/or the profession, and applying the right keywords for search engines. Translations have been obtained through online translation software (*Context.Reverso*, *Linguee*) and bilingual dictionaries. See in Zattra et al. (2018, 171) the table that represents the different translations of the term 'sound design' and 'sound designers' in the EU member states.

Translated results on sound design(er) were coupled with keywords such as 'university,' 'agency,' 'master,' 'course,' 'project,' 'profession,' 'website,' 'blog,' 'periodical,' etc. We also collected a list of more than 150 institutions – agencies, universities and academies – related to sound design production and sound design training in Europe. Lastly, a survey letter was sent to social networks (Facebook and LinkedIn), oriented mailing lists, personal contacts and other professionals (derived from conferences, associations, etc.) to enlarge the database, such as the ASD – Association of Sound Designers (sound designers in the UK theatre industry). The resulting list has collected 708 names (September 2018). However, several professionals cannot be considered as sound designers *per se*, as they rather qualify themselves, in many cases, as audio engineers or acousticians. For better results, those names were deleted from the list, depending upon choices made earlier while qualifying sound design and defining a typical sound designer activity (cf. head of section 3). All the skills/competences/facets that contribute to the conform definition needed to be taken into consideration (although in different proportions). After this skimming process, ASDP's database registered 560 professionals in sound design in Europe. Table 17.1 shows the geographical distribution of our preliminary list of contacts.

We launched our questionnaire at the end of July 2018 and ultimately sent it to 450 addresses (450/560). This definitive number is due to the fact that several messages bounced back to sender as undeliverable. We received 108 responses (108/450 = 24 percent participation rate). The final geographical distribution of participants to the questionnaire is shown in Table 17.2.

As expected, compared to the current inventory of sound designers in Europe (Table 17.1), Table 17.2 shows four representative countries: UK, Germany, Italy and France. This participation mirrors the number of professionals in our database

Table 17.1 Geographical distribution of our preliminary list of contacts for the questionnaire. In bold, the names of the countries from whence we received some answers to the questionnaire.

Country	N. of contacts	Country	N. of contacts
Austria	26	**Norway**	1
Belgium	4	**Poland**	3
Cyprus	1	**Portugal**	2
Czech Republic	2	**Romania**	1
Denmark	2	Serbia	2
Finland	2	Slovakia	1
France	115	**Spain**	4
Germany	99	Sweden	1
Greece	8	**Switzerland**	11
Hungary	2	**Turkey**	1
Iceland	2	**United Kingdom**	211
Ireland	3		
Italy	45		
Lithuania	1		
Malta	1		
Netherlands	9		
		Total	560

Table 17.2 Geographical distribution of participants to the ASDP questionnaire.

Country	Participants
Austria	3
Denmark	1
Finland	2
France	16
Germany	17
Greece	3
Iceland	2
Italy	15
Netherlands	3
Norway	1
Poland	1
Portugal	2
Romania	1
Spain	2
Switzerland	5
Turkey	1
United Kingdom	33
Total	**108**

and must be taken into account when analyzing the data (see section 3.4 on limitations). The disparity is due to several factors: the presence of well-organized associations (for instance, in the UK, the ASD – Association of Sound Designers for sound designers in the theatre industry), with well-structured (online) database which was easy to reach with our email campaign (and at the same time, on the other hand, an interest by such professionals in our questionnaire); the familiarity with certain names coming from the same country of provenance than those of the project members (France and Italy)– an acquaintance that drove the team to write and solicit a participation; and finally, the presence of recognized career paths in these countries, with schools, institutions, and Master's degree programs dedicated to sound design, which to our knowledge makes these countries leaders in this profession, with a growing number of professionals. Table 17.2 also represents the distribution of what was defined as 'place of business,' by means of the distribution of nations in which participants are working at the present time. However, there is also a percentage of migration among participants. In fact, a cross-analysis of nationality vs. working places shows that Germany, Denmark, Spain, Finland, France, Iceland and the Netherlands have received sound designers from abroad.

3.4 Limitations of the Current Approach

Even at this early stage of such a presentation, it is important to mention a few limitations to this study. The majority of professionals comes from 4 representative countries: UK, Germany, Italy and France (see section 3.3). This may well mirror the longevity and strength of sound design in those countries, but also conveys biases of underrepresentation of professionals working in other nations. To complete this picture, it should be necessary to relaunch or replicate the study from geographical areas far from the original institutions of the members of ASDP.

Another obstacle our research had to face was the scheduling. The online questionnaire was launched by email at the end of July 2018. The first deadline proposed by the team was September 30: we received 77 answers. The response was not satisfying due to poor timing. The deadline was therefore postponed and a second reminder was sent mid-September, with an extension of the deadline through October 31. As stated above, in total 108 answers were received – 108/450 = 24 percent participation rate – which is not an excellent survey response rate (although typical). This could be due to the timing (summer) and low motivations to complete a 39-question survey (approximately 20 minutes to complete). Among the participants there were persons who had personal relationships with our project team members, but we were pleasantly surprised that many participants were people unknown to us who showed interest in the topic of our questionnaire.

Our method (Creswell 1998) was positioned in the constructivist spectrum of research, knowing that the overall questions are likely to shape the outcomes of the study. One may also recognize that there isn't any absolute truth with such analysis, since any analysis can be criticized; at the same time, any human behavior – e.g., the participant's answer – can be biased or incomplete for several reasons (lack of time to answer; questions that are not clear enough; reluctancy to supply more information, etc.).

In retrospect, it has been noticed that the team did not ask questions addressing issues on ethnicity, gender identity, sexual orientation or inclusion, but rather let the participants choose what they could comfortably reveal in their comments. Nevertheless, to our knowledge, nobody has considered approaching these themes.

4 Key Findings

The analytical approach used was based on both quantitative and qualitative methods. The first analysis was based on the finite elements (closed questions). David Fierro was in charge of the analysis and creation of chart bars or numeric tables. For the open questions, Olivier Houix (Ircam STMS Lab) conducted a semantic analysis. Laura Zattra interpreted the results, comparing quantitative results with data and personal profiles of respondents, reading and interpreting semantic analysis, and making further qualitative analysis in case of totally open-ended questions. In this respect, she created a codification process for each participant – for example {2/75[m;56–65;TV,theatre,radio,music;DE; surname name]} – where the first number refers to the order of reception of the answers' submission, the second to our list of sound designers, then the gender, age group, professional area, nationality and name. In the following analysis (section 4), the names/code will be deleted. The acronyms SD (sound designer), SDs (sound designers), Q (specific question of the questionnaire) and P (the participant who responded to the questionnaire) will facilitate the reading.

4.1 Sound Designer: an Interdisciplinary Profile

Our survey corroborates the interdisciplinary profile of the profession, on the levels of training, day-to-day business and skills. As for training (Q.7: "What is your background and training?"), both qualitative and semantic analyses show recurrent terms in the list of skills learned during the training period (Table 17.3).

These data show that a combination of sound engineering and musical skills and practices (instrument/composition) is common. Such data also may be relevant to sound design program schools, pedagogical contents, etc. In fact, the word BA (Hons) appears in four answers, meaning that participants attend or have attended bachelor's degree courses with specialization in the third year. The analysis of personal profiles (according to nationality, age, current activity), stresses that younger

Table 17.3 List of skills learned during SDs training period (in decreasing order of occurrence)

Skill labels	N. of answers	Skill labels	N. of answers
Audio engineering	19*	Musicology	3
Musical instrument training	14	Self-taught in music/sound production (software)	3
Sound or electronic studio**	12	Architecture & urban design	3
Music composition	11	Physics	1
Recording/Production	9	Psychology	1
Film school and/or video production	8	Psychoacoustics	1
Self-taught in music performance***	7	Law school	1
Sonic Interaction Design / Auditory Display	6	Biology	1
Drama school and theatre training	4	Social Communication and Advertising	1
Arts & Humanities	4	DeeJaying	1

* three of whom are electronic engineers
** electronic music or computer music courses, e.g. in universities or conservatories of music
*** rock and popular music

SDs are the more trained in bachelor's and master's academic courses specifically designed for this profession. Older SDs, as expected, have more heterogeneous educations. One retired SD from France had been trained in architecture, music and acoustics; another from the Netherlands had studied biophysics. The UK is, as expected, a peculiar case because even older SDs can have been trained in institutional courses such as in Stage Management and Technical Theatre.

Regarding instead the day-to-day activity, the participants currently work especially in Design, Music composition and Artistic production. These are the answers more likely to be referenced as more attractive than other fields in Q.9 ("Which of the following fields characterize your areas of activity?"), while Architecture/Urbanism and activities such as Project Manager are less developed. Other fields emerge from the comments of several participants: they added categories such as creating sound devices, designing techniques, service design, interaction, cinema, theatre, storytelling or games. Q.10 ("During your career as a professional sound designer, please specify your areas of work") further examined the range of areas: Film industry (36 answers) and Theatre/scenography (56 answers) are very common (note that UK has its own Association in Theatrical Sound Design, should we remember); Music is the most common field (74 answers), a wide category that includes jingles and musical backgrounds; Sound Art (70 answers) and Soundscapes (59 answers) follow the 'Music' category in terms of popularity. In the final analysis, Music, Sound Art, Soundscape, Theatre (in this order) stand out against other categories. One may notice an interesting factor: inside the sound design field, 'Music' – as a conceptual framework – occupies a very important place. This depicts a clear profile of many SDs working in theatre and scenography (they do music but also soundscapes, drones, 'musical' situations, ambient, etc.). Among these respondents, many work also in the film or TV industry. However, 30 respondents did not check the 'Music' category at all, and selected the 'Soundscape' and 'Sound Art' boxes instead. They work in theatre (10), in film industry (8), video games (7), digital industry (auditory display/human-machine interfaces, sonic interactive devices) (6), museography (6), Foley/bruitage (5), transportation and/or medical industry (5), marketing (4), sound branding (6), architecture (2). It is as if there were two types of sound design, or at least the same concept is used within two perspectives: one sounds rather musical, and the other more sonic, although this assertion is only a hypothesis. There are also professionals among the 74, who chose music and a great number of boxes alongside it. One in particular is a very poliedric SD (14 boxes).

Skills in this profession (Q.9, 10, 37 –where the participants were asked to evaluate a list of sentences about SD's skills on a Likert scale) include: sound production, design, teaching, engineering, research (acoustics, computer science, etc.), digital know-how, marketing, audio/music creation for games and musical apps (smartphones, interactive musicals toys), music composition for events, websites, cinema, theatre, clubs and ships (one participant), food (one participant), music for education or therapy, service design, recording and/or field recording, research in sonification, instrument making, or more widely collaboration with other professionals when sound is involved.

The key finding here is that there are no significant ways of standardizing those skills; every participant indicates his/her own. This argument is supported also by the fact that (in Q.8: "How did you learn sound design?") the majority of participants say they are self-taught and/or have learned from other peers, meaning that sound design is a large discipline and that one has to follow his/her own path.

Moreover, there are other important skills in this profession. Q.37 was asked because, according to scholars on collaboration and cooperative creation, other psychological, interpersonal and cultural skills are important in creative professions. Then, in the next section (Figures 17.1 to 17.7), we parse answers made by SDs to some sub-questions of Q.37. We based these sub-questions on the methodology presented in Boldrini et al. (2015).

Figure 17.1 shows that most of SDs think the capacity for organization and planning is a fundamental skill, which means to establish "a systematic course of action for self or others to ensure accomplishment of a specific objective" (ibid, 62).

Even if not specifically related to the sound design profession, the working team wanted to know SDs' thoughts about the importance of a second language (Figure 17.2). Answers have been patchy and not every SD thinks that speaking a second language is important. Nevertheless, note that among those who answered from 1 (not important), to 3 (moderately important) many are English native speakers, according to our analysis of biographical profiles.

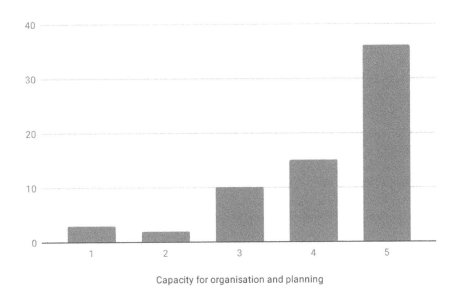

Figure 17.1 Capacity for organization and planning (part of Q.37). X-axis: Likert-scale marks (from 1 'not important at all' to 5 'very important'); Y-axis: number of participants' answers. Mean value: 4.2.

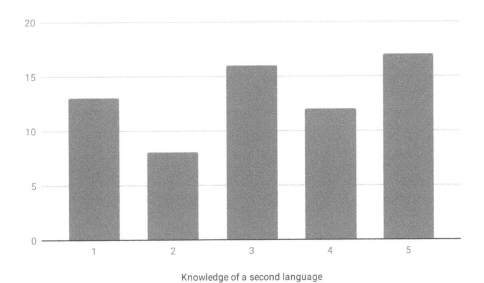

Figure 17.2 Knowledge of a second language (part of Q.37). X-axis: Likert-scale marks (from 1 'not important at all' to 5 'very important'); Y-axis: number of participants' answers. Mean value: 3.2.

Figure 17.3 indicates that SDs think that the knowledge of Information Technology is a necessity today. However, the scheme shows, unlike other questions, a significant diversity of opinions in the participants.

Figure 17.4 stresses that teamwork is important, but surprisingly almost half of the respondents think it is only moderately important (15) or not important (14). Hence, teamwork is important but can also be overlooked and replaced by personal skills and capacity to adapt (another sentence proposed with Q.37 asked about the ability to be quick to adapt and understand to new projects, situations and people, which requires social and emotional intelligence).

Figure 17.5 (Recognition of diversity and multiculturality) shows that this profession demands the ability to understand different cultural and personal characteristics in other collaborators. However, this is not an absolute and completely necessary skill.

Leadership is considered a lukewarm concept in SDs' mind (Figure 17.6). They collaborate, therefore they are not supposed to show strong leadership skills.

Figure 17.7 shows that participants' answers stand in the middle-far right of the scheme. This tells us that the economy of resources (we did not ask which ones, but these could be technological as well as human and physical) is important in this profession.

An interesting point is that there are sentences that have less importance for SDs, particularly those concerning the concept of leadership. The majority of

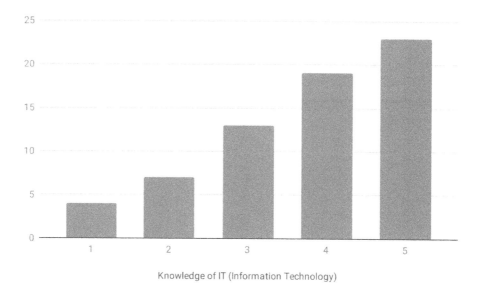

Knowledge of IT (Information Technology)

Figure 17.3 Knowledge of Information Technology IT (part of Q.37). X-axis: Likert-scale marks (from 1 'not important at all' to 5 'very important'); Y-axis: number of participants' answers. Mean value: 3.8.

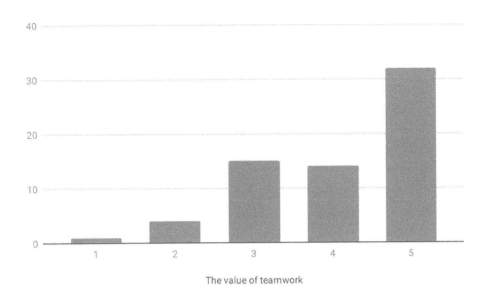

Figure 17.4 The value of teamwork (part of Q.37). X-axis: Likert-scale marks (from 1 'not important at all' to 5 'very important'); Y-axis: number of participants' answers. Mean value: 4.1.

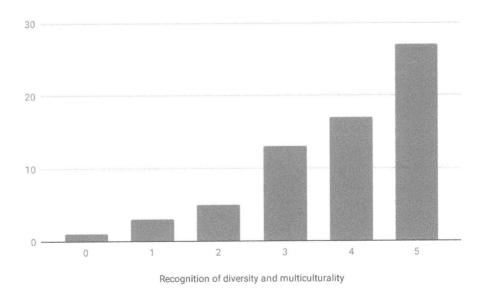

Figure 17.5 Recognition of diversity and multiculturality (part of Q.37). X-axis: Likert-scale marks (from 1 'not important at all' to 5 'very important'); Y-axis: number of participants' answers. Mean value: 3.9.

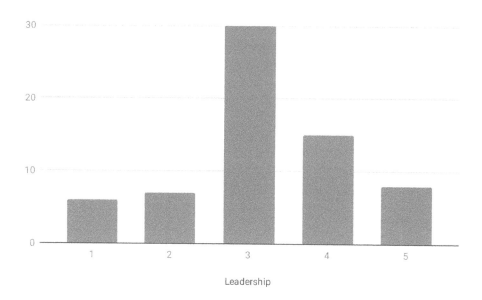

Figure 17.6   Leadership (part of Q.37). X-axis: Likert-scale marks (from 1 'not important at all' to 5 'very important'); Y-axis: number of participants' answers. Mean value: 3.2.

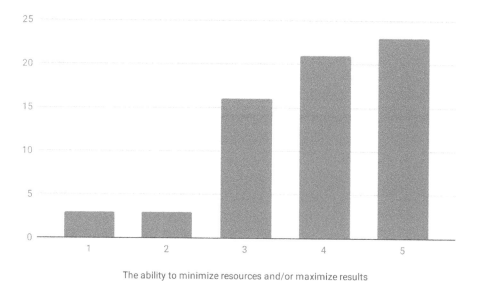

Figure 17.7   The ability to minimize resources and/or maximize results (part of Q.37). X-axis: Likert-scale marks (from 1 'not important at all' to 5 'very important'); Y-axis: number of participants' answers. Mean value: 3.9.

participants are neutral (median answer) to this concept. SDs are not meant to show leadership because their role is strongly based on collaboration.

4.2 Sound Design: a Self-taught Activity, Learned by Doing

Findings suggest that the sound design profession is in most cases self-taught (Q.8: "How did you learn Sound Design?") and learned by doing and working in the field with other professionals. This is true also for younger SDs from 26–35 years old (the second youngest age group). Some participants have also left comments such as: "by doing it while using skills from other fields of knowledge" writes P.18 who works in theatre, music and sound art (age group 26–35); "knowledge received by other professionals (industrial engineers, mix engineers, industrial designers, researchers, etc.)" writes P.66, whose range of activities covers films, video games, theatre, Foley, medical sound design, domestic electrical industry, digital industry, architecture, soundscape, sound branding, music, instruments, sound art (age group 56–65); in "seminars, workshops, books, etc.," writes P.44 (films, videogames, soundscapes; age group 26–35). Three participants even go as far as to name their mentors and colleagues from whom they have learned most.

The younger they are (up to 45 years old), the more they learn concomitantly via specialized diplomas (master's and/or specific training). This indicates that the sound design profession benefits from specialized training courses, increasingly arising in different countries. However, the majority feels that sound design is (and probably will be) a profession in which learning in the field remains necessary. The same is true when matching the training with years of experience. Even so, hands-on experiences are felt to be mandatory.

4.3 Sound Design Workflow: Short, Medium and Long Projects

Findings related to this section suggest that over time SDs have developed a common practice that goes beyond each domain's specificity. Analysis obtained with Q.23 ("What is the average duration of a sound design project?") and Q.24 ("Could you describe your typical time frame when working in a project?" – with some suggestions such as reflecting on the duration of a project/phases: first contacts, typical communication strategy, number of meetings, brainstorming, development, testing, revisions) helped us to understand the process leading to task completion, time frame, organizational structure, division of labor and coordination. Q.23 and Q.24 point out that only three SDs have difficulty generalizing – they answer: "It depends on the project." Actually, this is an answer the team tried to avoid, because as P.53 mentioned:

> The difficulty to generalize depends on the fact that some projects are strictly structured and others are more loose and informal with just a final deadline to

meet; some projects involve many people, others are restricted to one or two referents; and finally, some projects require many revisions, alternate versions, modifications; other projects are completed with a minimal number of iterations (the latter also depending on budget and time: the more time and budget, the more real-life testing, and therefore revisions).

But despite this apparent perplexity, our semantic (Figure 17.8) and qualitative analysis (Figure 17.10) have led to the following interpretations, detailed in sections 4.3.1 to 4.3.5 below.

4.3.1 General Approach

The first analysis examines the general workflow process which, according to qualitative analysis, usually develops in the following manner (Figure 17.10):

1. During the first contact/meeting (over the phone/online calls or in person) a lot of time is spent on getting to know each other's expectations (P.35, 66, 67, 73, 77). The SD shows his/her clients examples of past works when/where appropriate

Figure 17.8 Sound Design Workflow – Semantic Analysis of Q.24 (word cloud realized by Olivier Houix with TXM software – http://textometrie.ens-lyon.fr/)

(P.66) and/or discusses site, deadlines, costs, content (if theme driven) (P.76), brainstorming, creating the concept and discussion of technical feasibility.

2. The 'drafting time' corresponds to the first individual work in writing the project, planning phase, setting different strategies (P.35, 37, 67). Here, individual work may involve sonic browsing and mood boards to show to the client (P.66, 77), study of the acoustic environment or instruments (P.73). During this phase a series of email or other meetings (where appropriate) and other brainstorming (P.35, 67) may take place.

3. The third phase corresponds to a preliminary evaluation and validation (P.37, 66, 77). P.77 specifies that this is the tricky part because the sounds are often heard on systems that are very different from the final system. It is sometimes possible to make "sound simulations" of the final result, but it takes a lot of time and if the family of sounds is discarded, it is lost.

4. The fourth step involves the development and detailed design (P.37, 66, 67) with the setting of other meetings and exchange of emails or calls where appropriate (P.35, 67).

5. Testing is of crucial importance (P.67, 77) and may involve *in situ* testing and/or final sound system testing (P.77).

6. The sixth phase includes a series of revisions if needed (P.67), with more independent work and subsequent iterations of this working cycle (from four to seven), where appropriate (P.39).

7. Finally, during the last phase, the workflow ends with the submission, final implementation (in industry), diffusion (in cinema, theatre performance, marketing, etc.) (P.37, 66) or rehearsals, previews and opening night (P.35).

P.50 is one of the very few who stresses the importance of payment and financing. During the first contact P.50 talks about expectations and boundaries, prepares a demo to see if he understands the client clearly and comes up with ideas. At this point comes the first assurance payment (usually a third of the total payment). During the second meeting he works out any misunderstandings and proceeds with the bulk of the work (steps 5 and 6: Tests and Revisions) with the second assurance payment (another third of the total payment). And finally, "there are the finishing touches, test and revisions and the rest of the payment."

4.3.2 Typical timelines: three to six months and multiple parallel projects

According to the analysis of Q.23, the typical duration of a project is between one and six months, but the analysis of Q.24 also shows that, for most of the participants, an ideal project should last between three and six months (Figure 17.9). According to P.86, this is due to being involved in other projects (a common situation for other SDs): "I tend to work on projects up to 6 months because I am working on them in a

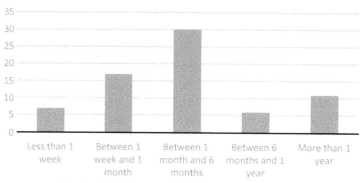

What is the average dura.on of a sound design project?

Figure 17.9 Average duration of a sound design project – qualitative analysis of Q.23.

part time capacity – 1 or perhaps 2 days per week." The second typical duration corresponds to shorter projects between one week and one month (again, freelancing and working in multiple parallel projects is the norm in this profession). P.72 also notes that projects have increasingly shorter time frames. This is confirmed by P.11 (theatre) whose average works develop over periods of days: "2 production meetings – 2 days of paperwork and programming – 2 days of equipment prep – approx. 4 or 5 days of rehearsal." However, the duration of a project depends also on the rate of repeatability; in music and theatre tours "performance period may range from 1 day for large outdoor ceremonies to a number of months»" writes P.11.

4.3.3 Three to Six Months Average Period (Movie/Theatre Industry)

It seems that in film and theatre industries (P.12, 18, 42) the labor organization is very standardized. The first contact starts 3–6 months in advance, or 3 months in case of short films/documentaries of 25–30 minutes (P.85). "In a 6-month project, the preparation and communication/meeting with the director starts during the 2nd month; the proper job (production) during the 3rd and 4th month; 1–2 months before the deadline there is a full involvement, with rehearsals and/or technical rehearsals, montage and a hard work in the final fortnight; press night during the 6th month" (P.18). When the project lasts three months the timeline is shortened accordingly (P.12, 32).

P.84 receives script/synopsis of project (1- first contact); then meets with principal team members (director, writer, director of music, lead engineer, etc.) (2- drafting time); engages with pre-production/production materials (view dailies, rushes, rehearsals, etc., as required) and constructs a holistic design plan, including dramatic interpretation through sound and/or music (3– first evaluation and validation). Phase 4 (development and detailed design) includes the audio production (recording, composition, etc.) and audio post (editing, mixing, etc.). According to P.26, phases 1, 2 and 3 are crucial for the success of the project:

> To discuss strategies in sound and to address technical elements of the shooting itself and of elements that might need to be included in the filming to help with the final sound is mandatory (e.g., making sure that set design has included elements that will appear in camera that might give extra layers of sound in the final Sound Design like different materials in flooring and other props that would justify a more complex sound experience; also, making sure that all measures will be taken to ensure a good dialog recording is crucial).

During the second phase (shooting of the film) P.26 prefers to

> do location sound myself. The third phase is additional sound recording on site or on other locations to record any elements that might be needed for the final film and not be available later on at the post production stage like seasonal ambient sounds [...]. Fourth Phase is after the first Draft Edit where I discuss with the Director and Editor if there are things concerning sound that need to be addressed in the film editing. Fifth Phase is Sound Editing usually after the Final Film Edit has concluded, where after cleaning the dialog and doing any needed ADR/Dubbing I start creating the extra sound elements of the film either with Foley in studio, or Sound Recording on location. Sixth phase is Sound Mixing and Mastering for the different media of the film like DCP, TV and Internet.

P.26 also stresses the difference between the work for films and for ads or TV. When doing sound design for ads and TV,

> most of the Location Sound details are discussed and resolved on site (if there is any location sound at all) and most of the post production is handled all at the same time after the Final Video Edit in one or more studio sessions with or without the director.

P.51 explains that the differences in film industry can be explained by the big variety of movie genres, which are demanding different types of soundwork:

> A typical cinema movie occupies me for about 12 to 15 weeks, in which I often do many parts of the work myself, including editing the location sound, dialogue, SFX and ambiance, ADR and music. I'm also present during the mixdown, until the film sound is finished completely, including also an international tape for dubbing. In the case of a normal TV movie of the same duration (90 min.) this same procedure takes me about 6 to 8 weeks.

4.3.4 Long Projects (Movie/Theatre Industry, Sound Art, Architecture)

The third typical duration shows projects of more than one year. These are typically based on art, sound installation, and architecture (P.62, 65). P.65 explains that for exhibition scenography and sound art/installation, the typical duration is one year, including first contacts, design draft and development, revisions, testing, realizations, with about three to six meetings distributed over all the phases. But for urban and architectural projects (again P.65) the project may last five to ten years, of which one year for the competition phase, one to two years for project development, one to two years of detail planning, one to three years of construction; and between those phases many meetings and sometimes also waiting times for the kick off of the next phase, because of revisions, client changes (especially, in the case of public space projects), etc.

4.3.5 Sound Design is a Multitasking Profession

The dimension of autonomy and entrepreneurship characterizing the sound design freelance activity mirrors the highly multitasking nature of this job and the need to be capable of organizing oneself between different time schedules. Older participants say they can count "more than 500 radio plays, 250 TV films, 500+ commercials, 100+ Music recordings and 200+ Live shows" (P 2, age group 56–65). Age matters when considering the number of projects, as it does for the extent and diversity. The analysis of this diversity (Q.10: "Specify your area(s) of work?") compared with SDs' age, shows that there is a relative tendency to accumulate different domains of activities once a SD grows older, and add new areas once their abilities, skills and professionalism increase. Mature SDs often work in marketing and create their own professional label or production company and/or work in legal aspects of the sound design activity, as in the acoustic regulation laws (P.24, age group 66+).

5 Discussion and Perspectives

The ASDP project had the ambitious objectives to build a consistent anthropological, historical, and geographical knowledge about the identity of the sound design profession and its actors, consequently promoting the understanding of creative processes underlying sound design practices. To do this – and address our issue: What is sound design? Ask sound designers! – digital ethnography (online questionnaire) has been the natural approach for capturing this community of practice (Lave & Wenger 1991; Wenger 1998), a method we already tested in the past.

Instead of trying to find an ultimate definition of sound design, a larger description based on sociological, operational, aesthetical, behavioral principles (section 3) has led us to identify 450 professionals (section 3.3) among whom 108 have replied to our questionnaire. Their answers (section 4) have helped us clarify the complexity

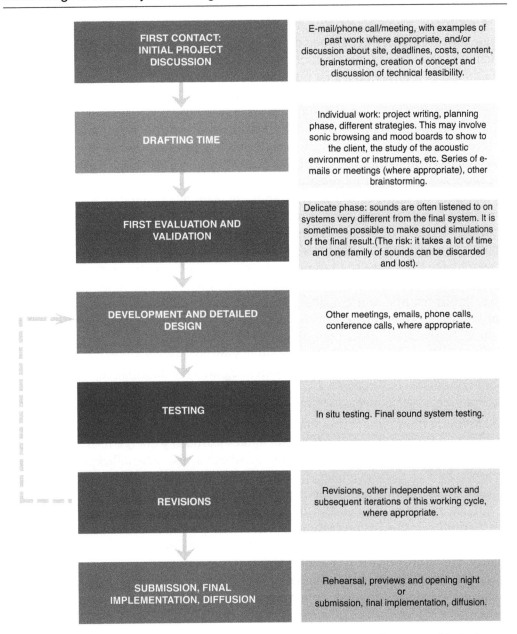

Figure 17.10 Typical workflow in sound design – qualitative analysis of Q.23–24 (block diagram made by Laura Zattra during the analysis phase of the study).

of this profession and advance both knowledge and definition of sound design. The advantage of the online questionnaire includes the ease of storage and retrieval, for future developments.

A solid discussion on our research methodology, research protocol and the first phases of the project (preparation of the questionnaire), as well as on the hypotheses and initial thoughts on our research topic, has characterized our project. Indeed, we can safely say that even our first publication (Zattra et al. 2018) and talks in conferences have represented a vehicle "to think [our] way through the research process" (Pryke et al. 2003, 2), because our "research [was] not a straightforward process" (ibid.). Writing about our methodological choices and our research protocol has helped us establish a more coherent framework for our outcomes. By sharing our research design and the protocol of this investigation, we also hope that such a learning process will help future research in this field.

We think the spirit of our interdisciplinary team has facilitated an attitude that avoided a top-down approach, which is essential when studying the behavior and actions (practices) of humans. Once the population was targeted, we needed to focus on the questions and then the answers. This is the quantitative/qualitative part of our research, where we have tried to rely as much as possible on the participants' views. However, aspects such as data typology and quantity, practical limits and imperfections in our method, our considerations and backgrounds, have inevitably shaped our knowledge (section 3.4). Every single choice of an open-ended question rather than a Likert scale or closed question may be debatable.

The ASDP method – which positions itself halfway between a concurrent procedure (we collected both forms of data and often integrated the information in the analysis) and a transformative procedure (even avoiding a top-down attitude, we used our background knowledge and theoretical lenses as an overarching perspective) – have led us understand, among our findings, that sound designers have differentiated skills and abilities learned in different ways (self-taught and/or learned by doing); their competences (knowledge, behaviors) seem to fit with one another, but also serve different tasks. Sound designers have an interdisciplinary profile, both on the levels of training and day-to-day practices, that outline a complexity typical of many creative (freelance) professions.

Clearly, this is not and cannot be the most comprehensive analysis of the sound design profession and its professionals. The ASDP project is limited to the European community, and the list of names and institutions we did develop is not exhaustive. One of the future goals is to promote our investigation in order to increase the professional network toward the sound design industry, and possibly open this research to the global sound designers' community.

Acknowledgments

This work was funded by the LabEx Creation, Arts and Heritage (France), as part of their 2018 call for projects, as well as the STMS Lab / Sound Perception and Design (SPD) research group. The authors thank all the respondents to the questionnaire for their contributions, as well as their colleagues from ASDP and STMS Labs and Claire

Richards. Special thanks to those sound designers who further discussed this study in its earlier stages: Roland Cahen, Saran Diakité, Jean Dindinaud, Nadine Schütz.

References

Archer, B. (1979). "Design as a discipline," *Design Studies*, 1 (1), 17–20.

Blattner, M.M., D.A. Sumikawa, R.M. Greenberg. (1989). "Earcons and Icons: Their Structure and Common Design Principles." *Human-Computer Interaction*, Volume 411–44.

Boldrini, F., Bracchini M.R., Fanti S. (2015). *ArtS – Skills for the Creative Economy. State of the Art and Mapping of Competences (Report WP: Defining Sector Skill Shortages and ECVET Strategy)*. Fondazione Centro Studi Villa Montesca, with the support of the Erasmus+ Programme of the European Union. http://cesie.org/media/R2.6.Mapping-the-Competences-of-the-Cultural-and-Creative-Sectors.pdf.

Bourdieu, P. (1979). *La Distinction: Critique Sociale du Jugement*. Éditions de Minuit.

Brown R. (2009). *Sound - A Reader in Theatre Practice*. Palgrave Macmillan.

Boussard, P., Dendievel C., Lachambre H. (2016). "Sketching Step in Sound Design: the Sound Designers' Point of View." *Proceedings of Internoise*, 4505–4513, Hamburg, Germany, 2016. http://skatvg.math.unipa.it/wp-content/uploads/2016/11/internoise2016.pdf.

Carron, M., Rotureau, T., Dubois, F., Misdariis, N. & Susini, P. (2017). "Speaking about sounds: a tool for communication on sound features." *Journal of Design Research*, 15 (2), 85–109.

Clarke, Thomas. (2016). "Emerging Professions." Professional Standards Councils. https://www.psc.gov.au/research-library/emerging-professions#

Collison, D. (2008). *The Sound of Theatre, A History: From the Ancient Greeks to the Modern Digital Age*. Eastbourne, Plasa Limited.

Creswell, John W. (1998). *Qualitative Inquiry and Research Design: Choosing Among Five Traditions*. Newbury Park, SAGE Publications.

Cross, Nigel. (2001). "Designerly ways of knowing: Design discipline versus design science." *Design Issues*. 17 (3), 49–55.

Cross, Nigel. (2007). *Designerly ways of knowing*. Springer Science & Business Media.

D'Amico, G. (2009). "La professione del sound designer: MioJob.it intervista sounDesign." March 3, 2009. Accessed February 9, 2021. https://www.soundesign.info/2009/03/03/la-professione-del-sound-designer-miojobit-intervista-soundesign/.

Dal Palù, D., C. De Giorgi, B. Lerma, E. Buiatti. (2018). *Frontiers of Sound in Design. A Guide for the Development of Product Identity Through Sounds*. Springer International Publishing.

Demazière, Didier & Gadéa Charles. (2009). "Introduction." In *Sociologie des groupes professionnels*. Demazière Didier & Gadéa Charles (eds.), 13–24. Paris, Éditions La Découverte.

Dent, Mike, Ivy Lynn Bourgeault, Jean-Louis Denis, Ellen Kuhlmann. (2016). *The Routledge Companion to the Professions and Professionalism*. 2nd ed. Oxford University Press.

Ebanista, Carlo. (2007). "Paolino di Nola e l'introduzione della campana in Occidente." *Dal fuoco all'aria. Tecniche, significati e prassi nell'uso delle campane dal Medioevo all'età Moderna*, Fabio Redi and Giovanna Petrella, (eds.), 325–353. Pisa, Pacini Editore,.

Farnell, A. (2010). *Designing Sound*. Cambridge, MA, The MIT Press.

Farnetani, N. Prodi, R. Pompoli. (2008). "On the acoustics of ancient Greek and Roman theaters." *The Journal of the Acoustical Society of America*, 124:3, 1557–1567.

Franinović, K. & S. Serafin (eds.). (2013). *Sonic Interaction Design*. Cambridge, MA, The MIT Press.

Gibbs, T. (2007). *The Fundamentals of Sonic Art and Sound Design*. Lausanne, Switzerland, AVA.

Houix, O., Monache, S.D., Lachambre, H., Bevilacqua, F., Rocchesso, D., & Lemaitre, G. (2016). "Innovative tool for sound sketching combining vocalizations and gestures." In *Proceedings of the Audio Mostly Conference*, 2016, 12–19.

Hug, D., & Misdariis, N. (2011). "Towards a conceptual framework to integrate designerly and scientific sound design methods." In *Proceedings of the 6th Audio Mostly Conference: A Conference on Interaction with Sound*, 23–30.

Humphrey, J.W., J.P. Oleson, A.N. Sherwood. (1998). *Greek and Roman technology: a sourcebook*. Routledge.

Kaye, D. & J. LeBrecht. (2009). *Sound and Music for the Theatre*. 3rd ed. Boston, MA, Focal Press. https://www.gamasutra.com/blogs/MarkKilborn/20130618/194567/Some_Advice_for_the_Aspiring_ Sound_Designer.php.

Lave, J. & Wenger, E. (1991). *Situated learning: Legitimate peripheral participation*. Cambridge, Cambridge University Press.

Louridas P. (1999). "Design as bricolage: anthropology meets design thinking." *Design Studies*, 20(6), 1517–535.

McEwen, Celina & Trede Franziska. (2014). "The Academisation of Emerging Professions: Implications for Universities, Academics and Students (2014)." In *Power and Education*, Volume 6, issue 2, 145–154.

Menger, Pierre-Michel (ed.). (2003). *Les professions et leurs sociologies: Modèles théoriques, catégorisations, évolutions*. New edition [online], Paris: Éditions de la Maison des sciences de l'homme. Available on the Internet: http://books.openedition.org/editionsmsh/5715.

Misdariis, N. (2018). "Sciences du design sonore Approche intégrée du design sonore au sein de la recherche en design." Habilitation à Diriger des Recherches (Diss.), Université de Technologie de Compiègne, Compiègne, France. https://hal.archives-ouvertes.fr/tel-02003043.

Misdariis, Nicolas & Cera Andrea. (2017). "Knowledge in Sound Design: The Silent Electric Vehicle—A Relevant Case Study." *Proceedings of the Conference on Design and Semantics of Form and Movement - Sense and Sensitivity, DeSForM 2017*, M. Bruns Alonso and E. Ozcan eds., 185–195.

Misdariis, N., Cera A., Gaxie S. (2015). "Creativity and Sound Design: a case study." TCPM – Tracking the Creative Process of Music Conference, (oral presentation) October 2015, Paris, France.https:// medias.ircam.fr/x3eca57.

Misdariis N., Pecquet F., Donin N. (2019). " Cartographier le design sonore? Qu'appelle-t-on design sonore? Première étape d'une recherche structurée sur la discipline." Sound Design Days (oral presentation), November 28–29, 2019. IRCAM, Paris, 2019 https://medias.ircam.fr/x8d4c89).

Pauletto, S. (ed.). (2014). "Perspectives on Sound Design." *The New Soundtrack* (special issue), (2).

Pecquet, F. (2017). "Sound Design or Musical Design, Provisional Problematization: Thoughts on James Murphy's *Subway Symphony* Project." *Les cahiers de la société Québécoise de Recherche en Musique*, Volume 18, (2), Automne 2017, 31–38. https://id.erudit.org/iderudit/1066438ar.

Pecquet, F. (2018). "Le nouvel ordre sonore - Recherche et création." Habilitation à Diriger des Recherches (Diss.). Université Paris1, EAS (Écoles des Arts de la Sorbonne).

Pecquet, F., Dupouey, P. (2021). *Design sonore : applications, méthodologie et études de cas*. Dunod editions.

Pryke, M., G. Rose, S. Whatmore. (2003). *Using Social Theory. Thinking Through Research*. Sage Publications.

Quagliarini, E. (2008). *Costruzioni in legno nei teatri all'italiana del '700 e '800: il patrimonio nascosto dell'architettura teatrale marchigiana*. Firenze, Alinea Editrice.

Rittel, H.W. and M.M. Webber. (1973). "Dilemmas in a General Theory of Planning." *Policy Sciences* 4 (2): 155–169.

Rocchesso, D. (2014). "Sounding objects in Europe." *The New Soundtrack*, 4 (2), 157–164.

Rocchesso, D. (2011). *Explorations in Sonic Interaction Design*. Berlin, Logos Verlag.

Rocchesso, D. & F. Fontana (eds.). (2003). *The sounding object*. Firenze, Mondo Estremo.

Rocchesso, D., Lemaitre, G., Susini, P., Ternström, S. & Boussard, P. (2015). "Sketching sound with voice and gesture." *Interactions*, 22 (1), 38–41.

Rocchesso, D., Serafin, S., Behrendt, F., Bernardini, N., Bresin, R., Eckel, G., Franinovic, K., Hermann, T., Pauletto, S., Susini, P., Visell, Y. (2008, April). "Sonic interaction design: sound, information and experience." In *CHI '08 Extended Abstracts on Human Factors in Computing Systems* (3969–3972).

Rodriguez, W. (2003). "Le design sonore, naissance d'une catagorie musicale." (MA Thesis). Mémoire de DEA, Ecole des Hautes Etudes en Sciences Sociales, EHESS, Paris.

Rogers, Justine, Dimity Kingsford-Smith, Thomas Clarke, John Chellew. (2016). "Modern Professional Practice and Its Future: The Promise of Professionalism in the 21st Century." Working paper presented at Modern Professional Practice and its Future, March 3, 2016. Retrieved from https://clmr.unsw.edu.au/

sites/default/files/attached_files/rogers_and_kingsford_smith_et_al_the_promise_of_professionalism_-_summary_from_march_2016_symposium.pdf.

Schafer, R.M. (1977). *The tuning of the world*. Alfred A. Knopf.

Sennett, R. (2008). *The Craftsman*. Yale University Press.

Sonnenschein, D. (ed.). (2001). *Sound Design: The Expressive Power of Music, Voice, and Sound Effects in Cinema*. Michael Wiese Productions.

Susini, P., O. Houix, N. Misdariis. (2014). "Sound design: an applied, experimental framework to study the perception of everyday sounds." *The New Soundtrack*, 4 (2), 103–121.

Susskind, Richard & Susskind, Daniel, (2016). *The Future of the Professions. How Technology Will Transform the Work of Human Experts*. Oxford University Press.

Vial, S. (2015). "Qu'est-ce que la recherche en design? Introduction aux sciences du design." *Sciences du Design*, (1), 22–36.

Vygotsky, L. (1978). *Mind in Society*. London, Harvard University Press.

Wenger, E. (1998). *Communities of practice: Learning, meaning, and identity*. Cambridge, Cambridge University Press.

Whittington, W. (2007). *Sound Design and Science Fiction*. University of Texas Press.

Wyatt, H. & T. Amyes. (2005). *Audio Post Production for Television and Film: An Introduction to Technology and Techniques*. 3d ed. Oxford, Focal Press.

Zattra, Laura. (2018). "Collaborating on composition: The role of the musical assistant at IRCAM, CCRMA and CSC," in *Live-Electronic Music. Composition, Performance and Study*. Friedemann Sallis, Valentina Bertolani, Jan Burle, Laura Zattra, eds., Routledge, 59–80.

Zattra, Laura & Nicolas Donin. (2016). "A questionnaire-based investigation of the skills and roles of Computer Music Designers." *Musicae Scientiae*, Special Issue "Tracking the creative process in music," September 2016, (3), 436–456.

Zattra, Laura, Nicolas Misdariis, Frank Pecquet, Nicolas Donin, David Fierro. (2018). "Analysis of Sound Design Practices [ASDP]. Research Methodology." In *Machine Sounds, Sounds Machine. Proceedings of the XXII CIM (Colloquio di Informatica Musicale)*, edited by Federico Fontana and Andrea Gulli, 20–23 November 2018: 168–175. Udine / Venezia, AIMI.

Zattra, L., N. Misdariis, F. Pecquet, N. Donin, D. Fierro. (2019a). " Cartographier le design sonore? Pratiques et praticiens: résultats du projet APDS (Analyse des Pratiques du Design Sonore)." *Sound Design Days*, (oral presentation, video), November 28–29, IRCAM, Paris, 2019. https://medias.ircam.fr/x73cb37.

Zattra L., N. Misdariis, F. Pecquet, N. Donin, D. Fierro. (2019b). "How do they work? An analysis of the creative process in Sound Design obtained through an online questionnaire." *Tracking the Creative Process in Music, International Conference*, 5th ed., Lisbon, Portugal, October 9–11, 2019.

Qualitative, Pragmatic and Hermeneutic Inquiries in Electroacoustic Compositions
Selected Case Studies

Maria Kallionpää

1 Introduction

Methods of electroacoustic music belong to the skillsets of most of the active contemporary classical composers of today. Furthermore, composers and performers alike seek inspiration from outside of the traditional, existing classical music concepts and musical forms. For example, as in the contemporary classical music culture composing a symphony or sonata is nowadays often considered somewhat outdated or uninteresting (and thus, with some exceptions, such works would less likely attract the attention of the gatekeepers, such as festival directors, conductors or intendents, potentially leading such musical works less likely of becoming successful and having a life after the very premiere), the music creators of today must define their compositional systems anew within their every composition project. Instead of resorting to existing forms and techniques, a composer sees an empty canvas ahead of them every time they start a new project: among other things, they need to find an answer to the question of how to define the overall form of the composition, as well as a method of selecting their rhythms and pitch material, as well as other components that should fly in their work. In some cases, this means that a musical work might not be just a creation of a single composer, but a joint effort of an entire research team that would contribute in generating the data on which the musical material would then be based, as well as on the process of designing and implementing a complex interactive system. Despite the technical nature of some of the composition projects discussed below, the desiderata for artistically valuable results prevails. Setting up an ambitious system just for the sake of it rarely satisfies the artistic goals of the composers or the musicians performing the music; nor would it usually appeal to the concert audiences, who still expect to be able to understand a musical performance without having access to a thorough analysis of the piece. The listeners still long for something they can relate to, be it either emotionally or intellectually. For example,

DOI: 10.4324/9780429356360-19

producing a concert piece simply describing the superficial characteristics of an ocean would not necessarily make sense, as Debussy and Ravel already did it much better more than a century ago. Instead, a composer of our times would probably apply a more systematic approach to it: for example, Olga Neuwirth's "Kloing!" utilizes the seismic data gathered during the Sumatran seaquake that caused the tsunami disaster in 2004. Another example to be discussed is "El Canto del Mar Infinito" (2020), composed by the author of this chapter. The work is based on computer-generated analysis of the sounds under and above the surface of an ocean.

Moreover, modern concert audiences have gotten used to expect innovation when listening to new works: they want to experience novel storytelling, immersive technologies, and all in all compositions that demonstrate relevance in the current cultural climate, potentially also contributing to the discourse on current societal topics and phenomena. The continuous re-definition is also one of the methods for contemporary classical music surviving in the concert programs, instead of becoming just a hobby for an elitist niche audience. As stated by Kallionpää et al., the successful implementation of interesting expressive ideas and technologies "may keep the contemporary classical music culture alive and fresh, also helping it to raise interest among the audiences that otherwise would not pay attention to it" (Kallionpää et al., 2018, p. 2)".

Undoubtedly, implementing new kind of techniques, technologies, aesthetics and narratives sets new challenges for the performers to respond to: new performative and technical skills must be acquired along the way (Kallionpää, 2014, p. 8). On the other hand, exploring such new musical territories may provide the performers more freedom to create their own interpretations, as the burden of tradition commonly plaguing the classical music culture (thanks to the modern recording industry), shines by its absence. Moreover, by challenging the concepts of traditional music forms, composers may acquire superior technical and aesthetic results that could not be achieved otherwise. This chapter approaches its topic from practice-based, qualitative and hermeneutic perspective, focusing on the cross-disciplinary artistic projects from the viewpoints of composers, performers and sound technicians. To shed some light on the research topic from a qualitative perspective, just a few representative composition project examples have been selected.

2 Analysis as a Foundation of a Musical Form

In arts research, it is not uncommon that an artist uses their actual artistic practice as means of collecting data and information for the purposes of their research. For example, Korhonen-Björkman mentions the potential of the music researchers simultaneously functioning as performers of their research materials. To illustrate this, she gives

Janet Schmalfeldt's work on Beethoven's "Bagatelles" op. 126 as an example. Schmal-feldt herself discusses her practice-led research by explaining how her system enabled her to analyze her research object from "inside," by simultaneously adapting the roles of a "performer self" and "analyser self." She regards these to be the two complement-ing sides of her performer identity (Schmalfeldt in Korhonen-Björkman, 2008, p. 3). Although Schmalfeldt's methodologies have been criticized (ibid.), the author of this chapter considers them to be promising, especially when investigating and analyzing repertoires that do not have a great deal of performance tradition or reception history. Furthermore, Borgdorff divides artistic research roughly into three different categories: "research on the arts, research for the arts and research in the arts" (Borgdorff, 2007 p. 5–6). However, the classifications suggested by Borgdorff are not carved in stone: the chapter author acknowledges the potential of applying all the above methods even within a single artistic research project. According to Borgdorff's second category, one of the main targets of arts research is to provide and create technical or artistic infor-mation that "may find its way into concrete (artistic) practices in one way or another" (Borgdorff, 2007, 6). A similar approach was adopted in the selection of compositions for this chapter, as some of the works discussed below were either composed or per-formed by the author herself.

When talking about musical works based on complex musical forms that may also include an electronic real-time system, discussing the analytical basis preceding the actual compositional work is inevitable. I argue that music analysis and composing go hand in hand: in order to build up an electroacoustic, communicative system that works in a coherent manner, a thorough pre-structuring phase must be conducted. The importance of this stage can be compared to that of preparing the drawings before constructing a building. Like a performer playing out a note on their instru-ment, a composer too (and in some cases their technical research team) needs to know beforehand what they intend to compose and how they plan to implement it. In research literature, there has been some debate about the relations and hierarchies between music making and analysis (or intellectual perception vs. creating or per-forming music). For example, Hämeenniemi argues that composers' and performers' *modus operandi* differ from each other: "whereas a composer approaches a musical composition in terms of its structure and composition techniques, a performing musician mainly approaches it, on top of the general sonic/aural outcome, from the perspective of performative physicality, namely in terms of motorics and instrumen-tal techniques" (Hämeenniemi, 2007, p. 34, as translated by the author). Nicholas Cook expressed a contrasting viewpoint, arguing that for a performer, learning a musical composition is primarily an analytical task (Cook, 1990, p. 77). Neuhaus, on the other hand, draws the reader's attention to the importance of mastering one's craft: the musician needs to know *how to* master their instrument, as the physical ability to play is the prerequisite for obtaining the most optimal artistic outcome (Neuhaus, 1986, p. 10–11).

Regarding on how rigidly the electronic system should follow the rules set by the composer, there might be some leeway to experiment with it in the process, as an electronic or electroacoustic composition may benefit from technical or artistic adjustments that were not foreseeable from the very beginning. Moreover, it may also take some time to spot the errors with regards to the functionality of such a system: often a fair amount of trial and error must be accepted as part of the process. In some cases, this may also come with the benefit of coming up with new technical or artistic ideas that might work better than the original plan: usually the artistic goals are prioritized in case they end up conflicting with the preset rules set by the composer.

One of the most important aspects to consider is, how much control the composer wants to have with regards to the final sonic outcome. For example, an interactive algorithmic system does not allow or require the composer to notate every detail that the system is supposed to play. As argued by Karlheinz Essl, although a composer has to give away a certain amount of control, the system "enables one to gain new dimensions that expand investigation beyond a limited personal horizon" (Essl, 2006, p. 1). Furthermore, according to the same author, preferring "process" over "object" (Essl, 1994b, p. 1) was one of the reasons why algorithmic composition techniques started to gain more popularity towards the end of the twentieth century. The same approach was also taken by Chadabe, who stated that one of the most significant steps in the development of contemporary music was the shift of emphasis from compositions being regarded as "fixed objects" to seeing compositions as living "processes" (Chadabe, 1996, p. 43). Moreover, within the same timeframe, a rising number of highly performing computer software were developed, including, for example, Pure Data and Max/MSP, the latter of which still is one of the main tools amongst the composers of today. In parallel, efficient computers that could be used to implement complex real-time computer-generated processes became even more widely available, thus enabling an increasing number of composers to experiment with such techniques in their compositions.

Algorithmic compositional techniques are usually connected to live recording and real-time processing of the gathered data. Essl calls them an "inspirational machine" that can be regarded as a powerful tool (Essl, 2006, p. 1). Essl applies these principles in his own compositional work. For example, his Sequitur series is a collection of works for a variety of instruments, based on the Sequitur Generator program that he designed himself (see Essl, 2008, and Essl, 2008–2010). There are also plenty of other composers, such as Volkmar Klien (Klien et al., 2010), who frequently apply such techniques in their work, but some others, such as, for example, Georg Friedrich Haas, have expressed a more critical stance on them (Haas, 2013). The author of this chapter believes that, depending on how the generative processes are being dealt with, this kind of techniques can provide an otherwise inaccessible richness of sonic, rhythmical and instrumental possibilities.

The reasons for composers choosing between the approaches of conventional and generative composition techniques differ. For example, as Laurie Spiegel described her work on generative systems: "I automate whatever can be automated to be freer to focus on those aspects of music that can't be automated" (Hinkle-Turner, 2006, p. 241).

Whereas generative systems are mostly based on the idea of flexibility, a thoroughly notated standard repertoire composition is normally based on a fixed form that is usually expected to remain the same in every performance, unless otherwise stated by the composer. The primary characteristic of algorithmic composition techniques resides in the fact that most of the composing work has been done at the programming stage, as the algorithm is in fact a code from which an entire musical composition can be derived. An algorithm in a compositional setting can be compared to a seed of a plant, in which all the DNA is already set, and which is just waiting for its turn to start flourishing. In such compositions the system controls the processes (and thus the compositional structure) that formulate the detailed sonic surface differently in each performance. Generative compositions leave room for flexibility in the performance situation by employing a real-time system that, along with the live performer, plays an equally substantial part of the contents of the actual composition. At their best, generative, algorithmic or dynamic music systems have a potential of combining the technical excellency of the computer with the expressiveness of the human imagination and intelligence.

2.1 History of Algorithmic Composition Techniques

Setting a rule from which an entire musical piece could be generated was a concept already known in the context of early Medieval music. The basic concept of the "organum" was presented by Hucbald of St Amande, who instructed how to improvise another voice to a given Gregorian chant. St Amande's considerations were the first steps on the path towards the development of the canon as we know it today (Essl, 2006, p. 2). Furthermore, the later introduced "Musikalische Würfelspiele," or musical dice games, can also be seen as early representations of algorithmic compositions (Edwards, 2011, p. 59). These games were especially popular among the composers of the late eighteenth and early nineteenth centuries. They were based on pre-composed excerpts that the participants would put in order by throwing a dice. For example, Mozart and C.P.E Bach both experimented with the form.

In the twentieth century, French composer Pierre Barbaud (1911–1990) proclaimed himself to be the original inventor of algorithmic music (Barbaud, 1960, p. 92). However, the concept of automatic musical composition was first patented by Rene-Louis Baron in 1998. Lejaren Hiller's (1924–1994) role as one of the pioneers of computer controlled algorithmic composition has also been widely acknowledged (Edwards, 2011, p. 61). On the other hand, Max Mathews' (1926–2011) contribution to the field has been regarded as very significant too, especially because of his involvement in the development process of the previously mentioned Max/MSP software, which is why the program was also named after him. However, one of the most famous composers working with algorithmic and generative music engines was Iannis Xenakis (1922–2001), who mentioned algorithmic composition techniques in his writings as early as 1957. Some researchers consider Xenakis' GENDY3 (1991) system to represent the very prototype of "a purely algorithmic composition" (Hoffmann, 2002, p. 121). According to Hoffmann, automatizing the compositional processes happens in this work on "the 'atom' of digital sound, the sample ('micro-composition')" (Hoffmann, 2002, p. 122). The GENDY program generates the micro- and macro-levels of the composition according to the composer's input (Hoffmann in Georgaki, 2005, p. 3). The terminology by which the composer himself referred to the process was "dynamic stochastic synthesis" (Xenakis, 1992, p. 289–293). It also later formed the basis of granular synthesis, as we now know it (Georgaki, 2005, p. 3). "The idea behind the GENDY program is that free, autonomous evolution of the music itself takes the place of the micro-formal choices usually made by the composer" (Gérard and Rebelo in Georgaki, 2005, p. 3). As demonstrated later in this chapter, the results of complex computer-generated music analysis can also fly into acoustic instrumentations. The same applied to Xenakis, who did not limit his techniques to be used in his electronics works only, but also made use of them in some of his acoustic pieces of music.

3 Case study 1. Olga Neuwirth: "Kloing!" (2008)

Olga Neuwirth's "Kloing!" (2008) for computer controlled piano, live video and human pianist found its inspiration from outside of the abstract realm of the classical music. Although the core material of the piece is based on the seismic data gathered in the Sumatran area during the seaquake that caused the tsunami disaster on December 26, 2004, it is simultaneously also constantly referring to known symbols of our Western culture. The composition is a collage of material produced by the composer, existing classical music literature and popular culture. "Kloing!" paraphrases famous excerpts from the classical virtuoso piano repertoire that include, for example, Franz Liszt's "La Campanella" and Chopin's "Raindrops" prelude, that all are played with a distorted player-piano sound that was specifically created by pre-recording samples on outdated, malfunctioning instruments. The work combines a plethora of technological solutions that are handled in a versatile manner: simultaneously, the human performer gets gradually stripped from their usual role of being in absolute control of their instrument. The work was realized on a Bösendorfer CEUS grand piano, which is capable of creating up to 1,000 velocity layers on each key. According to the music technology expert who translated the gathered data into musical material, the artistic idea behind "Kloing!" lies in the humans' fight against a machine (Plessas, 2008, p. 4): the concert instrument is simultaneously played both by the live musician and the computerized system. The role of the computer-generated data gradually increases its importance, whilst that of the human performer decreases: eventually the rapid automatized movements of the piano keys simulate in the concert room some natural forces that cannot be controlled or tamed by human intervention.

The original data was transformed into an audible form with the help of mathematical modeling and then converted into rhythms and pitches by using a specific computer software (Plessas, 2008, p. 1–3). The material based on the seismic data gets introduced gradually until it takes over: at the end there is barely any space for the pianist to play on the keyboard. At the same time, a live video is projected onto the screen, which gives the performance a theatrical and dramatic touch, underlining the superpowers of the piano and the nature. The projected visuals highlight the exaggerated superhuman movements of the automatically controlled piano keys to the audience. The simultaneous video presence of the cartoon characters Tom and Jerry, unpredictably reacting to the piano sounds, brings some lighter, humorous and ironic characteristics to the musical composition that bears quite some dark undertones.

By analyzing the music score and a recording of "Kloing!," it becomes evident that "Kloing!" is an individual concert project which comes with its very own setup and rules. For example, Neuwirth's ensemble work "Hooloomooloo" (1996–1997) strongly differs from it. Contrasting with the lavish musical gestures of the former, the latter work is structured around microtonal tensions between different instrumental groups. A continuing dialogue with the sine tones produced by an Ebow is omnipresent. An Ebow is an electronic bow that one uses to resonate the strings of a musical instrument, most often those of an electric guitar. In the case of "Kloing!" an Ebow is placed on the low keys of the piano to produce a sine tone of "d." The material realized with the help of this tool is paired with the fixed media background in order to maintain the tightly-knit harmonic cell on which the whole piece is based (Drees, 2001, p. 2).

"Kloing!" is a composition that reinterprets into music certain data that was gathered during an actual natural disaster. Thus, instead of just writing a generic piece describing a tsunami, the composer forces the listeners to take her work more seriously. By using authentic data from the event, the music manages to touch its audiences on a deeper level, providing them an immersive interdisciplinary experience that calls for more discussion and critical thinking on today's environmental issues.

3.1 Case Study 2. Maria Kallionpää: "Climb!" For Yamaha Disklavier Grand Piano and Interactive System

"Climb!" is a nonlinear composition that was inspired by the generative music systems of modern-day computer games. On top of the composition itself, "Climb!" includes an interactive system, real-time interactive visuals, an online archive and a smartphone application that helps the concert audience to follow the route that the pianist takes in the virtual environment. "Climb!" is intended to be both a narrative game as well as a virtuoso concert piece. However, the audience members do not necessarily have to follow the game structure in order to perceive the composition as a meaningful artistic experience. In this work, "the pianist navigates through a nonlinear musical adventure within the virtual space with the help of an interactive system" (Kallionpää & Gasselseder, 2019, p. 180). The concept design was first introduced by Kallionpää and Gasselseder in 2016, and the project was then hosted by the Mixed Reality Laboratory of the University of Nottingham. The music score and the interactive engine were developed simultaneously. The live system uses the Muzicodes software as a navigational tool (the program was developed in Nottingham, in parallel with this practice-based artistic research project of the composer). By playing musical motifs, the performer will be steered to another spot in the interactive music score. These particular motifs that are embedded in the interactive system, function the same way as any regular codes: a MIDI message consisting of a selection of pitches will trigger different tasks.

When working on the piece together with the technical research team in Nottingham, the composer constantly had to revisit the score from the performer's perspective, too, as the roles of the composer and performer overlapped in the process (Kallionpää and Gasselseder, 2019, p. 193). As pointed out earlier in this chapter, designing a complex technical work might require a lengthy process of trial and error: this also applied to "Climb!," as some of the problems in the system were often only detectable within the concert performances showcasing various development stages of the system. Improvements were made both to the music engine as well as in the music score itself.

3.1.1 Analysis as Part of the Composing Process of "Climb!"

Like in Janet Schmalfeldt's model, the composer of "Climb!" had to simultaneously divide herself into a "composer self" and "analyzer self." This was due to the branching structure of the work, which required the composer always to be in full control of the evolution of the piece's constantly changing musical form. "Climb!" is a large-scale work that takes around 30 minutes to perform, and the music score consists of approximately 150 pages. As the mosaic-like short movements of "Climb!" get organized differently in every performance situation, there are dozens of possibilities how the overall form may turn out. Furthermore, some branching options between different phases of the piece had to be added or changed along the way, which required the composer to reconsider which components of the piece could be combined linearly, and which ones not.

The originality of "Climb!" mostly lies in its form that resembles a kaleidoscope: as the work is primarily a concert composition, every possible version of it has to work in a coherent and meaningful manner. Thus, the first step before composing any music was to draw a map of its structure. This enabled the better control of the musical material in the actual composing phase of the project.

3.2 Case Study 3. Maria Kallionpää: "El Canto Del Mar Infinito" (2020)

I've lived near the sea all my life, and it's always been an important element for me. Although I've now left the waters of Southwest Finland and Satakunta that I love for the shores of Hong Kong, both are equally deserving of protection. "El Canto del Mar Infinito" is one of several works of mine exploring this theme. The presence of the sea is apparent in the musical material and instrumentation, some of which was generated from an analysis of underwater sound worlds.

Maria Kallionpää (the program note for "El Canto del Mar Infinito")

As was the case with Xenakis' generative techniques landing into some of his acoustic works, "El Canto del Mar Infinito" is also a purely acoustic composition based on musical material leaning on computer-based analysis. "El Canto del Mar Infinito" was commissioned by the Tampere Biennale 2020 festival (scored for fl, cl in B, pno, 2 vl, 1 vla, 1 vlc). It was tailormade for the Uusinta Ensemble, which is considered to be one of the finest contemporary music ensembles in Finland. The theme of the 2020 edition was protecting the nature and the global sea environments, which also called for technical solutions and artistic concepts that would refer to it in an insightful manner. As pointed out above, rather than just composing a contemporary music piece that would reiterate already established standard composition techniques, the composer intended to find a medium that would bring on some novel narratives, as well as technological and aesthetic perspectives that would make the work pay homage to the important societal discourse that the festival topic was linked to, as well as raise attention among the audiences that would not necessarily be familiar with the contemporary music repertoires. As pointed out in the program note, the goal was to give part of the authorship for the ocean itself. To do this, the composer ended up putting together a selection of sounds that would normally be associated with a sea environment. These included 1. the screaming of seagulls, 2. whales singing and 3. the sounds of ocean waves. However, the composer's intention was to keep a certain distance from the original sound sources, to avoid having them appear in a too obvious manner in the general soundscape. Although the audience would know from the beginning on that the ocean sounds would be part of the music, the sounds were not supposed to be clearly recognizable.

The sound samples were translated into music by using the ORCHID software by IRCAM. Before any actual music could be composed, the sonic excerpts had to be analyzed. After getting them in a notated form from the software, the composer would then craft the other musical materials of the piece. The composed materials had to be compatible with the rhythmic and intervallic shapes and structures of the music based on the ocean sounds, as the latter formed the "main themes" of the composition.

Although computer-based data analysis was used as an aid for creating the pitch material, rhythmical structures and instrumental combinations, it needs to be

Figure 18.1 An excerpt from the finished score of "El Canto del Mar Infinito." The sound sample "screaming seagulls" was translated into musical material, after modifications performed by the composer.

understood that, in order for "El Canto del Mar Infinito" to work as an artistically viable piece of music, the resulted fragments still needed quite some adjustments. A computer software does not think whether the rhythmical structures are playable, or if the instrumental textures are idiomatic: it just transforms the sonic spectrum into notes. In addition to the actual music textures, the composer decided to include two of her poems from 2006 (in Spanish and in English) that would contribute into the overall sonic world of the piece. "El Canto del Mar Infinito" is not vocal music, and the audience is not supposed to understand all the words of the text. The spoken, whispered, uttered or sung sounds mainly function as "a background noise" that melts into the texture, enriching the music that was produced by the sea itself.

4 Conclusions

This discussion has adopted a qualitative and hermeneutic approach to reflect compositional case studies based on different technical solutions. Moreover, to highlight the research topic from "inside," some parts of the discussion were conducted from the composer's perspective. The common denominator between the selected work examples presented in this chapter lies in the importance of analysis as a primary source of musical material: without a thorough analytical pre-construction, none of the works discussed could exist. The approaches taken by the composers depends on their technical and aesthetic goals: a spectral composer might use a different set of techniques than a composer working on generative or algorithmic systems, or when focusing on an individual composition project translating externally gathered data into a compelling musical work.

On top of the aspects mentioned above, modern concert audiences expect novelty when listening to contemporary music performances as part of festival and concert programs. Instead of engaging themselves in an abstract realization of the composer's artistic vision, they want to experience innovative storytelling, immersive technologies and musical language that demonstrates relevance in the current cultural climate, potentially also contributing to the discourse on current societal themes and topics. Music has always been composed for the audiences of its time: the continuous redefinition of its aesthetics, as well as implementation of interesting scoring techniques and technologies may help the classical music culture to appeal to a wider pool of listeners, and to defend it from becoming a museum for limited elitist audiences.

References

Barbaud, P., 1960. La Musique Algorithmique, in *Esprit*.

Berweck, S., 2012. *It worked yesterday: On (re-)performing electroacoustic music.* Doctoral diss., University of Huddersfield.

Borgdorff, H., 2007. The debate on research in the arts. *Dutch Journal of Music Theory*, Vol. 12 No 1.

Chadabe, J., 1996. The History of Electronic Music as a Reflection of Structural Paradigms. *The Leonardo Music Journal*, Vol. 6. The MIT Press.

Chadabe, J., 1997. *Electric Sound: The Past and Promise of Electronic Music.* Upper Saddle River, NJ: Prentice.

Cook, N., 1990. *Music, Imagination, and Culture.* Oxford: Oxford University Press.

Cook, N., 1999. Analysing Performance and Performing Analysis, in N. Cook and M. Everist, (eds.), *Rethinking Music*. Oxford: Oxford University Press.

Drees, S., 2001. Perspektivenwechsel: Tonraum-Deformationen durch Instrumentalklang- Verstimmungen bei Olga Neuwirth. http://www.stefandrees.de/publikation/hooloomooloo.html (accessed 9.10.2020).

Edwards, M., 2011. Algorithmic Composition: Computational Thinking in Music, in *Communications of the ACM*, Vol. 54 No. 7.

Essl, K., 1994a. Klangkomposition und Systemtheorie, in *Darmstädter Beiträge zur Neuen Musik*, Bd. XX, Ulrich Mosch and Gianmario Borio with Friedrich Hommel (eds.). Mainz: Schott.

Essl, K., 1994b. New Aspects of Musical Material. Translated by Joyce Shintani. (Lecture at Darmstädter Ferienkurse für Neue Musik, 1994.) http://www.essl.at/bibliogr/material.html (accessed 10.10.2020).

Essl, K., 2006. Algorithmic Composition. *The Cambridge Companion to Electronic Music*. Cambridge: Cambridge University Press.

Essl, K., 2008. *Sequitur V*. http://www.essl.at/div/scores/sequitur-V.pdf (accessed10.10.2020).

Essl, K., 2008–2010. Sequitur Generator. Software for realtime-processed live-electronics for Karlheinz Essl's "Sequitur" compositions. http://www.essl.at/works/sequitur-generator.html (accessed 9.10.2020).

Georgaki, A., 2005. The *Grain* of Xenakis' Technological Thought in the Computer Music Research of Our Days, in *Definitive Proceedings of the "International Symposium Iannis Xenakis."* Makis Solomos, Anastasia Georgaki, Giorgos Zervos (eds.) (Athens: 2005) http://www.iannis-xenakis.org/Articles/Georgaki.pdf (accessed 10.10.2020).Greenhalgh, C., Benford, S., and Hazzard, A. 2016. Muzicode$: Composing and Performing Musical Codes. *Audio Mostly 2016. Norrköping, Sweden, October 4–6, 2016, Proceedings.*

Hinkle-Turner; E., 2006. *Women Composers and Music Technology in United States: Crossing the Line.* Aldershot, England: Ashgate.

Hoffmann, P., 2002. Towards an "Automated Art": Algorithmic Processes in Xenakis Compositions, in *Contemporary Music Review*, Vol. 21. Nos. 2/3.

Hämeenniemi, E., 2007. *Tulevaisuuden musiikin historia.* Helsinki: Basam Books.

Kallionpää, M., 2014. *Beyond the Piano: The Super Instrument. Widening the Instrumental Capacities in the Context of the Piano Music of the 21st Century.* DPhil thesis, University of Oxford.

Kallionpää, M., Chamberlain, A., Gasselseder, H.P., 2018. Under Construction: Contemporary Opera in the Crossroads Between New Aesthetics, Techniques, and Technologies, in *Proceedings of the Audio Mostly 2018 on Sound in Immersion and Emotion (AM'18).* Association for Computing Machinery, New York, Article 14, 1–8.

Kallionpää, M. and Gasselseder, H.P., 2016. The Imaginary Friend: Crossing Over Computer Scoring Techniques and Musical Expression, in J.P. Bowen, G. Diprose and N. Lambert (eds.), *Electronic Visualisation and the Arts (EVA 2016)*, Jul 12–14. London: BCS Learning and Development Ltd.

Kallionpää, M., Gasselseder, H.P., 2019. "Climb!" – A Composition Case Study. Actualising and Replicating Virtual Spaces in Classical Music Composition and Performance, in Filimowicz, M. (ed.), *Foundations in Sound Design for Interactive Media.* New York: Routledge.

Klien, V., Grill, T., Flexer, A., 2010. Because We Are All Falling Down. Physics, Gestures, and Relative Realities, in *Proceedings of the International Computer Music Conference (ICMC'10).* New York.

Korhonen-Björkman, H., 2008. Eko som tema: En betraktelse av musiker- och lyssnarpositioner i Mon ami av Betsy Jolas. Helsinki: Musiikki 3–4.

Neuhaus, H., 1986. *Pianonsoiton Taide.* 2nd ed. Translated from German by Arja Gothóni. Juva: WSOY:n graafiset laitokset. Translated from Finnish by Neuhaus, H., 1969. *Die Kunst des Klavierspielens.* Leipzig: VEB Deutscher Verlag für Musik.

Plessas, P. 2008. *Arbeitsbericht: Olga Neuwirth – "Kloing!" für selbstspielendes Klavier, Live-Pianist und Live-Film (Uraufführung).* Institut für elektronische Musik und Akustik IEM. https://iem.kug.ac.at/fileadmin/media/iem/altdaten/projekte/composition/neuwirth/kloing/bericht.pdf (accessed 10.10.2020).

Schwartz, E., 1989. *Electronic Music: A Listeners Guide.* New York: Da Capo Press.

Xenakis, I., 1992. *Formalized Music*, revised edition. Stuyvesant, NY: Pendragon Press.

Discussions and lectures

Essl, K., 2014. email discussion. 6.8.2014.

Haas, G.F., 2013, *I Am Not a Spectral Composer.* Lecture at the Impuls festival, 15.2.2013. Graz, Austria.

Plessas, P., 2013, private discussions during the Impuls festival. 11.2.2013–20.2.2013. Graz, Austria.

Further reading

Kallionpää, M., Greenhalgh, C., Hazzard, A., Weigl, D., Page, K. and S. Benford, 2017. Composing and realising a game-like performance for disklavier and electronics. *New Interfaces for Musical Expression (NIME'17), Copenhagen, Denmark, May 15–18, 2017.*

Moscovich, V., 1997. French Spectral Music: An Introduction, in Calum MacDonald (ed.), *Tempo. A Quarterly Review of Modern Music*, No. 200.

The Bold and the Beautiful Methods of Sound Experience Research

An Introduction to Mixing the Qualitative and Quantitative Study of the Subjective Experience of Sound and Music in Video Games

Hans-Peter Gasselseder

1 Introduction

When you scream out loud, how do the people in your surroundings react? Surprised, confused, intimidated? How does that make you feel in exchange? Will you blush or not care? Would not either option represent a response in its own right? Notice, we are dealing with an interaction. And that is the basic quality of all that sounds: by triggering a response, it interacts with its listeners. Some of these interactions may be easier to observe than others. Stealing a candy from a small child may trigger an outcry, which in turn will have you return your loot to its rightful owner. Listening to a fast-paced music track while writing an essay may activate or distract you, leading to the continuation or change of music program and, alternatively, turning you to seek solace in social media, instead. The reader will likely have encountered and become familiar with the outcome of such situations. One may be even tempted to assume that familiarity with the task at hand contributes to a better understanding how sound affects performance as well as its effect on attention and emotion. For the modern sound designer, however, the challenge may be more complicated. Without immediate responses from an audience during the design process, all that sound designers and/or engineers are left with is their imagination of the aforementioned interaction patterns. For psychologists, this kind of understanding of others' mental states represents a fundamental aspect of social behavior, a theory of mind and empathic capacity (for a further review of the relevance of the 'theory of mind' (ToM) for music see Livingstone & Thompson 2009).

But ToM and empathy may not be of much help if we cannot ascertain whether sound and music impact the experience of situations in the virtual (as in interactive media) as much as in the physical. We may wonder whether any potential effects triggered by auditory stimuli depend on the motivational and attentional affordances

DOI: 10.4324/9780429356360-20

encountered in a specific situation. Depending on how pleasing you may find a specific task, you may expect a different kind of music program accompanying it. Most importantly, however, we may all agree to disagree upon what kind of sounds and music fit to performing a task or describing a specific situation. Indeed, while one user may feel distracted by a nerve-wrecking beep tone of a user interface or a nervously pounding music track, another user of the same program may consider these elements as beneficial to performance.

In the ongoing discussion on the use of sound and music in interactive media the above raised issues take various significance within the different forms of scientific inquiry. Whereas theory of mind provides us with a generic and rather specific framework of understanding the inner workings and logic of an outcome (i.e., 'agency,' 'intentionality'), affordances pose a wider range of situational factors that may call for a subjective account on how the environment affects the mental state of the user and vice versa (i.e., 'situation'). Rather than theoretically representing a model of reality, affordances act as a conditional parameter whose exact place in and effect on user experience seems indefinite as we cannot account for all possible scenarios a user may encounter. Thus, we may inquire about the range of situations and their associated experiential artefacts by interviewing users within the realm of qualitative inquiries, and/or we agree to consider a representative subset of situations that will limit the diversity of experience. This will allow us to look into a specific mode of interaction that we can further test within a quantitative experiment as to gain insight into the range of experiences that we would expect to find. The purpose of this sequential, mixed method paradigm is not just to cross-validate the outcome of one form of scientific inquiry. More so, it is meant to capture different dimensions of the same phenomenon so that we gain a more thorough understanding of the one experience we are interested in. For this chapter, this interest will be nothing less than the manifold faces of immersive experience of sound and music in video games whose methodological intricacies we will examine as part of a qualitative and quantitative investigation.

2 The Qualitative Inquiry: Exploring the Scope of Situation

Research into experience-related aspects of video games is characterized by a variety of methodological facets. Typically, these facets focus on a set of predetermined variables that are believed to represent the core aspects of the video game experience (cf. Klimmt & Trepte 2003). For the most part, these variables refer to tangible, particular features of the stimulus material in isolation rather than considering their contextual integration or potential effects of interaction. This particularization inevitably leads to study designs that favor prerequisites, such as the implementation of its pre-tested stimulus material, but may miss out on experiential artefacts that only arise when stimuli appear under different conditions. A good example of this may be found in what shall be called from this point on as the 'additive effect logic.' The additive effect logic assumes that contrasting the presence of a condition to the absence of the same condition allows us to determine its effect onto our outcome variable. While this logic may suit more stationary stimuli (such as visuals or game difficulty that change at a relatively slow pace), sound and music are temporary by nature. In fact, the processing of sound takes place at the highest temporal resolution in our perceptual system (Guttman et al. 2005). The additive logic, as in comparing an experimental condition with music to a condition without music, may not yield the same kind of ecologically valid results as in the other stimuli domains due to music experience being dependent on concurrent factors that affect its temporal processing (e.g., sound effects and dialogue for auditory streams; sudden bursts of light for visual streams). In other words, any experience accompanied by sound and music is more prone to be altered by concurrent stimuli or events. Thus, one may argue that comparing the effect of sound and music in isolation, as per the additive logic, will lead to game experiences that have little in common with the actual video game experience in real life settings, resulting in poor ecological validity and a potential skewing of effects on the one hand. On the other hand, by adding and removing the presence of a modality, we gain better control over interaction effects between conditions (that may well be in our interest but could also lead to less defined responses made by subjects). The question to what extent sound and music stimuli in video games influence immersive experience, requires a procedure that takes into account the properties of sound effects and music equally. Thus, even if counter-intuitive, the presented design will draw on the additive logic for the sake of ensuring that changes in stimuli are either consciously processed or more easily recalled by subjects.

2.1 Qualitative Inquiry: the Entanglement and Reference of Theory and Method

In view of the issues raised above, it seems sensible to determine a respective basis of reference and assessment of user responses in the context of a qualitative procedure that incorporates both ecologically valid environmental and behavioral prerequisites against the backdrop of keeping the constancy of conditions intact. This can be ensured, for example, by making use of a targeted stimulus presentation concept. In the case of surveys, it is advisable to use a questionnaire which, with the help of open item formats, is intended to enable subjects to incorporate aspects beyond the theoretical framework that led to the construction of the questionnaire. However, opening up this option harbors a risk of misunderstandings, which can arise both on the part of the respondent through incorrect interpretation of a question and on the part of the evaluation through misinterpretation of an answer. The sequence in which the questions are received and processed by test subjects can only be influenced to a limited extent, so that the risk of influencing the response behavior through context information remains (cf. Scholl 2003). In fact, this risk cannot be completely ruled out in any survey situation, but in the case of direct face-to-face communication of an interview, it can be avoided, albeit to a limited extent, by checking and monitoring the reflective behavior of the subject. This makes it possible to estimate the effect of a stimulus from the information provided by subjects without them immediately becoming aware of it, and to encourage subjects to reflect on the causes of effects only at a later point in time. The additive logic of our study design is thus used to avoid confusion, whether any change of experience should have occurred due to the change of conditions. This way, subjects are expected to recall the circumstances of their experiences and, thus, allow for the exploration of experiences that are linked with changes to sound and music in video games. With this in mind, our data collection intends to incorporate experiential facets that follow the concept of the "cognitive unconscious" (Kihlstrom 1987), according to which image and sound find their way into perception as partial effects that are hardly perceived individually. This suggests that the experience and performance of a video game is dependent on the entanglement of image and sound. Similarly, a common notion found in film music research asserts that cognitive processing plays a more passive role in consciousness, despite its noted significance for narrative interpretation and the experience of absorption (cf. Kreuzer 2009). Emotional responses, however, are typically attributed with higher saliency towards the spectrum of conscious experience. Either assignments are brought forward when subjects were specifically asked about the impact of (mostly atmospheric cues) of underscore music in a linear media program. When subjects are presented with more decisive musical cues that appear to prepare audiences for a forthcoming event within the narrative, the aforementioned allocation of conscious processing becomes more blurred, with structural aspects of the music material often being (consciously) recalled as the decisive measure of media experience (see Boltz 2004; Gasselseder 2015).

2.2 Qualitative Inquiry: Focus Interview

Our initial qualitative inquiry assumes that subjective involvement, depending on the player's prior experience and performance, allows conclusions to be drawn about the respective partial effects of image and sound in a video game. Since our research question aims to determine the effects of game audio on immersive experience, subjects are presented with stimuli that differ in terms of sound and music. In the context of interviews, they are expected to provide information about whether and to what extent a change took place in their experience. As a survey method we make use of focused interviews as proposed by Merton and Kendall (1946). By devising a structured interview guide, we intend to enable cross-case comparison on the one hand and avoid suggestive question-and-answer tendencies on the other.

The focused interview is a method of qualitative social research, which is considered to be somewhat closer to the side of the positivistic research logic on a continuum between qualitative and quantitative methodology (Lamnek 2005). Through Merton and Kendall (1946) the focused interview has developed into a quasi-independent scientific method and belongs to the older forms of qualitative interviews. The outset of the focused interview sees all subjects being exposed to an identical stimulus before the beginning of the interview, from which a content analysis will derive criteria that support the construction of an interview guide. "The persons interviewed are known to be involved in a particular situation: they have seen a film, heard a radio program (or) read a pamphlet" (Merton & Kendall 1946, p. 12). Thus, the focused interview is a combination of undiscovered observation and qualitative questioning. According to Lamnek (2005), the method is only conditionally open and process oriented. The presentation of stimuli as well as the application of an interview guide restrict the flexibility of the instrument. In so doing, a far-reaching concept fulfills the theoretical requirements and also delivers an explication. The focused interview is not only about developing and generating hypotheses, but also about testing hypotheses. To support this claim, the interviewer must observe the following three instructions in the specific interview situation (see Lamnek 2005):

1. non-influence: the researcher gives the interviewed person the opportunity to express about issues that are of central importance to them. This presupposes "that the hypotheses made by the researcher are not mentioned as such in the interview, because this would represent a predetermination of the course of the conversation and its findings [...]."
2. specificity: elicit specific reports in which subjects not only mention feelings and behaviors about the stimulus situation but also interpret and relate them to one another. This corresponds to the principle of explication. In order to prevent a predetermination and to obtain specific statements, the interviewer may temporally ignore the guideline.
3. depth: help subjects to describe affective, cognitive and evaluative meanings of situation and the degree of their involvement in it. Rather than superficial statements and raw descriptions, the interviewer should evoke in-depth and 'self-revealing' comments from the subject (cf. Merton & Kendall 1946).

2.3 Qualitative Inquiry: Interview Guide and Further Data Collection

The content as well as the course of the focused interview are prepared by a set of guidelines that endeavors to represent a sufficiently complete collection of information along the responses of subjects. Likewise, the theme-centered orientation aims to avoid rigid inquiries in the sense of questionnaires and allows the subject to add new content to the theoretical framework. The interviewer is urged to enrich the interview with a comfortable attitude in order to create an atmosphere of openness (cf. Kubinger 2006). An example of this are introductory questions that act, in a sense, as icebreakers. These questions introduce the topic and aim to integrate its references to the specific interview situation between the subject and interviewer. Their function is not limited to the initiation of the conversation, but also allow for insights into motivation, for example, when asked about the intention to continue the video game. The interview guide gains particular relevance when recalling episodes of experience and behavior with the simultaneous presentation of potentially unconsciously processed sound stimuli. By sketching out a basic sequence, we can reduce positional effects of question items that would otherwise trigger a premature awareness of the change in sound stimuli and thus avoid a potential influence on subject response. Therefore, the interview guide of our initial inquiry is based on a rough sequence of items. The structure of our problem-centered guide reflects this fact by giving precedence to items that do not make mention of sound effects or music as our manipulation. This reduces the risk of suggestion, which easily may occur if a conversation is conducted too intuitively (cf. Kubinger 2006).

While previous efforts in game studies typically adopt a free association paradigm of overall game experience (e.g., Calleja 2007), the present inquiry opts for a theory-based construction of its interview guide in an attempt to better identify responses that were brought forward by latent effects on involvement/motivation and emotion experience. Following the preceding theoretical considerations, our intention is to reformulate any potential outcome experience within broad categories. The first category, 'motivation,' entails observations that locate auditory-dramaturgical aspects of player experience in the vicinity of motivational attractors in a video game. This is indicated by findings where tactile elements of gameplay can grow into short-term game foci in combination with synchronized, interactive musical elements (cf. Collins 2007). Previous considerations have also found the implementation of music as well as its quality a facilitator of interest into game content (cf. Waterworth & Waterworth 2003; Zielinski et al. 2003; Gasselseder 2015). It therefore seems reasonable to understand motivation as a fundamental aspect of experiencing subjective involvement. In the qualitatively derived "Digital Game Involvement Model" (Calleja 2007), motivation is regarded as a constituent factor of macro-involvement in that its concept incorporates elements aimed at both gameplay or difficulty and perceived success. In the interview guide, this circumstance is addressed by means of question items such as "What level of difficulty from 1 (lowest) to 5 (highest) would you assign to session (A, B, C)," or "How did you cope with the video game during session (A, B, C)." The subjects are also encouraged to introduce additional motivational components into the survey, for example, in the question "Did the game in session (A, B, C) motivate you to master the task and challenges in the best possible way?" followed by "What contributed to it and what did not?" Sufficient space should be provided for the motivational aspect of the experience. The second category, 'emotion,' is represented by the two-dimensional emotion space of the circumplex model, which distinguishes between felt intensity/arousal and valence/ pleasure. Within the interview guide, these components act as a reference of reflection as exemplified in the questions "Did you like playing session (A, B, C)? "What did you like/not like?" as well as "Did you find playing in session (A, B, C) as being exciting or not? How did you find it exciting/not exciting?"

While seemingly basic, the questions listed above are designed to be easily understood but also wide enough in scope to include spontaneous responses/expressions on the part of subjects. Nonetheless, sufficient care must be taken in ensuring that the contentual question-response relationship remains unambiguous. A common misconception of such ambiguity may be exemplified by the argument between perceived and felt experience in emotion research. Whereas the former describes the identification of an emotion, the latter refers to the actual sensation of the same. However, many studies (both qualitative and quantitative) fail to distinguish between either facet in their item wording (and sometimes even theoretical foundation) (Vuoskoski & Eerola 2011). Overall, the recall of an event may be affected by suggestive wording of an item, but so also may an item be misunderstood when the interviewer inquires about one state while the subject may actually respond to another. Hence, it must be ensured that both interviewer and subject have a common understanding of the item in question. If in doubt, it is advisable to provide sufficient explanations ahead of data collection. In our present case, we explicitly ask subjects to provide information on their actual felt emotional response. However, this recall takes place as part of a reflective process that may be biased by social desirability. Despite being potentially rich in content, an oral response to the question "how do you like X" may foster more favorable accounts of the experience than not. Moreover, these accounts are typically characterized by a reflective process that involves subjects reviewing their experience, rather than giving an indication of how they actually felt right after finishing the game. To facilitate the latter in our data collection, our explorational endeavor takes on a semi-experimental notion by implementing an alternative item response format in the disguise of an intuitive visual interface.

The software application EMuJoy (Nagel et al. 2007), which is described as an "emotion-effect-response" survey instrument, is a procedure that attempts to compensate for the aforementioned deficits of free recall and rating procedures. In most cases, these measurement models merely refer to a target variable of perceived emotion that is based on the process of reflection (cf. Nagel et al. 2007). Instead, the concept pursued in EMuJoy aims at the feeling of emotions and operationalizes the circumplex model of emotion (Russel 1980). The circumplex model makes recourse to the two-dimensional approach of emotion space, where felt emotions are represented as a coordinate system between degree of valence (pleasure-displeasure, X-axis) and arousal (Y-axis). Along these coordinates, ratings are given by moving a cursor and pressing a button. The instrument has demonstrated good test criteria as found by high retest and construct correlations of about r>0.8 as well as high consistency between continuous and distinct measures (Nagel et al. 2007). Even though EMuJoy supports continuous measurements, for the purposes of our initial inquiry, we make use of distinct measures immediately before and after game presentation as to avoid interference with game experience.

Apart from these more general questions, the interview guide also includes items that specifically address the change of sound and music. If subjects do not express their observations of a change of conditions independently during preceding questions, the interviewer may point out the differences in stimulus presentation by drawing on questions such as "There was no music in session (A, B, C). In your opinion, does the background music have an influence on how you experience the game?" or "Can you determine an influence of sound effects on the game experience; what are the advantages and disadvantages?" This gives subjects a chance to interpret their previous responses within the confines of the study design, and thus may give rise to subjective findings that would otherwise remain unmentioned when comparing previous statements with regard to motivation and emotion.

Finally, the interview concludes with general questions about the amount of time spent playing action video games, genre preference and the motives and feelings before and after using genre video games. Importantly, at this point we address the extent of familiarity with genre-typical conventions in video games but also in film and music. In this regard, previous theoretical considerations for linear (Cohen 2001) but also non-linear media (Crathorne 2010: Gasselseder 2015) could have an impact on the strength of associative ties between the perception of dialogue, image and sound, with a particular emphasis put on the genre-specific music domain.

2.4 Qualitative Inquiry: Procedure

An information phase introduces subjects to the procedure and obtains consent for data collection/recording of the interview. The procedure involves subjects moving through three gameplay conditions (session A, B, C) of the third-person action-adventure video game "Batman: Arkham City" (Rocksteady 2011) in its "Penguin Museum" challenge map. The interviews are conducted after the completion of each gameplay condition.

During the trial phase, subjects were introduced to the interface of EMuJoy and the study's requirement to provide a rating immediately before and after each session. Prior testing, subjects were asked to familiarize with the game and usage of EMuJoy within 3 trials distributed over the course of 30 minutes (each rating on EMuJoy provided at a marked time reference of gameplay within a 10-minute period). Throughout the study, the game was displayed on a 15.6-inch notebook running at 1366 x 768 pixels, 32-bit, 60Hz, with second highest graphic settings. Sound was provided on closed stereo headphones (AKG K270 Studio) at 30 percent volume as set on an audio interface (MOTU 828 mkl). For controls, the standard assignments of a game controller (Xbox 360 for PC) were used. The trial phase involved a scenario taken from the same game that was designed to introduce players to its gameplay mechanics, logic and narrative. Notably, during the trial phase subjects were not wearing headphones and followed narrative development and dialogue via visuals and subtitles.

Next, over the course of the testing phase, subjects were not made aware of the manipulation of stimuli involving session condition (A) with sound effects present and music absent, (B) music present and sound effects absent and (C) both sound effects and music present. Each condition lasts ten minutes to ensure sufficient time for immersive experiences to evolve (cf. Örtqvist & Liljedahl 2010). In order to minimize information bias and positional effects, subjects were randomly assigned to session conditions. The dialogue as well as the overall volume output remained unaffected by the change of condition. Each condition started at the same outset involving a confrontation scenario to ensure the comparability of results. Before and after each session condition, subjects were asked to provide ratings of their current emotional state on EMuJoy. Ratings were provided within an average time period of nine seconds while the game and sound were paused. The subsequent interview lasted an average of 13 minutes and was recorded using a Dictaphone.

Following the completion of all conditions, subjects were debriefed within a final interview that invited them to further reflect on their experience throughout the testing phase. If stimuli manipulation had not been apparent to subjects at this point, they were informed and asked to give their thoughts on how sound and music may have impacted their experience. Finally, data was obtained on socio-demographics, genre preferences for music, movies and video games, as well as knowledge on current trends in the media. All data collection took place in a scarcely furnished, isolated room without windows at a university campus.

The analysis of responses was being conducted with the help of a computer-aided coding system (MAXQDA). The interviews were structured as codeword trees derived from inductive and deductive inferences on the basis of theoretical premises as well as existing data material.

2.5 Qualitative Inquiry: Subjects

The sample of the qualitative inquiry consists of seven female and eight male subjects aged between 21 and 26 years ($\bar{x}=22.73$, SD=1.79). According to Dill (2009), the majority of regular users of action video games are between 14 and 23 years of age. Anderson and colleagues (2010) point to the increasing use of violent action video games in young adulthood, citing the age range of 22 to 30 years. Although meta-analyses of video game experience show tendencies of an age-related effect, according to which children show a higher susceptibility than young adults to empathic and aggressive behavior in both affective and immersive terms, our data is limited to subjects of young adulthood (see Anderson et al. 2010). This is due to ethical as well as legal concerns with regards to violent depictions of fistfights and abduction in the video game used in this study. Furthermore, we may presume that subjects of younger adulthood possess higher experience and reflection skills than what would be expected of adolescents (see Anderson and colleagues 2010). While Dill and Dill (1998) note that the research literature at the time did not show any gender-specific effects of the affective experience of violent video games in quantitative studies, the authors draw attention to gender differences in isolated surveys. Twelve years later, Anderson and colleagues (2010) tend to agree with this finding in quantitative terms, but with regard to qualitative studies no gender differences were found as for the affective evaluation and aggressive tendency following playing violent video games. In view of the observations referred (and for the sake of brevity in this chapter), it seems reasonable to analyze results of both gender groups homogenously within the context of the present study.

It should be noted that genre familiarity holds a special place within the context of the cognitive and emotional experience of sound and music. Theoretical paradigms of music perception with a view on semantic processing, as proposed by Dowling and Harwood (1986) and Cohen (2001), have repeatedly stressed the importance of genre preference and familiarity by means of association and schemata recall. Hence, it was assured that subjects exhibited proficient knowledge of genre conventions.

The recruitment of subjects took place within the social media circles of LAN (Local Area Network) parties organized by student interest groups. These on-site events offer multiple players the chance to compete in video game tournaments. The participants often organize in teams and subsequently prepare the ground for holding LAN parties both in the private and public sphere (cf. Steinmetz 2004). Due to the competing scenarios addressed, the action genre lends itself particularly well as the content of LAN parties and subject recruitment appeared plausible. All subjects are in the final stages of their studies and were compensated for their participation with study credits. Due to the popularity and, in the same vein, representativeness of our stimulus game, it was stressed in the advertisement as well as during point of contact in the interview that the subject had not had any previous experience with the respective game. Furthermore, as with most qualitative endeavors, our subject count, while relatively small, focuses on optimal sample selection and, thus, rather serves representativeness across an expert population on the matter of interest. Along these lines, our findings may be interpreted within a framework that prioritizes a depth-first rationale, as it is typical for an initial exploratory study.

2.6 Qualitative Inquiry: Cross-case Results

Overall, the subjects of our qualitative inquiry form a largely homogeneous set in terms of socio-demographic characteristics as well as their use of video games for three up to ten hours per week depending on study workload. With regard to other leisure activities, subjects show a wide spectrum ranging from sports, cooking, computers and movies to drawing and media art. Though, music preferences vary primarily across rock-oriented as well as pop and techno/dance genres, seven subjects can name at least one game belonging to the action-adventure genre whose music score they found memorable. In addition, five subjects rank film and game soundtracks in their top three genre preferences. All subjects share an interest in action video games, with males showing a preference for shooters (both first-person and third-person perspective) and females slightly gravitating towards fantasy genres, such as role-playing games, while otherwise being open to other subgenres, too. As a motive for regular playing of action video games, both male and female subjects state boredom and the need for competition and experience of success. When asked about how they typically feel following the use of an action video game, both male and female subjects name states that indicate increased tension or relief, depending on the challenge as well as success in gameplay. In this regard, five subjects mention mixed states that, after a follow-up question by the interviewer, could be attributed to a disorientation/reorientation period following extended periods of gameplay. All but two subjects agree that they sometimes intentionally use video games for changing their current motivational and emotional state.

Moving on to the data gathered after each of the three conditions of gameplay in "Batman: Arkham City," we summarize responses to the following experiential categories: dramaturgy, motivation, gameplay/performance, spatial and emotion.

2.6.1 Qualitative Inquiry: Cross-case Results – Dramaturgy

Music creates atmosphere and characterizes relationships with the playing environment. At least this is the credo of our sample in relation to the experience of dramaturgy. While four subjects directly address the immersive qualities of music, as exemplified by the statement "without music, I'm not sure if I'm in the game" (subject 4), eight subjects refer to the situational reference brought about by the music score, naming emotional qualities but also elements that are beneficial to the experience of the narrative. The latter is illustrated by statements referring to a more differentiated perception of characters and a more credible depiction of the same under the influence of music. The same group of subjects describe attacks by opponents as more active, more emotionally charged and fights as longer lasting. For sound effects, the focus of responses gravitates towards the staging of confrontations. In this connection, 12 subjects point out sound effects as an essential ingredient of gameplay experience. Fighting scenes are described as more directly and positively experienced. Four subjects emphasize the reference to reality by understanding the range of interaction between the avatar character and its surroundings. Overall, subjects feel more immersed and recount confrontations more physically charged when sound effects are turned on. Specifically, five subjects address character traits and sensitivities conveyed by sound effects, using the example of enjoying the perceived strength of the avatar character 'Batman' but also learning about his vulnerabilities when taking blows from opponents.

2.6.2 Qualitative Inquiry: Cross-case Results – Motivation

Music promotes motivation, curiosity and free movement in the game environment. Subjects believe that long-term tasks, as framed by the narrative, are more worth pursuing when being accompanied by meaningful music. Overall, subjects believe that music motivates and encourages more successful fighting strategies but takes special precedence during exploration in the beginning stages of the game, while at a more advanced stage the constant presence of music can be distracting. Five subjects specifically mention their appreciation of adaptive music for motivating actions that align with the current demands/challenges as well as spatial properties of scenarios depicted in the game. Eleven subjects note how music supported their quest to explore the environment, but also to keep hidden from opponents. This is illustrated by statements of nine subjects alluring to the signaling function of (adaptive) music when changing from one underscore to the next as soon as opponents notice the avatar approaching. Music thus conveys the mysteriousness of the game environment, so that it sets incentives for the exploration of new worlds and rooms through distinguishing markers. Finally, in the case of mentions of adaptive music, subjects find themselves investing more effort in defeating opponents and enjoying success with the music (arrangement) changing depending on their progress. For sound effects, eight subjects take positive note on motivation by providing a more direct link to physical sensations that promote a stronger "anchoring" (subject 11) within gameplay. Six subjects point out that sound effects motivate more situation-specific behavior accompanied by a sense of achievement during confrontations.

2.6.3 Qualitative Inquiry: Cross-case Results – Performance

As for gameplay/performance, we encounter similar statements for music, which is described as arranging events, informing about tasks and upcoming dangers in the game. While six subjects believe that they experienced the entire session more successfully under the sole influence of music, five subjects describe the experience as "fair." It is particularly noticeable here that music increases the difficulty of confrontation scenarios and this impression does not diminish even when combined with sound effects. This assessment is supplemented by the observations of six subjects, in which the duration of a confrontation under the influence of music is found to be longer, whereas three subjects estimate confrontations in the absence of music as being shorter. Nonetheless, 12 subjects consider sound effects as more important for avatar control due to receiving immediate feedback of initiated actions. Through sound effects, subjects would tend to focus on themselves rather than on accomplishing mid- to long-term tasks, as implied by the narrative. Music and sound effects combined would generally have led to the most successful game experience, all but one subject agree.

2.6.4 Qualitative Inquiry: Cross-case Results – Spatial Experience

Music makes spaces appear more complex and differentiated. With music, eight subjects find movements more expansive and the environment larger while recounting places they had visited over the course of the game. Of these subjects, seven mention the search for and exploration of opponents and trophies as more targeted, fluent and successful under the influence of music, while in the absence of music, orientation problems would have somewhat prevailed. Notably, five subjects stress that fighting movements appear faster when music is paired with sound effects, but slower when only sound effects are active. Simultaneously, two subjects believe this to apply to situations where music accompaniment switches between different states (i.e., adaptive music). Furthermore, three subjects mention the impact of sound effects on estimating the size of rooms by means of reverberation but also atmospheric soundscapes. Sound effects intensify body awareness by informing about the actual occurrence and consequences of the actions taken and "make it easier to understand the characters' actions" (subject 5) as well as increasing attention towards the avatar persona.

2.6.5 Qualitative Inquiry: Cross-case Results – Emotion

Music structures sensations and elevates the game towards a more oppressive atmosphere, although overall eliciting a somewhat positive attitude towards gameplay in its entirety. These observations are shared by seven subjects, of which five imply a situation-dependent intensification of emotional response. All but one subject stated that they attributed a higher degree of sensed arousal to sound effects rather than to music. Music, however, would convey tension by announcing events, but also provides a sense of assurance. The majority of subjects find an emotional component mediated by sound effects particularly in confrontations. All subjects report a higher degree of emotional arousal after completing the game with sound effects as well as with music and sound effects turned on.

The results obtained with EMuJoy largely follow the above-mentioned statements. Discrepancies appear in statements of three subjects that attribute highest ratings on arousal to sound effects, whereas ratings in EMuJoy ascribe these to music. A further difference is found in ratings of valence by two subjects, whose interview responses indicated preference for music, whereas ratings in EMuJoy indicate a more positive emotional experience with sound effects. Again, it should be noted that different scaling of items as well as potential interactions between perceived and felt emotions may contribute to a contrast between a reflected and immediate rating position. Therefore, different aspects of game experience may be addressed and should be considered in interpretation. Notably, six subjects state that they did not recognize the changes between conditions until debriefing and subsequent reflection.

2.7 Qualitative Inquiry: a Review and Short Discussion

In general, the obtained findings indicate rather situation-specific influences of music and sound effects across experiential modalities. Of particular note are character-related mentions of music experience. Subjects appear to attribute more nuanced characterizations and relations to surroundings when music is active. Of interest, this notion was particularly stressed by a group of subjects that later had indicated a preference for film and game soundtracks as part of their top three genres. Thus, genre familiarity and preference may play a role in how subjects operationalize music as an empathic device in a video game.

Another notable observation relates to mentions of adaptive music and, as a consequence, the balance with sound effects. While the importance of adaptivity was specifically stressed with regards to functions ranging from motivation (i.e., motive, effort), performance (difficulty, signaling, time perception), and spatiality (marking places) by mentioning its implementation explicitly, most comments on music made reference to its alternating qualities, as exemplified in the perceived increase of difficulty during confrontation but also ease during phases of exploration. Thus, a further investigation of situational factors of music in game experience seems warranted.

As for sound effects, a group of subjects took note of a link to physical properties in both the avatar and the environment. From this, we may infer a proneness for perceptual realism, a construct that holds seemingly realistic depictions as the main effect of presence experience (cf. Hartmann et al. 2005; Gasselseder 2015). Here it is of interest that, again, subjects indicating a preference for film and game soundtracks make more frequent attributions to music with regards to spatial experience, whereas the remainder of subjects clearly accredit these functions to sound effects. Previous literature has linked music preference to personality, while similar results have been obtained in a more limited manner for immersive tendency (cf. Sandstrom & Russo 2011; Weibel et al. 2010). Accordingly, as we identify this as an overall trend within our data set, it appears worthwhile to investigate further within the realm of quantitative study.

3 The Quantitative Study: the Differential Processing of Situation

The perceptual sensitivity for music expression has been a recent theme in music psychology's endeavor to investigate the relationship between personality/genre preference and music experience. This makes reference to the construct of empathy, which in its original conception by Davis (1980) comprises a multidimensional definition as a cognitive and emotional process of perspective adoption. Notably in this view, cognitive empathy describes the ability to project one's thoughts and actions into fictional settings and derive mental models thereof. Its emotional counterpart draws on structures of automatic processing and is thus discussed in proximity to bottom-up pathways, such as emotional contagion. Both cognitive and emotional components of empathy have been shown to vary along interindividual characteristics (cf. Kreutz et al. 2008).

Perceptual sensitivity per se marks another personality trait that takes a more generic focus on immersive tendency. Slater (2003) suspects that subjects apply varying interindividual weighting of perceptual modalities during presence experience. However, rather than content per se, Slater's statement relates to form, as in audio fidelity. When combined with the above notion of empathy, we arrive at a dual process model that incorporates both expressive features taken from content as well as form/fidelity at the intersection of the physical and virtual. A related concept may be located within the empathizing-systemizing distinction, where the former takes on the role of a decoding capacity of sonorous quality of expression, while the latter focuses on structural aspects of musical syntax (cf. Baron-Cohen 2003; Kreutz et al. 2008). Garrido and Schubert (2011) associate empathizing with decoding accuracy of emotions and preference of music style, which puts the personality trait closer towards a consideration for potential effects on appreciating dramaturgic reenactment and situation in music (cf. Sandstrom & Russo 2011). Hence, analogues to Davis' "fantasy-empathy" construct, it appears justified to control for a trait structure responsible for a user's sensitivity towards the expressive qualities of music. These are believed to play a crucial role during the initial stages of immersive experience by influencing the formation of hypotheses of perception that are subsequently contrasted with incoming perceptual stimuli (for a short summary of the situational context model see Gasselseder 2019). In addition to a decoding structure akin to cognitive empathy, recent investigations have reinvigorated the need to control for a distinct processing structure that passes along decoded expression as emotional response (cf. Shamay-Tsoory et al. 2009). This process of emotional contagion is believed to come into play in a two-step manner, where the first marks the synthesizing of decoded expressive cues from multi-sensory streams into a holistic response, and its second step activates emotional involvement when congruence between hypotheses of perception and decoded response has been reached (cf. Gasselseder 2019; Weibel et al. 2010).

In order to investigate above claims empirically, the following sections incorporate the personality traits "music empathizing" and "emotional involvement" as well as the states "spatial self-location" and "emotional response" into a quantitative experimental design. Specifically, we are interested in how personality traits affect the immersive experience of sound and music in video games within a range of different situational settings. Furthermore, this study intends to examine the methodological usefulness of video games as an adequate research environment to study differential effects of music accompaniment over a series of situations within interactive media.

A total of 60 subjects (23 female, 37 male) aged 18–30 years (M=23.72, SD=3.4) answered self-report questionnaires of experiential states each time after playing the third-person action-adventure video game "Batman: Arkham City" (Rocksteady 2011) for ten minutes in one out of three randomized conditions. Experimental manipulation affected emotional arousal and structural-temporal alignment of non-diegetic music independently. This is achieved by contrasting conditions featuring adaptive music (AM), static music with low arousal potential (LAP) and static music with high arousal potential (HAP). Besides determining the experience of situation by measuring spatial presence self-location, data collection also included emotional valence and arousal. For the purpose of examining interindividual differences, our analysis will focus exclusively on personality traits related to the decoding/appraisal and synthesizing of expressive cues in music empathizing and immersive tendency/emotional involvement.

3.1 Quantitative Study: Subjects

Subjects spend on average 2.37 hours (SD=1.68) at 2.81 days per week (SD=1.78) with playing video games. As in our qualitative inquiry, none of the included subjects had played the stimulus game or any of its predecessors before. As verified during debriefings, no subjects took notice of the experimental manipulation. This is likely due to the present design's discarding an additive logic in favor for a gradual approach (as per different qualities) of contrasting stimuli conditions.

3.2 Quantitative Study: Materials

As in our qualitative inquiry, we made use of the challenge map "Penguin Museum" from "Batman: Arkham City" (Rocksteady 2011). The instruction for the player indicated ten minutes of time (as shown on a countdown) to distract opponents from chasing escaping hostages before challenging them in a final battle. Again, this time period was set for immersive experiences to manifest (see Örtqvist & Liljedahl 2010; Gasselseder 2015). The orchestral score of "'Batman: Arkham City," written by Nick Arundel and Ron Fish, makes use of a horizontal mechanism that reflects calm and confrontational situation changes by musical expression of LAP and HAP (see 're-sequencing' in Gasselseder 2019). In addition, the score utilizes a vertical mechanism that reflects dramaturgic aspects ranging from danger to task progress by adding and removing four orchestral stems to the mix relative to the actions and performance of the player (see 're-orchestration' in Gasselseder 2019).

3.3 Quantitative Study: Instruments

The following section gives a short overview of the content and test parameters of the scales used as part of the quantitative inquiry. Pursuant to our interest in the effect of personality traits on immersive experience, we distinguish between experiential 'state' and personality 'trait' measures.

3.3.1 Quantitative Study: Instruments – Self-location (State)

The dimension 'self-location,' taken from the "MEC Spatial Presence Questionnaire" (Vorderer et al. 2004), refers to a sense of physical projection when interacting with the game. The scale contains four items and is rated on a Likert-type scale scored from 0–4. Studies undertaken by the questionnaire's authors show good internal consistencies between $\alpha=.80$ to $\alpha=.92$ as well as construct validity of about $k>.70$.

3.3.2 Quantitative Study: Instruments – Emotion (State)

As in the qualitative inquiry, EMuJoy (Nagel et al. 2007) was used for recording emotional responses of subjects. (Please refer to the description found in section 2.3 of this chapter.)

3.3.3 Quantitative Study: Instruments – Music Empathizing (Trait)

'Music empathizing' is a dimension taken from the "Music Empathizing-Music Systemizing Inventory" (Kreutz et al. 2008), that aims to assess the ability to identify and respond to expressive content in music. The ME-MS Inventory presents music empathizing with nine items in randomized order on a Likert-type scale ranging 0–3. Its German translation in use here, is characterized by an internal consistency of $\alpha=.71$, which compares well to the $\alpha=.69$ of the original version in English.

3.3.4 Quantitative Study: Instruments – Emotional Involvement (Trait)

The dimension 'emotional involvement' emerged from a factor analysis in Weibel and co-workers' (2010) German adaptation of the "Immersion Tendency Questionnaire" (ITQ). The dimension aims to describe the tendency for emotional reactions during media usage and (day-) dreaming. The subscale consists of five items presented in randomized order and rated on a Likert type scale scored 0–4. Results of the present study show acceptable internal consistency of $\alpha=.71$.

3.4 Quantitative Study: Procedure

Display and sound output were analogous to our initial qualitative inquiry. Sound effects were fed to the monitoring input of a DAW (Apple Logic Pro set at 128 samples buffer) and, for static conditions, mixed with the pre-recorded original music tracks (A-weighted volume matched). Prior testing, subjects took a 30-minute training session involving game mechanics and EMuJoy. Before starting the game, EMuJoy ratings on current emotional state were recorded. Next, the game excerpt is presented in three sessions of ten minutes length, each reflecting one out of three music modalities contrasting adaptive/static mechanisms and arousal potential characteristics in randomized order. At the end of each condition an animation of five seconds signaled successful completion, which marked the point when sound had faded out gradually. Following this, subjects were asked to give ratings on their current emotional state on EMuJoy and measures of self-location. After having completed all sessions, subjects finished the experiment by filling out the personality trait questionnaires ME-MS Inventory and ITQ.

3.5 Quantitative Study: Results

For the following correlations, attention is drawn to prior studies in music and social psychology with mean effect sizes ranging typically between $r = .21$ and $r = .40$ (see Sandstrom & Russo 2011). Investigating the relationship of trait empathy and emotional experience with Spearman ranks, music empathizing and pre-post arousal measures were correlated only in the LAP condition, $r = .26$, $p = .04$, while no correlations manifested in conditions presenting AM, $r = .01$, $p = .96$, and HAP, $r = .06$, $p = .63$. Comparable results are found for emotional involvement for which a moderate correlation with pre-post arousal appeared only in LAP, $r = .33$, $p = .01$, whereas non-significant results were obtained from the HAP, $r = .22$, $p = .09$, and the AM condition, $r = .10$, $p = .46$. When changing over to MEC-SPQ measures of self-location, music empathizing correlates roughly moderately after playing in the AM, $r = .28$, $p = .03$, and LAP conditions, $r = .26$, $p = .04$, albeit results from the HAP variant remain non-significant, $r = .21$, $p = .10$. Interestingly, emotional involvement and self-location showed relatively strong correlations only in the HAP condition, $r = .39$, $p = <.01$, despite non-significant results for AM, $r = .13$, $p = .33$, and LAP conditions, $r = .13$, $p = .34$.

3.6 Quantitative Study: Discussion

Compared to the tendency towards general emotional involvement, it is expected that the ability to decode expression in music will correlate more strongly with pre-post arousal measures in the adaptive music condition. While this hypothesis was not supported in terms of music empathy and emotion measures, the expected outcome for music empathy appeared in reported self-localization during adaptive music and static conditions with low arousal potential (LAP), although without any observable difference between the two conditions. The absence of the same correlation reaching significance in the high arousal potential (HAP) condition might indicate an inferior role of musical decoding structures in the experience of spatial presence when exposed to higher volume/intensity of sensory stimulation. This view agrees with the observation that the trait 'emotional involvement,' a construct closer to emotional contagion, and self-localization correlated only in the high arousal potential condition (HAP), indicating the presumed role of the former in the synthesis of decoded expressive cues from multiple channels and their transformation into a holistic experience. Furthermore, when the expressive parameters of music are examined separately for each condition, the interpretation presented above finds support in the limited dynamic range of the high arousal potential music score (HAP) in comparison to the more complex, far-reaching dynamics and ambivalent expression that characterizes the music with low arousal potential (LAP) (see Gasselseder 2015). This may also play a role in why music with low arousal potential (LAP) remains the only condition where both the traits 'music empathizing' and 'emotional involvement' correlate with pre-post arousal measures. In line with previous reports of predictive validity for the enjoyment of sad music (cf. Garrido & Schubert 2011), we find that the correlation between the trait 'music empathizing' with pre-post arousal may also be due to higher demands for the decoding of music with low arousal potential (LAP). In the case of 'emotional involvement,' the result pattern indicates the participation of a distinct structure that could be responsible for the vicariousness of the decoded expressive stimuli. This may also indicate a link to perceptual realism as characterized by feedback from the environment found in the low arousal potential (LAP) condition which features a higher prominence of sound effects due to the relative lower volume of the more atmospheric music score.

4 The Beautifully Bold: Combining the Qualitative and the Quantitative

Qualitative and quantitative methods both shine light on a research subject; however, each with a different shade and character. Where the former compels with depth, content and ecological validity, the latter excels in objectivity, generalization, as well as uncovering latent phenomena and constructs, such as personality traits. Our journey from a theory-driven to a qualitative and quantitative approach has revealed a transformation from a holistic consideration of a phenomenon towards its particularities. The beginning of the chapter introduced readers to the basic notion of theory of mind, situational context and its ties to understanding other acting entities within music experience (albeit, in a simplified manner for the purposes of this chapters' prevailing topic of presenting qualitative and quantitative methods in practice). From here we went on to further explore by seeking answers to the question of the meaning of sound effects and music in game experience, its situational ties as well as its relation to aspects of immersive experience, such as spatial presence. As we have reached the qualitative realm, while condensing the rich accounts collected from the interviews into a clearer structure, we nonetheless tried to depict a wide spectrum of subjective experience so as to appreciate its contextual implications within situational as well as persona-related constraints, such as genre preference. In doing so, we found that a subset of subjects experience music to alter the perceived difficulty of gameplay and set the dramaturgic tone of the narrative stronger than others. These insights then gave rise to an investigation of personality-related effects on immersive experience that vary across situational settings, as realized by contrasting adaptive music conditions on the grounds of its implied effect on perceived difficulty (amongst other factors, see Gasselseder 2015). By consulting our theoretical framework in conjunction with previous findings, we adopted a dual processing model of music experience in the form of music empathy (expression decoding) and emotional involvement (expression synthesis and emotional contagion). Rather than applying our previous study design by contrasting the presence and absence of sound modalities (additive logic), we learned to embrace the effects of their interaction, as illustrated by the relationship between emotional involvement and spatial self-location during conditions of pronounced sound effects in adaptive and low arousal potential music.

So much for the method. But what do our results entail for research on sound and music in video games? Some studies on music perception have suggested a weak correlation between trait music absorption, empathy and pre-post arousal measures. Subsequently, these personality traits appeared to be a less than ideal choice when investigating interindividual differences of reported intensity changes in (linear) music dramaturgy and seemed a better fit for post-measures only (see Sandstrom & Russo 2011). However, our study may offer a reconsideration of this discrepancy when looking at experiential measures that include cognitive components such as spatial presence. This finding is consistent with an appreciation of trait music empathizing as a decoding capacity akin to cognitive empathy. Similarly, our study provides insights into a division of trait emotional involvement where expression synthesis and emotional contagion take on the roles of a cognitive-perceptual capacity contrasted by an affective capacity promoting decoded expression towards the sensation of emotion. In this vein, we have provided initial insights on the cognitive-emotional processing styles in the experience of music in video games. As would be expected with a pilot endeavor, the scales in use will require further evaluation and adaptation to determine more precisely the impact of sound and music on game experience. However, our results indicate that components of spatial presence, such as self-location, correlate with domain-specific decoding structures whose effect vary depending on music condition (see Gasselseder 2015).

As for a more pragmatic perspective towards the implications of these findings for sound design, we may assume that self-localization depends on the decoding of expressive dynamics of sound and music and the projection of these dynamics onto the game situation. Supposing that a player has below-average decoding capabilities, sound designers are recommended to provide arousing stimuli, such as pronounced sound effects, at context-sensitive volume levels, as realized by adding/removing instrument layers using vertical techniques in adaptive music (see Gasselseder 2019). This way, we maintain high stimulation for emotional involvement while preserving audibility of critical sound effects for the purpose of establishing perceptual real- ism. Likewise, situational settings and attention focus may be altered by utilizing music in a differential manner for decoding and involvement structures. Accord- ingly, as indicated in our qualitative inquiry, a new situation may be proposed to players by implementing horizontal or sequential changes in the music score where one cue fades into another depending on gameplay context. Here, a variation in the sequence, as found in vertical layering, may be less desirable, since stable cues, such as those expressed by the valence potential of music, are more important for establishing initial stages of immersive experience than abrupt changes as realized by vertical layering. In other words, subjects with low decoding capacity may benefit in immersive experience when introducing changes of musical structure at a slower and more moderate pace. Vice versa, subjects with high decoding capacity profit from more articulate and clear distinctions in the arrangement of situational markers conveyed by adaptive music. These aspects of decoding and emotional involvement illustrate how situational dispositions of users may influence game experience.

As with the bold and the beautiful, our insights will gain on relevance with every further season of innovative implementations of game audio. Specifically, decod- ing structures as well as situational trait paradigms, as exemplified in differential behavioral tendencies across situations, may perhaps one day find their way into user-aware adaptive soundtracks and game engines. Until then, this author asks the reader to keep watching or rather *listening* for further methodological advances in the study of the subjective experience of sound and music in video games.

References

Anderson, C.A. et al., 2010. Violent Video Game Effects on Aggression, Empathy, and Prosocial Behavior in Eastern and Western Countries. *Psychological Bulletin*, 136, no. 2, pp. 151–173.

Baron-Cohen, S., 2003. *The essential difference: Male and female brains and the truth about autism.* New York: Basic Books.

Boltz, M.G., 2004. The Cognitive Processing of Film and Musical Soundtracks. *Memory & Cognition*, 32, pp. 1194–1205.

Calleja, G., 2007. Digital Game Involvement. A Conceptual Model. *Games and Culture, 2,* no. 3, pp. 236–260.

Cohen, A.J., 2001. Music as a Source of Emotion in Film. In: Juslin, P.N. and Sloboda, J.A. (eds.), *Music and Emotion: Theory and Research* (Series in Affective Science). New York: Oxford University Press, pp. 249–272.

Collins, K., 2007. An Introduction to the Participatory and Non-Linear Aspects of Video Games Audio. In: Hawkins, S. & Richardson, J. (eds.), *Essays on Sound and Vision*. Helsinki: Helsinki University Press, pp. 263–298.

Crathorne, P.J, 2010. *Video game genres and their music*. Master Thesis at the Department of Music, University of Stellenbosch, Stellenbosch, RSA. Viewed Oct. 10, 2020. https://citeseerx.ist.psu.edu/viewdoc/download?doi=10.1.1.883.1120&rep=rep1&type=pdf.

Davis, M.H., 1980. A multidimensional approach to individual differences in empathy. *JSAS Catalog of Selected Documents in Psychology*, 10, pp. 85.

Dill, K.E., 2009. *How Fantasy Becomes Reality*. New York: Oxford University Press.

Dill, K.E. and Dill, J.C., 1998. Video Game Violence: A Review of the Empirical Literature. *Aggression and Violent Behavior,* 3, no. 4, pp. 407–428.

Dowling W.J. and Harwood, D.L., 1986. *Music cognition*. New York: Academic Press.

Garrido, S. and Schubert, E., 2011. Individual differences in the enjoyment of negative emotion in music. A literature review and experiment. *Music Perception*, 28, pp. 279–296.

Gasselseder H.P., 2015. Re-sequencing the Ludic Orchestra. In: Marcus, A. (ed.), *Design, User Experience, and Usability*: Design Discourse. Lecture Notes in Computer Science, vol. 9186. Cham: Springer.

Gasselseder, H.P., 2019. Composed to Experience: The Cognitive Psychology of Interactive Music for Video Games. In: Filimowicz, M. (ed.), *Foundations in Sound Design for Interactive Media*: An Multidisciplinary Approach. New York: Routledge.

Guttman, S.E., Gilroy, L.A. and Blake, R., 2005. Hearing what the eyes see. Auditory encoding of visual temporal sequences. *Psychological Science*, 16, no. 3, pp. 228–235.

Hartmann, T., Böcking, S., Schramm, H., Wirth, W., Klimmt, C. and Vorderer, P., 2005. Räumliche Präsenz als Rezeptionsmodalität. Ein theoretisches Modell zur Entstehung von Präsenzerleben. In: Gehrau, V., Bilandzic, H., Woelke, J. (eds.), *Rezeptionsstrategien und Rezeptionsmodalitäten*. Munich: Fischer, pp. 21–37.

Kihlstrom, J.F., 1987. The cognitive unconscious. *Science,* 237, pp. 1445–1452.

Klimmt, C. and Trepte, S., 2003. Theoretisch-methodische Desiderata der medienpsychologischen Forschung über die aggressionsfördernde Wirkung gewalthaltiger Computer- und Videospiele. *Zeitschrift Für Medienpsychologie*, 15, pp. 114–121.

Kreutz, G., Schubert, E. and Mitchell, L.A., 2008. Cognitive Styles of Music Listening. *Music Perception*, 26, no. 1, pp. 57–73.

Kreuzer, A.C., 2009. *Filmmusik in Theorie und Praxis*. Konstanz: UVK Verlag.

Kubinger, K.D., 2006. *Psychologische Diagnostik. Theorie und Praxis psychologischen Diagnostizierens*. Göttingen: Hogrefe.

Lamnek, S., 2005. *Qualitative Sozialforschung*. Ein Lehrbuch. Weinheim: Beltz.

Livingstone, R.S. and Thompson, W.F., 2009. The emergence of music from the Theory of Mind. *Musicae Scientiae,* 13, pp. 83–115.

Merton, R. and Kendall, P., 1946. The Focused Interview. *American Journal of Sociology*, 51, no. 6, pp. 541–557.

Nagel, F., Kopiez, R., Grewe, O. and Altenmüller, E., 2007. EMuJoy. Software for continuous measurement of perceived emotions in music. *Behavior Research Methods*, 39, no. 2, pp. 283–290.

Örtqvist, D. and Liljedahl, M., 2010. Immersion and Gameplay Experience. *International Journal of Computer Games Technology*, no. 3, pp. 1–11.

Rocksteady, 2011. *Batman. Arkham City* (video game). Warner Bros. Interactive.

Russell, J.A., 1980. A circumplex model of affect. *Journal of Personality and Social Psychology*, 39, no. 6, pp. 1161–1178.

Sandstrom, G.M. and Russo, F.A., 2011. Absorption in music. Development of a scale to identify individuals with strong emotional responses to music. *Psychology of Music*, 11, pp. 1–13.

Scholl, A., 2003. *Die Befragung*. Konstanz: UVK.

Shamay-Tsoory, S.G., Aharon-Peretz, J. and Perry, D., 2009. Two systems for empathy. *Brain*, 132, no. 3, pp. 617–627.

Slater, M., 2003. A note on presence terminology. *Presence-Connect,* 3, no. 3, pp. 1–5.

Steinmetz, W., 2004. *LAN Party. Hosting the Ultimate Frag Fest*. Hoboken, NJ: Wiley.

Vorderer, P. et al., 2004. *Development of the MEC Spatial Presence Questionnaire, MEC-SPQ*. Report for the European Commission, IST Programme 'Presence Research Activities.'

Vuoskoski, J.K. and Eerola, T., 2011. Measuring music-induced emotion. A comparison of emotion models, personality biases, and intensity of experiences. *Musicae Scientiae*, 15, no. 2, pp. 159–173.

Waterworth, J.A., and Waterworth, E.L., 2003. *The Core of Presence: Presence as Perceptual Illusion*. Presence Connect. Viewed Oct. 10 2020. https://www.academia.edu/881161/The_core_of_presence_Presence_as_perceptual_illusion.

Weibel, D., Wissmath, B. and Mast, F. W., 2010. Immersion in mediated environments. The role of personality traits. *Cyberpsychology, Behavior, and Social Networking*, 13, no. 3, pp. 251–256.

Zielinski, S., Rumsey, F., Bech, S., de Bruyn, B. and Kassier, R., 2003. Computer Games and Multichannel Audio Quality - The Effect of Division of Attention between Auditory and Visual Modalities. 24th AES Conference, June 20th. Viewed Oct 10 2020. https://www.researchgate.net/publication/30929471_Computer_games_and_multichannel_audio_quality_-_the_effect_of_division_of_attention_between_auditory_and_visual_modalities.

Index

Note: Page numbers followed by "n" denote endnotes.

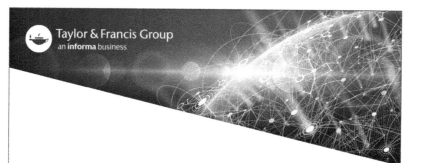

Milton Keynes UK
Ingram Content Group UK Ltd.
UKHW052016071024
449327UK00027B/2293